# *Service Science, Management, and Engineering*

# *Intelligent Systems Series*

# Service Science, Management, and Engineering

## *Theory and Applications*

Gang Xiong
Zhong Liu
Xi-Wei Liu
Fenghua Zhu
Dong Shen

AMSTERDAM • BOSTON • HEIDELBERG • LONDON
NEWYORK • OXFORD • PARIS • SANDIEGO
SANFRANCISCO • SINGAPORE • SYDNEY • TOKYO

Academic Press is an imprint of Elsevier

Academic Press is an imprint of Elsevier
The Boulevard, Langford Lane, Kidlington, Oxford OX5 1GB, UK
225 Wyman Street, Waltham, MA 02451, USA

First edition 2012



**British Library Cataloguing in Publication Data**
A catalogue record for this book is available from the British Library

**Library of Congress Cataloging-in-Publication Data**
A catalog record for this book is availabe from the Library of Congress

ISBN–13: 978-0-12-397037-4

For information on all Academic Press publications
visit our web site at *books.elsevier.com*

Printed and bound in the US

12 13 14 15 16 10 9 8 7 6 5 4 3 2 1

*Typeset by*: diacriTech, Chennai, India

# Contents

# Preface

With the fast development of the service sector in the national economy, it is observed that the employees in this sector account for a great proportion of the labor force. For example, in some developed countries such as the USA, the labor force in the service sector is over 50% nowadays. Generally speaking, service has played an important role in our daily lives and work. We even could say everyone is surrounded by kinds of services. However, there still lacks in-depth research on this newly-emerged sector. This motivates academics to pay more attention to service sector research, which further gives impetus to the subject Service Science, Management and Engineering (SSME). As an emerging discipline, SSME has no unified theory as yet. In general, SSME focus on the related research around services, such as service system design, improvement of specific industry services, software and hardware implementation of service, and many other issues. In short, the research of service is aimed at building a general theory of service including problems, tools, methods and practical applications.

The concept of SSME was first introduced by IBM, based on research of service marketing, service quality, service innovation etc. This research has been well-developed and provided the foundations for SSME. Unlike that research, SSME emphasizes the multi-disciplinary aspects. However, it is still an open question as to which disciplines are involved in SSME. There are many kinds of science related to SSME, such as natural science, social science, and technology etc. In order to promote the development of SSME, there are some conferences and workshops, for example the IEEE International Conference on Service Operations and Logistics, and Informatics ( SOLI ) is one of them.

This book is a follow up effort after the publications of IEEE SOLI 2011. The seventeen chapters in this book are written by experts from different research fields and addresses problems and methods of SSME from different viewpoints. Most of them are extensions of the papers published at IEEE SOLI 2011.

This book includes two parts, namely theory and applications. The former includes five chapters while the latter consists of twelve chapters.

This book starts with a review chapter on SSME, given by Zhen Shen, Dong Shen, Gang Xiong and Fei-Yue Wang. Some important topics are addressed in this chapter. The topics include: what is SSME, why is it important and how do you apply it? Chapter 2 by Yong Lin, Yongjiang Shi and Shihua Ma presents servitization strategy based on the analysis of two PC industry cases and secondary documentation research. Two types of servitization strategy are compared from different points of view. In Chapter 3, Miao He, Changrui Ren, Qinhua Wang and Jin Dong discuss the concepts and modeling for supply chain finance, where approximate dynamic programming (ADP) is used as a basic tool to deal with multi-period problems under uncertainty. In Chapter 4 by G.R. Gangadharan, Anshu N. Jain and Nidhi Rajshree, a methodology for participatory service design for emerging markets is addressed, based on real world case studies. The issues of lack of transparency and the existence of corruption are the primary focus. In Chapter 5, Xinxin Bai, Jinlong Wu, Haifeng Wang, Meng Zhang, Jun Zhang, Yuhu Fu, Xiaoguang Rui, Wenjun Yin and Jin Dong develop recommendation algorithms for implicit information. A new similarity measure and rating strategy for neighborhood models are suggested to obtain better recommendation accuracy.

In Chapter 6, Dennis Güttinger, Eicke Godehardt and Andreas Zinnen compare several online approaches for optimizing the emergency supply after a major incident. The authors show that the combination of a greedy strategy and a subsequent application of a workload adapted version of Simulated Annealing works well for the given online assignment problem. Fenghua Zhu, Zhenjiang Li and Yisheng Lv in Chapter 7 evaluate traffic signal control systems based on Artificial Transportation Systems (ATS). The effectiveness of the evaluation method is verified by two practical applications, which may be difficult to carry out by traditional methods. In Chapter 8, the problem of optimizing police patrol activities is addressed by Li Li, Zhongbo Jiang, Ning Duan, Weishan Dong, Ke Hu and Wei Sun. The authors integrate a spatial pattern identification approach with an efficient route optimization algorithm to produce randomized optimal patrol routes. A case study is provided to illustrate the proposed approach. In Chapter 9, Wei Wei and Changjian Cheng propose a novel emergency management framework of Parallel Emergency Management, based on an ACP (Artificial System, Computational Experiments, Parallel Execution) approach for the problem of how to insure the effectiveness of emergency rescue. Two case studies are given to show the rationality and feasibility of the ACP approach. In Chapter 10, Feng Li, Hongbin Lin, Yu Yuan, Changjie Guo, Wanli Min and Lei Zhao discuss the problem of bus arrival prediction and trip planning for better user experience and services. The bus travel pattern is first classified into eight clusters based on a linear model, and then a system and several algorithms for bus trip planning services are proposed. Chapter 11, by Gang Xiong, mainly proposes a kind of mass customization manufacturing solution, which has been applied by a global mobile phone manufacturer successfully. The four main phases (Marketing, R&D, Production and Purchasing), especially the customized order processing, process quality assurance, statistical process control and the solution's architecture, are proposed in detail. In Chapter 12, Xiwei Liu, Xiaowei Shen, Dong Fan and Masaru Noda give plant human machine interaction evaluation methods based on ACP theory. It is shown that Fault Detection and Isolation (FDI) performance can be improved by comparing the evaluation results of different plants' human machine interaction design schemes. Timo R. Nyberg, Gang Xiong and Jani Luostarinen in Chapter 13 address topics on "cloud of health" for connected patients. The health problems of the 21st century lifestyle are analyzed and three drivers are identified that will change the health services landscape and the mindsets of patients, public and clinicians. Then the concept "cloud of health" is introduced to describe the new and emerging application of the Internet, mobile and wireless technologies to connect patients with expert advice, other patients and devices. Also, the Short Messages System (SMS) is also discussed in-depth. In Chapter 14, Gang Xiong, Xisong Dong and Jiachen Hou present the problem of how to construct artificial power systems based on an ACP approach. As an application case study, the artificial power grid model is constructed with actual data from North China Power Grid, and its vulnerability is tested under random, dynamic and static attacks. Thus the proposed approach could provide theoretical guidance and practical support for the security and stability, quality and economical operations research of power grids and the smart grid. In Chapter 15, Sven Schulze, Christian Engel, and Uwe Dombrowski discuss the influence of electric vehicles on the after sales service. Specifically, the chapter analyzes after sales service in the automotive industry, changes due to the increasing market share of electric mobility and the impact on stakeholders. In Chapter 16, Sheng Liu, Gang Xiong and Dong Fan present service modeling optimization and service composition QoS analysis. An easy-to-use BPEL4WS modeling method and tool is designed to encapsulate computer terms and convert business models to BPEL4WS models directly. Then the number of modeling elements can be cut by more than 85 percent, the modeling time can be saved by 80 percent, and the model running speed can be accelerated by more than 40 percent. So it enables the enterprise to create and run its business processes more quickly than other methods. In the last chapter by Dong Shen and Songhang Chen, a basic framework of an urban traffic management system based on ontology and a multi-agent system is proposed. The ontology of an Urban Traffic Management System (UTMS) is discussed in detail, and a three-level architecture is proposed for an agent-based distributed traffic control system.

All the editors would like to thank our contributing authors for their outstanding research, hard work and great patience with our tedious editing process. We thank Prof. Fei-Yue Wang for his support and guidance for this book. We also thank Prof. Yanqing Gao for her great help and Ms. Jiaying Xu of Zhejiang University Press for her patience. The publication of this book is supported by project 60921061, 70890084, 90920305, 60974095, 61174172, 60904057, 61101220, and 61104054 of The National Natural Science Foundation of China (NSFC); project 2F09N05, 2F09N06, 2F10E08, and 2F10E10 of the Chinese Academy of Science.

*Gang Xiong, Zhong Liu, Xi-Wei Liu, Fenghua Zhu & Dong Shen*

CHAPTER 1

# Overview of Service Science, Management, and Engineering

**Zhen Shen, Dong Shen, Gang Xiong, and Fei-Yue Wang**
*State Key Laboratory of Management and Control for Complex Systems, Institute of Automation, Chinese Academy of Sciences, Beijing 100190, China*

The economies of the world are shifting labor from agriculture and manufacturing sectors to service sectors, as measured by percentage of labor (jobs) (Maglio, Srinivasan, Kreulen, & Spohrer, 2006). This makes researchers pay more attention to the service sectors, and the concept of "Service Science, Management, and Engineering (SSME)" becomes more and more popular. IBM regards SSME as the next trend in college and professional education (International Business Machines Corporation [IBM], 2008). SSME is believed to play an important role in the future world. In this chapter, we try to explain the following: (1) what SSME is; (2) why it is important; and (3) how to apply SSME to several kinds of real world problems.

## 1.1 What Is SSME?

The term "Service Science, Management, and Engineering (SSME)" was introduced by IBM to describe an interdisciplinary approach to the study, design, and implementation of a services system to provide value for others by suitable arrangements of people and technologies (Hefley & Murphy, 2008; IBM, 2008; Spohrer & Maglio, 2008), where an elementary concept "service" is involved. There are various definitions of "service," for example, "…a result that customers want," "…sometimes referred to as intangible goods; one of their characteristics being that in general, they are 'consumed' at the point of production," "…consumer or producer goods that are mainly intangible and often consumed at the same time they are produced… service industries are usually labor-intensive," "intangible products," "a set of intangible activities carried out on [the customer's] behalf," "any act or performance that one party can offer to another that is essentially intangible and does not result in the ownership of anything," "invariantly and undeviatingly personal, as something performed by individuals for other individuals," "…a change in the condition of a person, or of a good belonging to some economic unit, which is brought about as the result of the activity of some other economic unit…," and "economic activities that produce time, place,

Service Science, Management, and Engineering.


DOI: 10.1016/B978-0-12-397037-4.00001-6

form, or psychological utilities" (Sampson & Froehle, 2006). We take the viewpoint of Vargo and Lusch and define "service" as the application of competences (such as knowledge and skills) by one entity for the benefit of another (Vargo & Lusch, 2004, 2006, 2008; Vargo, Maglio, & Akaka, 2008). Around the concept of "service," "science" means "what service systems are and how to understand their evolution," "management" means "how to invest to improve service systems," and "engineering" means "how to invent new technologies that improve the scaling of service systems" (Spohrer, Maglio, Bailey, & Gruhl, 2007).

In national economic statistics, the service sector usually refers to those that are not in the agriculture or manufacturing sectors. In contrast to providing goods in the agriculture and manufacturing sectors, knowledge and skills are provided from one party to another in the service sectors. Examples of service sectors are many, such as tailoring a suit for a customer, teaching a class, and consulting. It appears that the service sector plays an important role in the national economy while a country develops. According to the study by Spohrer and Maglio (2008), in the year 2003, 50% of the labor force of China was in agriculture, 15% in goods production, and 35% in services. When compared with the year 1978, the percentage change of the labor force of China in service sectors was increased by 191%. Nevertheless, in the United States, the labor forces in agriculture, goods production, and services were 3%, 27%, and 70%, respectively, in the year 2003. Over the past three decades, service sectors have become the largest part of most industrialized nations' economies (Spohrer et al., 2007). According to a report of the National Academy of Engineering of the United States (US National Academy of Engineering, 2003), the service sector accounted for more than 80% of the US GDP in the year 2003 (Spohrer et al., 2007).

With the rapid development of service economics, related topics on service deserve more and more study. This is why many researchers turn to develop a general theory for SSME. There is no doubt that SSME is important for our daily life. Here, we want to give more descriptions on some specific points, which may play a significant role in the research of SSME.

### 1.1.1 Information and Communication Technology

The first point that we want to emphasize here is about information and communication technology (ICT). It is believed that ICT will play an important role in SSME. ICT is often used as an extended synonym for information technology (IT), but it stresses the integration of telecommunications, computers, middleware, and necessary software, storage, and audiovisual systems. It is a term that includes any communication device or application, such as radio, television, cellular phones, and computer and network hardware and software. Because of the development of ICT, the spread of knowledge and skills becomes much faster and easier. The provision of service has low cost. One example is the search engine, which makes companies such as Google and Baidu very successful. Another example is the e-commerce systems for which Amazon is the most famous. These giant corporations were hardly known about 10 years ago, but now they play an important role in our daily life.

When talking about "service" and ICT, we have to introduce the concept of "cloud computing" (Armbrust et al., 2009; Buyya, Yeo, & Venugopal, 2008; Li, Chen, & Wang, 2011; Wang & Shen, 2011). Cloud computing is the delivery of computing as a service rather than a product, where resources, software, and information are provided to end users as a utility over a network. For cloud computing, the end users do not need to know the physical location and the configuration of the system or, rather, they do not need to have knowledge about the system. Usually, the end users only need a web browser to enjoy the service provided by the compute clouds. This is similar to the electricity grid in that end users do not need to understand the infrastructures or the devices that produce and transfer the electricity.

The three models of services in cloud computing are as follows: Infrastructure as a Service (IaaS), Platform as a Service (PaaS), and Software as a Service (SaaS). IaaS delivers the virtual IT resources such as storage and computing and networking capability to the end users. PaaS delivers a computing platform together with necessary software subsystems or components to the end users. PaaS may offer facilities for application design, development, testing, deployment and hosting as well as application services such as database integration, team collaboration, and application versioning. SaaS, as the name indicates, delivers software as a service, which makes end users free of the need to install and run applications on their own computers and simplifies maintenance and support.

There are many advantages of cloud computing. Cloud computing can provide on-demand service to end users. The users do not need to buy hardware or software. They only need to apply for resources according to their needs. The users can access the system using a web browser regardless of their locations or what device they are using. The maintenance work can be left to the provider. And the provider can distribute the servers at different places, and the reliability can be guaranteed. There are many successful stories of cloud computing including Amazon's Elastic Compute Cloud (EC2), Apple's iCloud, and Microsoft's Windows Azure.

### 1.1.2 ACP Theory

In the era of providing services by compute clouds, new methodologies are desired to solve complex problems in the real world. Besides the basic services of weather report, news report, finding a telephone number, and a suitable restaurant, we have more complicated requests such as finding a suitable route while driving a car and consulting for special knowledge on managing a factory. It is believed that ACP (Wang, 2004a, 2004b, 2004c, 2010a) is a suitable approach for providing more intelligent solutions for service requests. What ACP stands for is Artificial systems, Computational experiments, and Parallel execution. As the name indicates, the three steps of ACP are as follows: (1) modeling and representation using artificial systems,

(2) analysis and evaluation by computational experiments, and (3) control and management through parallel execution of the real system and artificial systems (Wang, 2010a). In the ACP approach, key characters of the real system are considered in modeling, and the artificial system can create an electronic copy of the real world; various methods may be used for the evaluation and optimization of the artificial system; for parallel execution, the real system and artificial systems interact with each other; the results of the computational experiments are applied to the real system, and the real system is used to calibrate and verify the artificial system.

The ACP approach has been applied to the transportation system (Z.J. Li et al., 2011; Shen, Wang, & Zhu, 2011; Wang, 2004c, 2010a; Wang & Shen, 2011; Xiong et al., 2010). In particular, it has been used for the transportation management and guidance service for the 2010 Asian Games.

Cloud computing and the ACP approach share the same idea of considering the system as a whole. With the development of Internet of Things (IoT), which can connect identifiable objects (things) together by their virtual representations in the computer, collecting real world information becomes easier. With the information, an artificial system can be built. With the powerful computing infrastructure in the compute clouds, the computing experiments can be done and the experiment's results can be delivered as services to the end users to guide the real systems.

## 1.2 Why Do We Need SSME?

The service sector usually becomes more and more important as a country develops. The research of SSME may make our world better and have a great impact on our life. We use the example of e-banks as an explanation. As we know, the bank is closely related to our daily life. People can deposit spare money in the bank for safe keeping and interest and get loan from the bank for expanding production or other needs. This is the main utility that the bank offers to us. However, with the development of ICT, e-bank service now plays an important role. E-banks, also known as web banks or online banks, can provide the end users many kinds of services by Internet, such as querying, reconciliation, transferring, credit, Internet securities, investment, and financing. In other words, e-bank can be regarded as a virtual bank counter but with more convenience. Therefore, e-banks are also called "3A banks," because they can provide financial service Anytime, Anywhere in Anyway. To the customers, they do not have to go to a bank anymore, as now they can deal with almost all kinds of business just by an Internet-available computer. Although the e-bank provides us services with more convenience, there are a lot of new problems to work on. For example, how to guarantee that the account is safe? How to organize the accounts data, so that the many kinds of queries can be easy? How to make the response time as short as possible, as when the customer invests in stocks or futures, the prices can be changing all the time? All these problems make people put more effort into SSME. SSME is an interdiscipline of computer science, operations research, management science, systems engineering, industrial engineering, applied mathematics etc.

It is also closely related to transportation, enterprise management, logistics, health care, finance system, and e-commerce system.

Here we want to emphasize the importance of SSME at the times where ICT has been well developed. As the e-bank example shows, financial services can be greatly improved when the latest ICT systems are applied. Nowadays, we can buy and sell things, track a package, attend a meeting via the Internet, and use GPS for route guidance. Without ICT, these things could be much more tedious.

The development of ICT makes the concept of a Cyber-Physical System (CPS) (Lee, 2008; Sha, Gopalakrishnan, Liu, & Wang, 2009; Wang, 2010b) popular. CPS refers to a system featuring a tight combination of and coordination between computational and physical resources (Wang, 2010b). A CPS is usually a network of interacting elements with physical input and output. Examples of CPS range from the smart phone system to the smart grid system. The term "smart phone" is usually used to describe computer-like phones with advanced computing ability and connectivity usually equipped with a camera, a GPS, and a web browser. A smart grid is a type of electrical grid, which attempts to predict and intelligently respond to the behavior and action of all electric power users to efficiently deliver reliable, economic, and sustainable electricity services.

People connect themselves to the physical world via CPS, and the service is delivered to people by the CPS. ICT makes CPS an indispensable part of human life and then makes service much faster and easier to provide and obtain. SSME is desirable to make better our life in this new situation.

With the human and social dimension added, a CPS becomes a Cyber-Physical-Social System (CPSS) (Wang, 2010b). The difference between CPS and CPSS lies in whether the human is considered to be outside or within the system. For the control and management of a smart grid, humans can be considered as outside the system. However, for the transportation system and the enterprise system, it is better if people are considered as a part of the system. For the ACP approach, the human and social factors can be considered as within the system, and this is why we use the term "artificial societies." The modeling process needs human intelligence and experiences, and at the step of computational experiments, we are in fact trading the computation for "intelligence." There is a trend that the world is shifting from a physical world to a CPSS and ACP approach, which combines human intelligence and computation intelligence, making it possible to provide smarter services.

## 1.3 How Do We Benefit from SSME?

In this section, we give some examples of SSME. Specifically, we review some related papers presented at the 2011 IEEE International Conference on Service Operations and Logistics, and Informatics, which is a recent conference closely related to SSME.

### 1.3.1 Transportation System

The transportations system is a typical CPSS. For the transportation system, the car position information and the demand of the driver can be reported to the compute clouds. On the basis of the ACP approach, the clouds can optimize the demands of all the drivers and then provide services of detailed daily plans to the drivers with minimal delays while balancing the traffic load of the road network (Wang & Shen, 2011). Moreover, the optimization of the traffic lights can be performed by considering all the traffic lights as a whole system. Also, a Graphics Processing Unit (GPU) can be used to accelerate the optimization algorithms at the compute clouds (Shen et al., 2011). GPU usually consists of many cores working together and is a popular device for parallel computing. It can be facilitated at the compute clouds to accelerate the algorithms. The end users do not need to know how the GPU works, although the GPU is used to provide better services to the users. This shows the advantage of the cloud computing architect. Also, by collecting all available information, traffic information such as bus arrival time ( Z.J. Li et al., 2011) can be better predicted. The predicted information is delivered as services to the interested users.

### 1.3.2 Logistics System

Logistics is the management of goods flow between the point of origin and the point of use to meet the requirements of customers and other constraints. Logistics involves the integration of information, transportation, inventory, warehousing, material handling, packaging, and often security. There are many research topics in logistics research. Wang et al. (2011) quantify the value of collaboration in supply chains by computational experiments on the business process. The authors study a typical case of a four-tier network, including suppliers, plants, warehouses, and end customers. And then two situations of information sharing and resource sharing are analyzed. The results show that with collaboration, the supply chain performance can be improved significantly. With this result, a collaboration mode may be adopted to improve the performance of the system.

Guettinger, Godehardt, and Zinnen (2011) compare several online approaches for optimizing the emergency supply after a serious incident. And by applying a workload adapted version of the Simulated Annealing (SA) algorithm subsequent to each greedy iteration, the authors show that the total damage yielded by the greedy strategy can be reduced by about 10% with an acceptable computation budget. This shows the power of computation when providing services to users.

### 1.3.3 Health Care System

Health care is the diagnosis, treatment, and prevention of disease, illness, injury, and other physical and mental impairments in humans. Nyberg, Xiong, and Luostarinen (2011a) propose the idea of a "connected health service" for which ICT is used to connect the patient to expert advice and information knowledge databases, to connect patients to each other in self-help groups, to connect the patient to monitoring devices for self-diagnosis, and to connect the patient physiological measurement data to the clinician. The devices used to connect the patient could be a smart phone such as an iPhone or even a dedicated Personal Health Assistant (PHA) device. This is a typical CPSS for which the patients, the experts, and the cyberspace of knowledge and data are connected. The information can be transmitted easily via this network and then there are many benefits for the patients. The expert knowledge can be easily delivered from a website as a service to patients, and the communication between patients is easier, which enables them to help each other, and short messages can be used as a reminder of appointments, and it is easy to obtain sexual health advice via the short messages (Nyberg, Xiong, & Luostarinen, 2011b).

### 1.3.4 E-Commerce System

The e-commerce refers to the buying and selling of products or services over electronic systems such as the Internet and other computer networks. E-commerce is closely related to cloud computing, because there is a strong need for data storage and computing for better selling strategies. Bai et al. (2011) focus on how to use implicit and hybrid information to produce efficient recommendations for e-commerce systems. They make improvements in the matrix factorization (MF) method to explore hidden information more efficiently and then extend the improved MF to integrate user or item information to obtain a new algorithm for the recommendation. This process involves a combination of human intelligence and computation intelligence to provide good services to end users.

### 1.3.5 Financial System

The financial system is the system that allows the transfer of money between depositors and lenders. The financial service is one of the most elementary services in our life. Nowadays, electronic financial systems are becoming popular. But there are still many things to do. He, Ren, Wang, and Dong (2011) study the Supply Chain Finance (SCF) for which the bank evaluates the credit risks from the perspective of the supply chain instead of a single buyer. They formulate a three-stage supply chain problem into a stochastic dynamic programming problem. The objective is to maximize the expected income for the bank under the supplier's credit risk, the buyer's credit risk, and the inventory-in-transit risk. This is a study on a new topic of the service system. Agarwal, Desai, Kapoor, Kumaraguru, and Mittal (2011) propose a system that uses voice as a medium to percolate knowledge through the thick layers of illiteracy and to overcome the barrier of reach. The service system is applied to a rural area in India where many farm households are illiterate and do not have access to any credit. The Self-Help Group (SHG)—Bank Linkage Program was developed to link the unbanked rural population to the formal financial system. IBM's Spoken Web Technology is used to communicate between SHG members. This is a successful application of ICT to deliver services.

## 1.4 Summary

We summarize our ideas and opinions as follows. Service Science, Management, and Engineering (SSME) is an important field in the service sector, which plays a more and more important role as a country develops. With the development of information and communication technology (ICT), especially cloud computing, the service usually can be delivered in a faster and easier way. A service system is a cyber-physical system (CPS) if the service is delivered via the Internet. The CPS provides a viewpoint of considering cyberspace and the real world as a whole. If social factors are considered, a CPS is a cyber-physical-social system (CPSS) for which human impacts are considered as part of the system. The artificial systems-computational experiments-parallel execution (ACP) provides an approach to analyze and optimize CPS and CPSS by combining human intelligence and computation intelligence.

In this chapter, we have provided some examples related to the topics mentioned earlier. In the future, we will see a whole integration of the topics. With further development of ICT and methodologies such as ACP, we see a promising future for SSME.

## References


Agarwal, V., Desai, V., Kapoor, S., Kumaraguru, P., & Mittal, S. (2011, July). *Enhancing the rural self help group–Bank linkage program.* Paper presented at the IEEE International Conference on Service Operations and Logistics, and Informatics, Beijing.

Armbrust, M., Fox, A., Griffith, R., Joseph, A. D., Katz, R. H., Konwinski, A., … Stoica, I. (2009). *Above the clouds: A Berkeley view of cloud computing* (Technical Report No. UCB/EECS-2009-28). Berkeley: EECS Department, University of California.

Bai, X., Wu, J., Wang, H., Zhang, J., Yin, W., & Dong, J. (2011, July). *Recommendation algorithms for implicit information.* Paper presented at the IEEE International Conference on Service Operations and Logistics, and Informatics, Beijing.

Buyya, R., Yeo, C. S., & Venugopal, S. (2008). Market-oriented cloud computing: Vision, hype, and reality for delivering IT services as computing utilities. *Proceedings of the 10th IEEE International Conference on High Performance Computing and Communications, Dalian*, China, Sept. 25–27, 5–13.

Guettinger, D., Godehardt, E., & Zinnen, A. (2011, July). *Online strategies for optimizing medical supply in disaster scenarios.* Paper presented at the IEEE International Conference on Service Operations and Logistics, and Informatics, Beijing.

He, M., Ren, C., Wang, Q., & Dong, J. (2011, July). *An innovative stochastic dynamic model to three-stage supply chain finance.* Paper presented at the IEEE International Conference on Service Operations and Logistics, and Informatics, Beijing.

Hefley, B. & Murphy, W. (2008). *Service science, management and engineering for the 21st century.* New York: Springer.

International Business Machines Corporation (IBM, 2008). *IDEAS from IBM.* Retrieved from http://www.ibm.com/ibm/ideasfromibm/us/compsci/20080728/resources/IFI_07282008.pdf

Lee, E. A. (2008). Cyber physical systems: Design challenges. *Proceedings of the 11th IEEE International Symposium on Object Oriented Real-Time Distributed Computing (ISORC), Orlando, FL,* May 5–7, 363–369.

Li, F., Yu, Y., Lin, H. B., & Min, W. L. (2011, July). *Public bus arrival time prediction based on traffic information management system*. Paper presented at the IEEE International Conference on Service Operations and Logistics, and Informatics, Beijing.

Li, Z. J., Chen, C., & Wang, K. (2011). Cloud computing for agent-based Urban transportation systems. *IEEE Intelligent Systems, 26*, 73–79.

Maglio, P. P., Srinivasan, S., Kreulen, J. T., & Spohrer, J. (2006). Service systems, service scientists SSME, and innovation. *Communications of the ACM, 49*, 81–85.

Nyberg, T., Xiong, G., & Luostarinen, J. (2011a, July). *Connected health services: Internet, mobile and wireless technologies in healthcare*. Paper presented at the IEEE International Conference on Service Operations and Logistics, and Informatics, Beijing.

Nyberg, T., Xiong, G., & Luostarinen, J. (2011b, July). *Short messages in health promotion: Widespread and cost effective two-way communication channel for health*. Paper presented at the IEEE International Conference on Service Operations and Logistics, and Informatics, Beijing.

Sampson, S. E. & Froehle, C. M. (2006). Foundations and implications of a proposed unified services theory. *Production and Operations Management, 15*, 329–343.

Sha, L., Gopalakrishnan, S., Liu, X., & Wang, Q. (2009). Cyber-physical systems: A new frontier. In J. J. P. Tsai & P. S. Yu (Eds.), *Machine Learning in Cyber Trust* (pp. 3–13). New York: Springer-Verlag.

Shen, Z., Wang, K., & Zhu, F. (2011, October). *Agent-based traffic simulation and traffic signal timing optimization with GPU*. Paper presented at the 14th International IEEE Conference on Intelligent Transportation Systems (ITSC 11), Washington, DC.

Spohrer, J. & Maglio, P. P. (2008). The emergence of service science: toward systematic service innovations to accelerate co-creation of value. *Production and Operations Management, 17*, 238–246.

Spohrer, J., Maglio, P. P., Bailey, J., & Gruhl, D. (2007). Steps towards a science of service systems. *Computer, 40*, 71–77.

US National Academy of Engineering (2003). *The impact of academic research on industrial performance*, US National Academy Press.

Vargo, S. L. & Lusch, R. F. (2004). Evolving to a new dominant logic for marketing. *Journal of Marketing, 68*, 1–17.

Vargo, S. L. & Lusch, R. F. (Eds.). (2006). *Service-dominant logic: What it is, What it is not, What it might be*, Armonk: M. E. Sharpe Inc.

Vargo, S. L. & Lusch, R. F. (2008). Service-dominant logic continuing the evolution. *Journal of the Academy of Marketing Science, 36*, 1–10.

Vargo, S. L., Maglio, P. P., & Akaka, M. A. (2008). On value and value co-creation: A service systems and service logic perspective. *European Management Journal, 26*, 145–152.

Wang, F. Y. (2004a). Computational theory and methods for complex systems. *China Basic Science, 6*, 3–10.

Wang, F. Y. (2004b). Artificial societies, computational experiments, and parallel systems: A discussion on computational theory of complex social-economic systems. *Complex Systems and Complexity Science, 1*, 25–35.

Wang, F. Y. (2004c). Parallel system methods for management and control of complex systems. *Control and Decision, 19*, 485–489.

Wang, F. Y. (2010a). Parallel control and management for intelligent transportation systems: Concepts, architectures, and applications. *IEEE Transactions on Intelligent Transportation Systems, 11*, 630–638.

Wang, F. Y. (2010b). The emergence of intelligent enterprises: From CPS to CPSS. *IEEE Intelligent Systems, 25*(4), 85–88.

Wang, K. & Shen, Z. (2011). Artificial societies and GPU-based cloud computing for intelligent transportation management. *IEEE Intelligent Systems, 26*, 22–28.

Wang, W., Chai, Y., Dong, J., Ding, H., Ren, C., & Qiu, M. (2011, July). *Evaluating the value of collaboration in supply chain through business process simulation*. Paper presented at the IEEE International Conference on Service Operations and Logistics, and Informatics, Beijing.

Xiong, G., Wang, K., Zhu, F., Chen, C., An, X., & Xie, Z. (2010). Parallel Traffic Management for the 2010 Asian Games. *IEEE Intelligent Systems, 25*, 81–85.

CHAPTER 2

# Servitization Strategy: Priorities, Capabilities, and Organizational Features

**Yong Lin*, Yongjiang Shi†, and Shihua Ma‡**

**School of Business, University of Greenwich, London, SE10 9LS, UK*

*†Institution for Manufacturing, University of Cambridge, Cambridge, CB3 0FS, UK*

*‡School of Management, Huazhong University of Science and Technology, Wuhan, 430074, China*

## 2.1 Introduction

Servitization has been widely recognized by manufacturers as a strategy to get more revenues and competitive advantages from service units (Baines, Lightfoot, Benedettini, & Kay, 2009a; Neely, 2008; Schmenner, 2009; Vandermerwe & Rada, 1988) instead of products. It is expected to have many financial and strategic benefits for manufacturers. Although the considerable potential of servitization strategy for manufacturing firms is not very apparent to most companies (Glueck, Koudal, & Vaessen, 2006), it has been reported that revenues contributed by service units represent an average of more than 25% of total revenues in a wide range of manufacturing industries (Koudal, 2006).

Almost 40 years ago, Levitt (1972) made a provocative assumption that everybody is in the business service. Bitner (1997) even asserted that all businesses are kinds of service businesses. Currently, a growing number of manufacturing companies from developed economies are actively pursuing the transition from providing physical products to offering services or combining services with products. The successful cases include Xerox (White et al., 1999), IBM (Mathieu, 2001), Rolls Royce (Johnstone, Dainty, & Wilkinson, 2009), Cannon and Parkersell (Martinez, Bastl, Kingston, & Evans, 2010), Nokia, ABB, and General Electric (Godlevskaja, van Iwaarden, & van der Wiele, 2011). They have all been successfully transformed from physical product production-focused companies to service-offering companies providing services or product–service combinations to customers.

Confronted with the fierce competition, manufacturers from both developed and developing economies focus on the servitization strategy as an efficient and effective way for their

Service Science, Management, and Engineering.


DOI: 10.1016/B978-0-12-397037-4.00002-8

sustainable development. However, the transition from a traditional manufacturer to a servitized manufacturer or a service provider correspondingly leads to priorities changing and more complex processes (Brax, 2005), and investments in service offerings may not always result in the expected benefits. Furthermore, this transition needs both cultural shift and organizational changes (Mathieu, 2001).

In particular, in the PC manufacturing industry in China, after several years, high-speed development with low profit rate becomes one of the bottlenecks of further sustainable development for both the industries as a whole and the PC manufacturers. A new source of profit and development is needed. After IBM successfully transformed to be a global IT service provider (Mathieu, 2001), servitization strategy attracts more and more attention from PC manufacturers in China. Some of the companies have implemented or started to apply servitization strategy in their operations. To overcome these practical challenges during servitization implementation and, actually, to derive benefits from the servitization strategy, new strategic priorities, capability requirements, and organizational features are required for a conventional manufacturer.

## 2.2 Background

With a fast speed of growth, but low profit rate, the PC industry is confronted with a difficult situation for sustainable development. Servitization strategy provides a potential option for conventional PC manufacturers pursuing further development.

### 2.2.1 Context of the PC Industry

The PC industry has become a fast-developing global service economy (Haynes & DuVall, 1992) for many years. Especially when IBM shifted to offering services, the emerging servitization strategy has attracted much attention in the global PC industry.

As the world leading IT company, IBM started its worldwide service strategy in 1991 with the goal of being "a world-class service company" by 1994 (IBM). A new unit, IBM Global Services (IGS), which develops and delivers services to customers worldwide, was formed in 1995. IBM subsequently became the largest IT service provider in the world. In 2001, service revenue surpassed hardware revenue for the first time. According to IBM's annual report, service contributed 56.9% to the total revenue in 2010 and 39.2% to the pre-tax income (IBM Annual Report, 2011). To some extent, IBM's service strategy has reached a cultural level (Mathieu, 2001), which means that IBM has successfully transferred from a product manufacturer to a service provider.

The Chinese PC market is expected to overtake that of the United States as the largest computer market in the short term, driven by factors such as rise in Internet usage, a huge consumer base, rising income, accelerating urbanization, and changing lifestyles (Netscribes, 2012). Despite being affected by the global economic recession, after a slowdown during the end of 2008 and early 2009, the Chinese PC industry returned to fast growth from May 2009 and continued its strong double-digit growth rate, 19.2% and 37.0% in May and June respectively (see Fig. 2.1). It is also interesting that PC shipments in the Chinese market exceeded those in the United States in the second quarter of 2011 (IDC, 2011).

Within such a fast-growing PC industry, there are over 560 companies in China producing desktops, laptops, notebooks, and computer accessories. They produced 320 million units in 2011 with a growth rate of 30.3% (Ministry of Industry and Information Technology of China, 2012).

Most of these Chinese PC manufacturers are still focused on offering physical products to the PC market; even for some leading companies, they are still strongly focused on the PC products and not on the services. Unfortunately, the PC industry has moved into an era of low profit. The average profit rate declined to 2–3% during 2007–2008 (Table 2.1), from a value of 10.1% in 2000, and some manufacturers even claimed their profit level is less than 1%. That is the reason why the competition in the PC industry has become much fiercer than 10 years ago in China. How to further develop the service economy environment has emerged as a strategic issue for Chinese PC companies.

During the last decade, servitization has emerged as one of the strategic options to gain competitive advantage for sustainable development. However, transferring from product

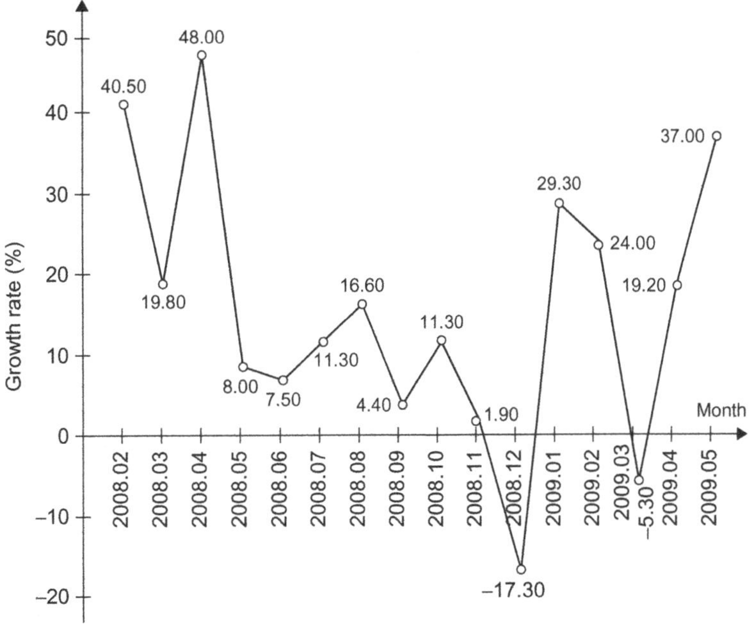


**Figure 2.1: Growth rate of Chinese PC industry.**

*Source: National Bureau of Statistics of China.*

**Table 2.1: Sales profit rate (%) for 2007–2008.**

| | Profit Rate (%) | |
|---|---|---|
| **Month** | **2008** | **2007** |
| January–February | 2.90 | 1.67 |
| January–May | 2.76 | 2.20 |
| January–August | 2.52 | 2.67 |
| January–November | 2.75 | 2.73 |

*Source: Ministry of Industry and Information Technology of China.*

manufacturer to service provider generates many challenges (Brax, 2005) for traditional manufacturers. That is the reason for this research to aim at better understanding in how to establish and implement a servitization strategy efficiently and effectively.

### *2.2.2 Definitions of Servitization*

The first use of servitization (sometimes referred to as servitisation or servicisation) can be traced back to the work of Vandermerwe and Rada (1988), which has been repeatedly referenced in the literature.

"Servitization is happening in almost all industries on a global scale. Swept up by the forces of deregulation, technology, globalization and fierce competitive pressure, both service companies and manufacturers are moving more dramatically into services" (Vandermerwe & Rada, 1988). This definition outlines the original premises of servitization strategy.

Currently, there are many different definitions in the literature with different perspectives.

Many of the definitions are grounded in the traditional theories of manufacturing and operations management. For example, Robinson, Clarke-Hill, and Clarkson (2002) viewed servitization (or servitisation) as a concept "which goes beyond the traditional approach of providing additional services but considers the total offer to the customer as an integrated bundle consisting of both the goods and the services" (Robinson et al., 2002). From a view of operations management, "companies … are becoming aware of the value of the servitization of their products. That is, marketing the capability that their products bring" (Slack et al., 2004). Following that, Slack (2005) thereafter proposed that "Servitization is the generic (if somewhat unattractive) term that has come to mean any strategy that seeks to change the way in which product functionality is delivered to its markets."

Obviously, transition from products to service becomes an emergent strategy to create more value and gain more competitive advantages. Hence, Åhlström and Nordin (2006) referred to servitization as strategic attempts in a manufacturing company "to establish service supply relationships to deliver product services to augment their physical products," and thus "differentiate themselves from the competition by offering a higher level of services than their competitors." Moreover, Lindberg and Nordin (2008) believed that servitization is a strategy trend where "firms move from manufacturing goods to providing services or integrating products and services into solutions or functions." Furthermore, Neely (2008) introduced servitization as the movement in which manufacturing firms "move beyond manufacturing and offer services and solutions, often delivered through their products, or at least in association with them." These definitions all have emphasized the strategic importance of servitization for a conventional manufacturer.

Within the different definitions of servitization, the focus is on the transition from product to service. But that is not to say transferring from a product manufacturer to a pure service provider. Most of the definitions are still focusing on bundling or combining services with products. For example, Brax (2005) defined servitization as a process that leads to "companies adding more and more value to their core offering through services experiencing a shift in their core business." Pawar, Beltagui, and Riedel (2009) not only viewed servitization as a phenomenon where "a transition has been recognized from an emphasis on the manufacture of products to the provision of service" but also characterized it as "the trend toward bundles of customer-focused combinations, dominated by service" (Pawar et al., 2009).

To achieve the benefits of servitization, innovation of service offering is essential to the servitization strategy. That is the reason why there are many literature definitions of servitization from an innovation perspective. For instance, Baines et al. (2009a) defined servitization as "the innovation of an organization's capabilities and processes to shift from selling products to selling integrated products and services that deliver value in use." This definition also shares basic principles of product-service systems (PSSs) (Baines et al., 2007). Based on this definition, Baines et al. (2009b) also proposed the concept of product-centric servitization as "the phenomena where a portfolio of services is directly coupled to a product offering" as well as the offering of "goods combined with closely related services (e.g., products offered with maintenance, support, finance, etc.)." Schmenner (2009) treated servitization as a term "coined to capture the innovative services that have been bundled (integrated) with goods by firms that had previously been known strictly as manufacturers" (Schmenner, 2009). These definitions have paid much attention to innovations of service offering.

Some of the definitions started to view servitization from the perspective of the whole supply chain, not the single manufacturer or service provider. Johnson and Mena (2008) identified that servitization "involves a customer proposition that includes a product and a range of associated services." In particular, from the view of the automotive supply chains, Lewis and Howard (2009) acknowledged servitization to be a strategy with "a greater emphasis on a whole range of novel product-service combinations." These definitions extended the scope from internal to external covering the whole supply chain.

In summary, servitization is defined at both strategic (innovation, strategy, trend, and phenomenon) and operational (process) levels as an innovative approach to create more value and to gain more competitive advantages. One essential point among those definitions lies in the understanding of combining and bundling service with product. Obviously, more and more companies will provide both products and services to customers. However, different companies will put different levels of focus on products and services. That is to say a servitized manufacturer will focus on either product or service. With the different focuses, their strategies will be different from each other. That is the reason why we are undertaking this research to identify their strategy differences when they have different focuses.

### 2.2.3 Benefits and Challenges of Servitization

Along with the fast development of the knowledge body of servitization, there are both benefits and challenges in using a servitization strategy.

#### 2.2.3.1 Benefits of Servitization

Transforming from physical product manufacturer to service provider leads to many financial benefits, including facilitating sales of product, balancing the effects of economic cycle on different cash flows, and responding to demands (Brax, 2005). Moreover, this could lead to many strategic advantages (Matthyssens, Vandenbempt, & Berghman 2006) such as increasing customer loyalty, creating opportunities for growth, in particular in mature markets (Brax, 2005).

Servitization is regarded as a business model innovation (Spring & Araujo, 2009), which worked as an innovative approach to facilitate growth (Canton, 1984; Sawhney et al., 2004), higher profitability (Cohen, N. Agrawal, & V. Agrawal, 2006; Neely, 2008; Oliva & Kallenberg, 2003), and enhance economic stability (Lele, 1986; Quin & Gagnon, 1986).

#### 2.2.3.2 Challenges of Servitization

However, when pursuing the benefits and advantages of servitization, several areas of operations are substantially impacted, which includes facilities, information and communication technologies, performance measurement systems, organizational processes, and human resources (Baines, Lightfoot, & Smart, 2011). As a result, there are also many challenges confronted by manufacturers who are trying to get a foothold in the service economy (Martinez et al., 2010).

The first challenge is to change the *strategy priority* from product focused to service focused. Because services are substantially different from physical products with the nature of perishable, complex, and multifunction (van Biema & Greenwald, 1997), servitization requires comprehensive changes in the whole company. Designing and managing a service business model is more challenging than for a manufacturing business model. Strategy priorities have to be transformed to be service focused not product focused.

Moreover, because the service delivery processes are much more complex than the product manufacturing processes, new *capabilities* are needed to be a service provider rather than a traditional product manufacturer (Oliva & Kallenberg, 2003). One reason for the high complexity of the service offering process is that customers are involved in the value-creation process. Hence, the company should focus on the process of creating value for customers (Grönroos, 2000) during service delivery, particularly its interactions with customers, which also makes the service delivery processes much more complex. As a result, new capabilities of a service business are a critical challenge to both conventional manufacturers and service companies from the service sector, particularly when the product is only a small part of the service.

Another challenge is that the servitization strategy requires different *organizational structure* (Brax, 2005) to support the service offering. Servitization strategy needs change not only in design, operations, and skills but also in cultural shift (Gebauer, Fleisch, & Friedli, 2005; Malleret, 2006; Oliva & Kallenberg, 2003) and organizational changes (Mathieu, 2001), which are the most important challenges. From this perspective, necessary organizational changes are essential to support the establishment of any new capabilities for servitized manufacturing.

#### *2.2.3.3 Conceptual Research Framework*

Making the change from a product manufacturer to a service provider is not that easy (Brax & Jonsson, 2009; Gebauer et al., 2005; Mathieu, 2001; Neely, McFarlane, & Visnjic, 2011), and the system is much more complex to control and manage. Consequently, investments in servitization do not always achieve the expected results (Gebauer et al., 2005; Lay, Copani, Jäger, & Biege, 2010).

On the one hand, a servitized manufacturer could create higher revenues, and on the other it also leads to lower net profits than pure manufacturing companies (Neely, 2008). Obviously, necessary strategic and operational skills for implementing a servitization strategy are needed.

This chapter focuses on the strategy priorities, capability requirements, and organizational features of the servitization strategy during the transformation from a product manufacturer to a service provider. These three essential elements constitute the conceptual framework of servitization strategy in this research (see Fig. 2.2).

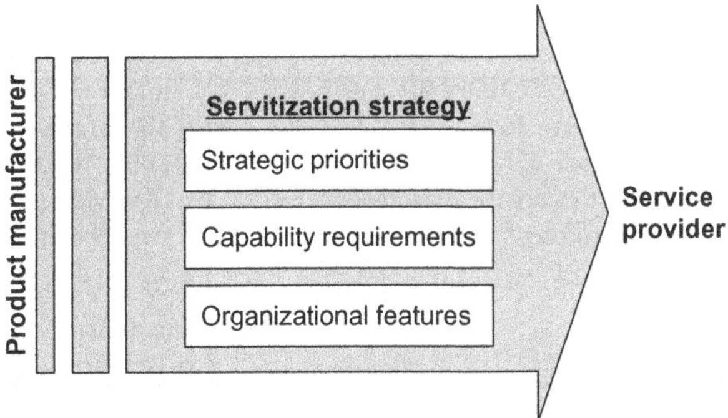


**Figure 2.2: Conceptual framework of servitization strategy.**

## 2.3 Research Methodology

To tackle the defined research question, case study is adapted in this research. Results are derived from the data analysis on the collected data in the two case companies.

### 2.3.1 Case Study

The research aims to investigate the development and the application of the servitization strategy in the PC industry. Specifically, two types of research questions are studied in this research, "why" (why a certain strategic option (priority) of the servitization strategy is chosen by the conventional manufacturer) and "how" (how a particular servitization strategy is supported by specific capabilities and organizational configurations). Based on the nature of the research questions and the research objectives, case study methodology is adopted in this research (Yin, 1994).

Two leading companies are selected from the Chinese PC industry, and the choice and administration are in accordance with replication logic (Eisenhardt, 1989). They reflect two different corporate strategies in their history. Data are collected through in-depth interviews, official company websites, and secondary documentations. The data of the industries are mainly collected from the National Bureau of Statistics of China and the Ministry of Industry and Information Technology of China, and the other data from the industry research companies and consulting companies.

Both these two cases are leading PC companies in China, and they are both spin-offs from the same group of companies (see summary in Table 2.2).

**Table 2.2: Sales profit rate (%) of 2007–2008.**

| | Case A | Case B |
|---|---|---|
| Strategic priorities | PC products as the core business; global company | Customer-focused & service-oriented (IT services, supply chain services) |
| Product/service revenue | • Product originates 98% of the total revenue of 2009/10<br>• Desktop: 35%<br>• Laptop: 63%<br>• Start to create a leveraged IT consulting services business | • Service contributes 22% of the total revenue of 2008/09<br>• 30% of the total revenue of 2009/10 Q1-Q3<br>• IT services: 21%<br>• Supply chain services: 9% |
| Organizational feature | Production and sales network | Service network (strategic partnership with over 100 leading IT vendors worldwide) |

*Source: Company A and B's annual reports.*

### 2.3.2 Case Company A

Company A, founded in 1984, has been the market leader in the Chinese PC industry since 1997 and is a world leading company that makes award-winning PCs for global customers.

With greater clarity and focus on product, the strategic priorities of company A are aimed at stabilizing its PC products (including laptop and desktop) as the core business and taking it to a higher level. Company A allocates about 80% of its resources to make innovative consumer PCs, which constitutes about 98% of its annual revenues for 2009/10, including 35% for desktop and 63% for laptop. With continuing sales growth, its No.1 market position is strengthened both in China and Asia and the top 5 in the world. Its success is mainly because of product innovation and operations efficiency. To enhance its R&D capability to ensure product innovation, company A has built major research centers not only in Mainland China but also in Japan and the US.

Furthermore, company A intends to expand geographic and organizational boundaries on a global scale. After several mergers and acquisitions, company A has become a successful global company. After the globalization strategy, company A is now providing PC products in more than 160 countries, covering not only American and European markets but also some emerging markets (accounting for 16% of the total annual revenues for 2009/10). Consequently, company A focuses on optimizing production processes and a global logistics network to meet customer demands and to respond quickly to market changes.

Recently, the growth of a new non-PC-related area is becoming important to company A. In fact, company A has started to create a leveraged IT consulting services business from its core product business. However, it is still difficult for company A to achieve competitive advantage in service business, particularly consulting services. Hence, the key business area in company A is still in the field of providing physical products, not services.

### 2.3.3 Case Company B

Company B is a leading IT services provider in China, providing high quality and comprehensive integrated services, particularly to the financial, telecommunications, and government sectors. It focuses on providing customers with sophisticated and applicable IT products and comprehensive system integration services in the Chinese market.

In fact, after its spin-off from the group company 8 years ago, company B strived to become a premier IT services provider and a customer-focused and service-oriented organization that can provide comprehensive services to customers in the PC market in China. Currently, the business segments of company B include IT planning, business process outsourcing, application development, system integration, hardware infrastructure services, maintenance, hardware installation, distribution, and retail. Furthermore, the innovation in company B means not only service innovation but also management innovation and human resource innovation, which actually facilitate the service innovation.

The high-quality and comprehensive services contribute a lot to the annual revenue. According to the recent annual report, the service revenue has increased to 30% of the total revenue for 2009/10 Q1-Q3, including 21% from IT services and 9% from supply chain services. Furthermore, the services business reported rapid and sustainable growth compared comparing to the corresponding period in the previous financial year. Such growth underpinned strongly vast improvements of the services strategy as it started to make profitable contributions to the company.

As IT systems become more and more complex, varied services are needed to support the efficient and effective operations of the IT system. From the beginning, company B has concentrated on designing innovative services to meet customers' individual requirements, with wholehearted support from top management. For example, company B proposed a service package covering the whole IT life cycle, which helps to sustain customers through periods of economic recession. The service package includes services like maintenance, operations outsourcing, system testing, professional services, consulting services, and training services.

Company B has established regional centers in 19 major cities in China and a network of over 10,000 re-sellers and agents across the country. Meanwhile, company B has built strategic partnerships with over 100 leading IT vendors worldwide. With such a strong regional and global network, a full range of the best and most convenient IT services are available to industry clients, large enterprises, SMEs, and individual consumers. Moreover, several years ago, company B opened an Enterprise Innovation Center to help customers experience IT services and to develop and design innovative services in the center.

## 2.4 Servitization Strategy for PC Industry

According to literature review and case study, most recently service has emerged as one of the strategic options to gain competitive advantage in the PC industry. However, there are few PC makers in China that are implementing a servitization strategy. Most of these companies continue to be traditional PC manufacturers focusing on physical product offerings. However, there is an obvious trend that more PC manufacturers have already begun to, or are planning to, place more emphasis on service provision.

Based on analyzing and comparing the results of the case study, a servitization strategy in terms of *strategic priority*, *capability requirement*, and *organizational feature* is summarized in Fig. 2.3.

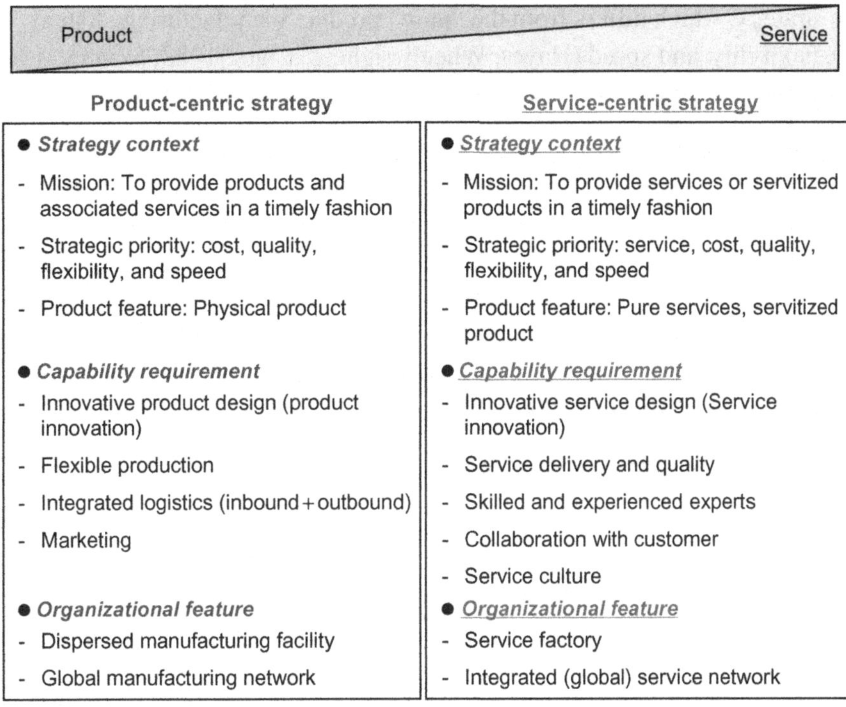


**Figure 2.3: Framework of servitization strategy.**

## 2.4.1 Strategic Priorities

First of all, servitization changes a company's strategy from a focus on products to an emphasis on service (Wilkinson, Dainty, & Neely, 2009); hence, two types of servitization strategy are defined here. One is product-centric servitization strategy with the mission of providing PC products and associated services (normally after-sale services) to the customers on time, and the other is service-centric strategy with the mission of providing IT-related services or servitized products in a timely fashion. The capability requirements and configuration features are different for these two types of servitization strategy. This classification is also based on the classical conception of a product service system (Baines et al., 2007), and it reflects the company's involvement in service.

### 2.4.1.1 Product-Centric Servitization Strategy

The results show that most Chinese PC manufacturers still focus on material transformations and the production of tangible products (Chase, 1991), but they have started to think about creating value through service offerings, which is different from the traditional PC manufacturer who only focuses on physical product offerings and the related after-sale services, particularly maintenance services. Hence, this research defined it as a product-centric servitization strategy, which differs from the "pure" product manufacturing strategy. Normally, cost, quality, flexibility, and speed (Hayes, Wheelwright, & Clark, 1988; Skinner, 1969, 1974) are regarded as the four priorities of the product-centric servitization strategy.

There are examples in other industries like General Electric and Rolls-Royce that followed a product-centric servitization strategy (Baines et al., 2009b). Their strategic priorities are focused on efficiently and cost-effectively delivering the integrated products and service to their customers. However, most of their values still come from the physical products. Company A is a classical example of a product-centric strategy. Due to the extremely fierce competition in the PC industry, company A is continuing to concentrate on its PC products, with only minor investments in services. As a result, cost reduction and high quality are their major strategic priorities for competitive advantages.

#### *2.4.1.2 Service-Centric Servitization Strategy*

For those companies with a service-centric strategy, their strategic imperatives have changed from product dominated to service oriented (Johnston, 1994; Martin & Horne, 1992; Oliva & Kallenberg, 2003; Quinn et al., 1990; Wise & Baumgartner, 1999). The concept of a service-centric servitization strategy is applied to represent the service as the fifth competitive priority in manufacturing strategy (Chase, 1991; Chase & Erikson, 1988; Chase & Garvin, 1989; Chase, Kumar, & Youngdahl, 1992; Garvin, 1993; Spring & Dalrymple, 2000; Voss, 1992; Youngdahl, 1996).

A company with a service-centric servitization strategy puts more emphasis on providing either pure services or servitized products. The pure service is normally a combination of information, problem solving, sales, and support activities to both internal and external customers (Chase, 1991), and these services are explicitly measured, monitored, and marketed (Chase, 1991). On the other hand, the servitized product refers to the bundle of products and services to the customers (Levitt, 1983). The main difference between a servitized product and the normal product is that most of the added values come from the bundled services in the servitized product, whereas for the normal product, the main value comes from the physical product part. That is the reason why companies with a service-centric servitization strategy concentrate on into service design and service delivery, whereas the companies with a product-centric servitization strategy are are primarily concerned with product design but have started to integrate service and production.

Only a few Chinese PC companies, such as company B, are fully involved in servitization strategy. Company B announced its strategic target is to be a leading IT service provider like IBM. That is to say, service has become an important competitive priority in its manufacturing strategy. In the Chinese PC industry, IT-related services are still quite new. Although consulting and training may add value to the core business, it still requires a paradigm shift to understand and accept the service-centric strategy. This is the reason why company A is still focusing on the PC products market and occupies only 2% market share in the field of IT services, but almost 30% market share in the PC products market in China. The products provided by company A are an integration of its desktop, laptop, and Netbook products with after-sale services.

However, for company B, even facing competition from leading companies such as IBM and Accenture in this market, becoming a professional IT service provider in the Chinese market is the key priority of its development strategy. As a result, company B focuses on service provision rather than hardware and PC products. The services provided by company B include supportive service to hardware and software, installation and configuration services, outsourcing services like desktop and data center outsourcing, professional services like business consistency service and system operation monitoring service, and testing service. Obviously, service is the key priority of the manufacturing strategy in company B.

### *2.4.2 Capability Requirements*

For different servitization strategies, different capabilities are required to support the success of the servitization strategies. The essential difference lies in the strategy focus, product innovation, or service innovation.

#### *2.4.2.1 Capabilities for Product-Centric Servitization Strategy*

For companies with a product-centric servitization strategy, most of the investment goes into innovative product design, particularly the capability development for R&D. Product innovation, integrating services with PCs, flexible production, marketing, and time-to-market and delivering value to the end users are critical factors to the success in the PC industry when adopting a product-centric servitization strategy.

##### 2.4.2.1.1 Product Innovation

The leading PC company A in China who adopted a product-centric servitization strategy is good at product innovation. For example, was the first company to introduce the Netbook into the Chinese PC market. Meanwhile, this company has also advanced in after-sale services. However, company A is not only committed to new product developments but also focused on developing integrated software solutions for its hardware products and support services along with the physical products. For the product offering, the efficiency, effectiveness, and flexibility of the production facility are fundamental requirements of this type of company.

##### 2.4.2.1.2 Logistics Integration

For product-centric companies like company A, efficiently and effectively operating the whole supply chain is the key to success. The logistics integration, particularly the integration between inbound logistics for the components supply and outbound logistics for the finished products delivery, plays an important role in ensuring product availability to satisfy customers. In the case of company A, there is a strong experienced team to manage the operations of the whole supply chain and its logistics.

##### 2.4.2.1.3 Marketing

Another main strength of company A is marketing, including its advantages on price, promotion, people, and the national retail channel, which is one of the important capabilities required to deliver the physical products to the market on time. Company A is good at segmenting customers into different categories and then develops complementary business models accordingly. These business models not only ensure that company A delivers the right PC product at the right price to the right customer segments, but also help the company to respond quickly to the fast-changing market and operate efficiently with an optimized cost structure.

#### *2.4.2.2 Capabilities for Service-Centric Servitization Strategy*

For a company with a service-centric servitization strategy, service innovation, service delivery and quality, skilled and experienced experts, and service culture are the key capability elements required to achieve a high-level performance.

##### 2.4.2.2.1 Service Innovation

The product-centric company focuses on innovative product design to gain competitive advantages, whereas in a service-centric company, service design emerges as a critical issue in its operations management (Johnston, 1999) and also is a vital part of the service experience (Zehrer, 2009).

Service design is the activity of planning and organizing all kinds of resources such as people, infrastructure, communication, and material components of a service to improve its quality and the interaction between service provider and customers, which were considered as part of the domains of marketing and management disciplines at an early stage (Shostack, 1982, 1984). Together with the more traditional methods used for product design, necessary methods and tools for service design are required to control and manage new elements of the design process (Morelli, 2006), such as the time and the interactions between service providers and customers.

First of all, for service design activities, service innovation is the fundamental requirement for a service-centric servitization organization (Gremyr, Löfberg, & Witell, 2010). Service innovation is first discussed as an approach to overcome problems associated with service characteristics like the difficulty in demonstrating the service to the client or the problems in storing and building up stocks of the service (Miles, 1993). It may also be referred to as innovation in service, service process, and service organization.

For a service-centric company, service innovation (Moller, Rajala, & Westerlund, 2008) is required to improve the customer experiences of receiving the servitized products or pure services. For example, the IT consulting services in company B demand state-of-the-art knowledge about a variety of software platforms, middleware, and applications that corporate customers demand in a one-stop-solution format. To become a serious contender in this area, company B needs to refine its software and applications services and its service delivery processes innovatively to provide customers with required knowledge for solving enterprise-wide challenges and issues.

For service innovation, the most important thing is that the designing service is a challenging job for the PC company as a service provider, because services always have many unique characteristics (Heizer & Render, 2008), which are totally different from the physical products. In 2008, company B redefined its service designs to enhance the service capability by setting up solution centers to provide product and application solutions that are specific to the enterprise-level customers, which finally resulted in 68.39% growth in turnover compared with 2007.

#### 2.4.2.2.2 Service Delivery and Experienced Experts

Furthermore, as a service provider, service delivery and its quality are the key to its success. Service quality is regarded as a measure of how well a service is delivered to match customer expectations (Lewis & Booms, 1983). The service delivery process should be well designed and well organized to ensure service quality (Furrer, 2005), which can help to increase customer value, profitability, and lower cost. First, continuous innovation is needed to support the design of the service delivery process. To some extent, a well-designed process is a vital component of service delivery innovation, and it will surpass customer expectation and increase customer satisfaction.

Second, specialist knowledge, experiences, and skills are mostly required for the service delivery process. That is the reason why skilled and experienced experts are fundamentally required for a successful service-centric servitization company. For example, IBM gathers this expertise during its merger and acquisition activities, to support its service business.

Finally, according to the service nature of customer involvement, effective and continuous collaboration with customers is one of the essential ways to create greater value for the customer and to keep them for the long term. One of the successful experiences of company B is that they always get their customers involved in the design process in different ways like Innovation Center. For one thing, company B can better understand customer requirements to improve service design. For another, customers can provide more suggestions after experiencing the services and the design processes.

2.4.2.2.3 Service Culture

Another important factor is the service culture within the company. The service culture implies different types of organizational culture and employee behavior, and different service culture leads to different levels of concern when serving customers. Development of service culture largely depends on the company's industry background, product/service characteristics, size, and business model. Obviously, it is not an easy job at all. Service culture can be built only by a sustained and consistent effort over an extended period, and it cannot be introduced by just top management or by any consulting company. After 10 years, focusing on IT services, company B has built its own service culture that is largely different from the product-oriented culture in company A.

When changing from product provider to service provider, the organization structure and the operations management will be changed dramatically. It is impossible to achieve a high quality of service delivery without a service-based organizational culture. Successful companies like IBM have achieved cultural level as a service firm (Mathieu, 2001). Successful examples in the Chinese PC industry like company B show that even the leadership plays a critical role in the application of a servitization strategy, and it also leads the whole organization to a high level of service offering.

### 2.4.3 Organizational Features

Organizational innovation is a critical part of service innovation, and many important service innovations involve organizational changes together with specific technologies. When applying servitization strategy, organizational principles and organizational structures (Oliva & Kallenberg, 2003; Voss, 1992) present more challenges to PC companies, especially if they have a globalization strategy now or in the future. Based on the case study results, global manufacturing network and global service network are proposed respectively as the organizational options for companies who adopt different product-centric or service-centric strategies.

#### 2.4.3.1 Global Manufacturing Network (GMN)

For product-centric companies, dispersed manufacturing facilities (plants) push them to extend the operations view from internal to external, which transfer the factory into an organizational form as an extended enterprise (Bititci, Mendibil, Martinez, & Albores, 2005; O'Neill & Sackett, 1994). With a globalization strategy applied in the company, an extended enterprise could be further extended to be a Global Manufacturing Network (GMN).

After leading for several years in the PC market in China and Asia, company A launched its globalization strategy and has started to establish its GMN to develop its global competitive capabilities (Shi, 2003; Shi & Gregory, 1998) and to enhance its competitive advantages (Bolisani & Scarso, 1996; Toni, Filippini, & Forza, 1992; Young, Kwong, Li, & Fok, 1992).

Along with production bases in China, company A has opened plants in Southeast Asia, North America, and Eastern Europe to form a global manufacturing network, which follows its globalization strategy to be "A New World Company." For the global operations of its core products, the customer research and product planning have been done in the United States, hardware design in Japan, and the products are assembled in its global manufacturing network, and finally, they are shipped to every region of the world.

In general, GMN is regarded as a network of manufacturing factories (Ferdows, 1989); thus location decisions (Canel & Khumawala, 1996; Meijboom & Vos, 1997; Vos, 1991) for various manufacturing system and factory designs become the vital strategic issues in GMN. The most important and difficult task of managing the GMN is to manage the dispersed plants network to respond quickly to increasingly competitive and volatile environments (Colotla, Shi, & Gregory, 2003), particularly within a global context. Currently for company A, the most difficult thing is to manage and control the global dispersed plants and facilities to follow the same product quality and management standards. One of the critical barriers is to overcome cultural conflicts to achieve the same high efficiency in different countries.

#### *2.4.3.2 Global Service Network (GSN)*

For the service-centric company, service facilities also need to be well organized. It will compete with other competitors under an organizational form of service factory (Chase, 1991; Chase & Erikson, 1988; Chase & Garvin, 1989) at an early stage, bundling services with products, forecasting, and responding to customer requirements. After successful mergers and acquisitions with several software and consulting companies, company B successfully transformed itself from a leading PC retailing company to an IT service provider in China. These service and software companies are regarded as the service factories of company B. As a result, the former operations' focus on the retail channels is shifted to organization and operations of these service factories.

Along with the development of a globalized economy, an emerging trend for service factories is the extension to a *Global Service Network* (GSN; Lin, Shi, & Zhou, 2010), which is a network consisting of service providers or service firms (Freeman & Sandwell, 2008; McLaughlin & Fitzsimmons, 1996; Kathuria, Joshi, & Dellande, 2008; Sharma & Loh, 2009) not only providing servitized products (Baines et al., 2009a; Spring & Araujo, 2009; Wilkinson et al., 2009) but also offering professional pure services to global customers.

A national service network for company B has been built up over a 10-year period, consisting consisted of a network of more than 10,000 re-sellers and agents, and several regional centers in 19 major cities in China. In the future a, global service network is planned that uses global resources and serves customers worldwide. For the first target, company B has built their global IT partner network with over 100 leading IT vendors. For the second target, it is still in the early stage of becoming fully involved in the global competitive market. To this end, a global service network for company B has not been finalized as yet, but work to do so is in progress.

### 2.4.3.3 Evolution from Factory to GMN or GSN

In terms of organization, a traditional manufacturing factory could evolve into either a GMN or a GSN through two pathways (see Fig. 2.4, adapted from Lin et al. (2010)): servitization and globalization. Furthermore, a GMN also could be extended to a GSN.

The traditional factory is focused on manufacturing visible physical product (Chase, 1991), and it can be expanded to a new organizational form of service factory when the company is trying to create more value through bundling services to products (Chase & Erikson, 1988; Chase & Garvin, 1989; Chase et al., 1992; Youngdahl, 1996), for example adopting a service-centric strategy. Within the globalization tide, a service factory could be further extended to a service-driven global supply chain (Youngdahl & Loomba, 2000) and then to a global service network (Lin et al., 2010) mainly driven by servitization and globalization strategies. On the other hand, a GMN could be developed into a GSN following the servitization strategy, particularly when the GMN enters a mature stage seeking opportunities for sustainable development through service offerings.

Contrary to the globalization strategy in company A, company B retains its focus on the Chinese PC market. One of the reasons for this is that the government sets a long-term plan to vigorously develop the Chinese IT industry, which boosts the market of IT services in China. Consequently, company B aims to deploy its service network to provide a full range of IT services to industry clients, large enterprises, SMEs, and individual consumers in China. However, from a long-term view, it is necessary to include the GSN development

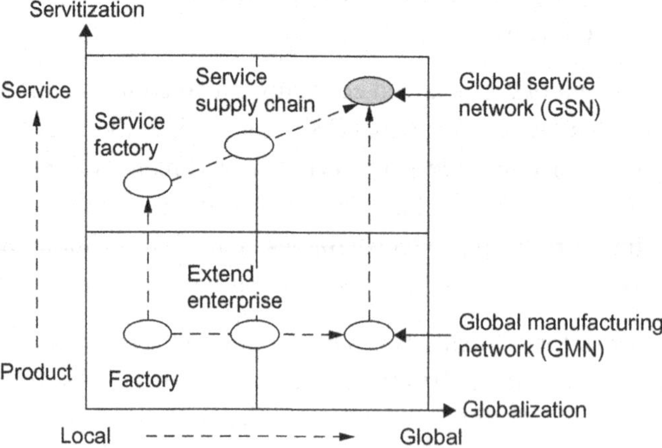


**Figure 2.4: Servitization-globalization matrix of the evolution of global service network (Lin et al., 2010).**

into its corporate strategy for future development. During the past 3 years, company B has established collaborative relationships with more than 100 professional service firms around the world. GSN is a necessity for the development of service-centric servitization companies such as company B.

## 2.5 Managerial and Practical Implications

The results show that core competence, leadership and top management support, and industry policy substantially affect the adoption and development of servitization strategy in the PC industry in China.

### 2.5.1 Strategy Priority Depends on Core Competence

The company strategy mainly depends on the core competence that the company has established. For a company involved in product innovation with advanced R&D capability, production facilities, and manufacturing network, a product-centric strategy will be the first option; whereas for a company concerned with service innovation that has advanced capability in service design and service delivery, a service-centric strategy will be a reasonably good choice. The establishment of the core competence decides the strategy priority in a company, which also means that not all the PC companies should follow the servitization trend, even though it is very successful in some leading companies such as IBM.

Furthermore, integrating services with physical products does not create advantages in all circumstances. That is why not all manufacturers are successful in the transition from product producer to service provider and not all manufacturers will be able to quickly integrate their manufacturing with a service offering, especially if they have developed significant manufacturing productivity (Schmenner, 2009). For example, DEC refused to provide services because they regarded computer design as their core competency (Oliva & Kallenberg, 2003).

In the Chinese PC industry, the most important thing is to identify and establish its core competence to gain competitive advantages. Servitization is increasingly regarded as a value-creation strategy in the PC industry, but not all Chinese PC companies need to follow it. If the company chooses a service-centric strategy, the first thing is to find a suitable starting point to develop its service-based competences and its supported capabilities and configurations.

### 2.5.2 Leadership and Top Management Support

One interesting thing is that leadership and top management support play essential roles in servitization strategy. To some extent, different types of leadership lead to different manufacturing strategies in a PC manufacturer. The leader of company B is more creative and innovative to pursue new targets and new strategies. After spin-off from the group company, the leader made his decision to lead company B to be a leading IT service provider in China, whereas the leader in company A is more firm and steady, and quietly works hard. After few failures in service, the leader finally decides to stay focused on physical products. This is one of the major reasons why there are two leading companies with two different strategies.

Another important thing is that top management support is critical to the success of the manufacturing strategy. No matter to which strategy, product-centric or service-centric, top management support can help to get necessary resources (facility, capital, IT, and human resource) and then to achieve the benefits of different strategies. In both the companies, all the top managers paid sufficient attention to the servitization strategy. Also, the service innovation and establishment of the service culture are substantially dependent on support from top management.

### 2.5.3 Industry and Government Development Policy

The PC industry is at the end of the IT industry chain. Consequently, its development largely depends on the development of other IT industries, including communication, telecom, and Internet. For example, the number of Internet users in China had reached 420 million by June 30, 2010 (China Internet Network Information Center, 2010). With the fast growth of the number of Internet users, there arises great opportunities to the PC industry. Recently, the Netbook used for the Internet has become one of the most popular products on the Chinese PC market, which is the result of the increased number of Internet users. As a result, the government or industry associations should put more focus on the inter-relationships between the IT-related industries, and a big picture should be established to focus on all the related IT industries and not just one.

Meanwhile, government policy is vital to the development of the Chinese PC industry. After several years of fast development, low profit becomes one of the important bottlenecks restricting the further development of the PC industry in China. Balancing the trade-off between low labor cost and high value-added service is the critical issue faced by the Chinese government. Furthermore, to increase profit and gain more competitive advantage, finding new sources of profit growth is the urgent problem. Fortunately, service as a value-added strategy attracts more attention from the Chinese government and the PC industry. For the sustainable development of the Chinese PC industry, well-designed policy environment is very important to the application of servitization strategy in the Chinese PC industry.

## 2.6 Conclusions

After several years of fast development at two-digit rate, low profit has become the bottleneck of sustainable development for the Chinese PC industry. Seeking new sources of profit, growth and value adding are critical issues for both the Chinese PC makers and the government. Servitization as a value-adding strategy attracts more attention from both practitioners and academics.

Based on the case study and industry analysis, this research defines two types of servitization strategies: product centric and service centric. The purpose is to propose these strategy options to manufacturers in the PC industry and to help them better understand the trend of servitized manufacturing. Strategy priorities, capability requirements, and organizational features for different servitization strategies are summarized to give potential directions to the PC makers in China for their sustainable development.

The results show that there is no common strategy suitable for every manufacturer as regards servitization. The strategy choice largely depends on the core competences that they have established in the past. Moreover, the success of the servitization strategy is also affected by the leadership style and top management support. Furthermore, the government and industry associations play a critical role in the success of implementing servitization strategy. Well-designed policy systems and well-organized industry environments will ensure a healthy and sustainable development of the PC industry in China.

Further study will focus on the capability identification, establishing organization structure, and designing operating principles. More in-depth case studies in the Chinese PC industry are needed to facilitate these findings. In particular, cross-case studies between China and other developed countries will be helpful to enhance the research results. The most important thing is that policy development needs to be further studied at both the government and industry level to help the Chinese PC industry create a healthy environment for its sustainable development.

## Acknowledgments

This research was supported by the project RAE-BUS-007/10 funded by the Research & Enterprise Investment Programme 20010/11 of the University of Greenwich, the Project 70672040 funded by the National Natural Science Foundation of China (NSFC).

## References

Åhlström, P. & Nordin, F. (2006). Problems of establishing service supply relationships: Evidence from a high-tech manufacturing company. *Journal of Purchasing and Supply Management, 12*(2), 75–89.

Baines, T., Lightfoot, H., Benedettini, O., & Kay, J. M. (2009a). The servitization of manufacturing: A review of literature and reflection on future challenges. *Journal of Manufacturing Technology Management, 20*(5), 547–567.

Baines, T., Lightfoot, H., Evans, S., Neely, A., Greenough, R., Peppard, J., … Michele, P. (2007). State-of-the-art in product-service systems. In *Proceedings of the Institution of Mechanical Engineers – Part B, Journal of Engineering Manufacture, 221*(10), 1543–1552.

Baines, T., Lightfoot, H., Peppard, J., Johnson, M., Tiwari, A., Shehab, E., & Swink, M. (2009b). Towards an operations strategy for product-centric servitization. *International Journal of Operations and Production Management, 29*(5), 494–519.

Baines, T., Lightfoot, H., & Smart, P. (2011). (Research note) Servitization within manufacturing: Exploring the provision of advanced services and their impact on vertical integration. *Journal of Manufacturing Technology Management*, *22*(7), 1–8.

Bititci, U. S., Mendibil, K., Martinez, V., & Albores, P. (2005). Measuring and managing performance in extended enterprises. *International Journal of Operations and Production Management*, *25*(4), 333–353.

Bitner, M. J. (1997). Services marketing: Perspectives on service excellence. *Journal of Retailing*, *73*(1), 3–6.

Bolisani, E., & Scarso, E. (1996). Operation International manufacturing strategies: Experiences from the clothing industry. *International Journal of Operations and Production Management*, *16*(11), 71–84.

Brax, S. A. (2005). A manufacturer becoming service provider—challenges and a paradox. *Managing Service Quality*, *15*(2), 142–155.

Brax, S. A. & Jonsson, K. (2009). Developing integrated solution offerings for remote diagnostics: A comparative case study of two manufacturers. *International Journal of Operations and Production Management*, *29*(5), 539–560.

Canel, C. & Khumawala, B. M. (1996). A mixed-integer programming approach for the international facilities location problem. *International Journal of Operations and Production Management*, *16*(4), 49–68.

Canton, I. D. (1984). Learning to love the service economy. *Harvard Business Review*, *62*(3), 89–97.

Chase, R. B. (1991). The service factory: A future vision. *International Journal of Service Industry Management*, *2*(3), 60–70.

Chase, R. B. & Erikson, W. J. (1988). The service factory. *Academy of Management Executive*, *2*(3), 191–196.

Chase, R. B. & Garvin, D. A. (1989). The service factory. *Harvard Business Review*, *67*(4), 61.

Chase, R. B., Kumar, K. R., & Youngdahl, W. E. (1992). Service-based manufacturing: The service factory. *Production and Operations Management*, *1*(2), 175–184.

Chinese Internet Network Information Center (2010). Internet fundamental data. Available at http://www.cnnic.cn/en/index/0O/index.htm (accessed March 1, 2012).

Cohen, M. A., Agrawal, N., & Agrawal, V. (2006). Winning in the aftermarket. *Harvard Business Review*, *84*(5), 129–38.

Colotla, I., Shi, Y. J., & Gregory, M. J. (2003). Operation and performance of international manufacturing networks. *International Journal of Operations and Production Management*, *23*(10), 1184–1206.

Eisenhardt, K. M. (1989). Building theories from case study research. *Academy of Management Review*, *14*(4), 532–550.

Ferdows, K. (1989). *Managing international manufacturing*. Amsterdam: Elsevier Science Ltd.

Freeman, S. & Sandwell, M. (2008). Professional service firms entering emerging markets: The role of network relationships. *Journal of Services Marketing*, *22*(3), 198–212.

Furrer, O. (2005). Service quality: Research perspectives. *International Journal of Service Industry Management*, *16*(4), 408–410.

Garvin, D. A. (1993). Manufacturing strategic planning. *California Management Review*, *35*(4), 85–106.

Gebauer, H., Fleisch, E., & Friedli, T. (2005). Overcoming the service paradox in manufacturing companies. *European Management Journal*, *23*(1), 14–26.

Glueck, J., Koudal, P., & Vaessen, W. (2006). *The Service Revolution. Manufacturing's Missing Crown Jewel*, Deloitte Review, Deloitte Research.

Godlevskaja, O., van Iwaarden, J., & van der Wiele, T. (2011). Moving from product-based to service-based business strategies. *International Journal of Quality and Reliability Management*, *28*(1), 62–94.

Gremyr, I., Löfberg, N., & Witell, L. (2010). Service innovations in manufacturing firms. *Managing Service Quality*, *20*(2), 161–175.

Grönroos, C. (2000). *Service management and marketing: A customer relationship management approach*. Chichester: John Wiley & Sons.

Hayes, R. J., Wheelwright, S. C., & Clark, K. B. (1988). *Dynamic manufacturing: Creating the learning organisation*. New York: Free Press.

Haynes, R. M. & DuVall, P. K. (1992). Service quality management: A process-control approach. *International Journal of Service Industry Management*, *3*(1), 14–24.

Heizer, J. & Render, B. (2008). *Principles of operations management* (7th ed.). New Jersey: Pearson Education.

IBM (2011a). IBM Global Services: A brief history. Available at http://www-03.ibm.com/ibm/history/documents/pdf/gservices.pdf (accessed December 12, 2011).

IBM (2011b). IBM Annual Report 2010. Available at ftp://public.dhe.ibm.com/annualreport/2010/2010_ibm_annual.pdf (accessed October 1, 2011).

IDC (2011). China surpassed the United States to become the largest PC market in the world in the second quarter of 2011. Available at http://www.idc.com/getdoc.jsp?containerId=prUS22997711 (accessed March 1, 2012).

Johnson, M. & Mena, C. (2008). Supply chain management for servitised products: A multi-industry case study. *International Journal of Production Economics*, *114*(1), 27–39.

Johnston, R. (1994). Operations: From factory to service management. *International Journal of Service Industry Management*, *5*(1), 49–63.

Johnston, R. (1999). Service operations management: Return to roots. *International Journal of Operations and Production Management*, *19*(2), 104–124.

Johnstone, S., Dainty, A., & Wilkinson, A. (2009). Integrating products and services through life: An aerospace experience. *International Journal of Operations and Production Management*, *29*(5), 520–538.

Kathuria, R., Joshi, M. P., & Dellande, S. (2008). International growth strategies of service and manufacturing firms: The case of banking and chemical industries. *International Journal of Operations and Production Management*, *28*(10), 968–990.

Koudal, P. (2006). *The service revolution in global manufacturing industries*, A Deloitte Research Global Manufacturing Study. Available at http://www.deloitte.com/assets/Dcom-Turkey/Local%20Assets/Documents/The%20Service%20Revolution%20in%20Global%20Manufacturing%20Industries(1).pdf (accessed October 1, 2011).

Lay, G., Copani, G., Jäger, A., & Biege, S. (2010). The relevance of service in European manufacturing industries. *Journal of Service Management*, *21*(5), 715–726.

Lele, M. M. (1986). How service needs influence product strategy. *Sloan Management Review*, *28*(1), 63–70.

Levitt, T. (1972). Production-line approach to service. *Harvard Business Review*, *50*(4), 41–52.

Levitt, T. (1983). After the sale is over. *Harvard Business Review*, *61*(5), 87–93.

Lewis, R. C. & Booms, B. H. (1983). The marketing aspects of service quality. In L. Berry et al. (Eds.), *Emerging perspectives on services marketing*. New York: AMA.

Lewis, M. & Howard, M. (2009). Beyond products and services: shifting value generation in the automotive supply chain. *International Journal of Automotive Technology and Management, 9*(1), 4–17.

Lin, Y., Shi, Y. J., & Zhou, L. (2010). Service supply chain: Nature, evolution, and operational implications. In G. Huang, K. Mak, & P. Maropoulos (Eds.), *Advances in intelligent and soft computing* (pp. 1189–1204, Vol. 66), Berlin/Heidelberg: Springer-Verlag.

Lindberg, N., & Nordin, F. (2008). From products to services and back again: Towards new service procurement logic. *Industrial Marketing Management*, *37*(3), 292–300.

Machuca, J. A. D., González-Zamora, M.d. M., & Aguilar-Escobar, V. G. (2007). Service operations management research. *Journal of Operations Management*, *25*(3), 585–603.

Malleret, V. (2006). Value creation through service offers. *European Management Journal*, *24*(1), 106–116.

Martin, C. R. & Horne, D. A. (1992). Restructuring towards a service orientation: The strategic challenges. *International Journal of Service Industry Management*, *3*(1), 25–38.

Martinez, V., Bastl, M., Kingston, J., & Evans, S. (2010). Challenges in transforming manufacturing organisations into product-service providers. *Journal of Manufacturing Technology Management*, *21*(4), 449–469.

Mathieu, V. (2001). Service strategies within the manufacturing sector: Benefits, costs and partnership. *International Journal of Service Industry Management*, *12*(5), 451–475.

Matthyssens, P., Vandenbempt, K., & Berghman, L. (2006). Value innovation in business markets: Breaking the industry recipe. *Industrial Marketing Management*, *35*(6), 751–761.

McLaughlin, C. P. & Fitzsimmons, J. A. (1996). Strategies for globalizing service operations. *International Journal of Service Industry Management*, *7*(4), 43–57.

Meijboom, B. & Vos, B. (1997). International manufacturing and location decisions: Balancing configuration and co-ordination aspects. *International Journal of Operations and Production Management*, *17*(8), 790–805.

Miles, I. (1993). Services in the new industrial economy. *Futures*, *25*(6), 653–672.

Ministry of Industry and Information Technology of China (2012). 2011 Statistical Report of the Electronic and Information Technology Industry. Available at http://www.miit.gov.cn/n11293472/n11293832/n11294132/n12858462/14475184.html (accessed March 1, 2012).

Morelli, N. (2006). Developing new PSS, methodologies and operational tools. *Journal of Cleaner Production*, *14*(17), 1495–1501.

Moller, K., Rajala, R., & Westerlund, M. (2008). Service innovation myopia? A new recipe for client-prodier value creation. *California Management Review*, *50*(3), 31–48.

Neely, A. (2008). Exploring the financial consequences of the servitization of manufacturing. *Operations Management Research*, *2*(1), 103–118.

Neely, A., McFarlane, D., & Visnjic, I. (2011). Complex service systems – identifying drivers characteristics and success factors. In *Proceedings of the 18th international annual EurOMA conference: Exploring interfaces, 3–6 July 2011* (p. 74). Cambridge, UK: Cambridge University Press.

Netscribes (2012). Personal computer market in China 2012. Available at http://www.marketresearch.com/Netscribes-India-Pvt-Ltd-v3676/Personal-Computer-China-6746140/ (accessed March 1, 2012).

O'Neill, H. & Sackett, P. (1994). The extended manufacturing enterprise paradigm. *Management Decision*, *32*(8), 42–49.

Oliva, R. & Kallenberg, R. (2003). Managing the transition from products to services. *International Journal of Service Industry Management*, *14*(2), 160–172.

Pawar, K. S., Beltagui, A., & Riedel, J. C. K. H. (2009). The PSO triangle: Designing product, service and organisation to create value. *International Journal of Operations and Production Management*, *29*(5), 468–493.

Quin, J. B. & Gagnon, C. E. (1986). Will services follow manufacturing into decline? *Harvard Business Review*, *64*(6), 95–103.

Quinn, J. B. (1992). *Intelligent enterprise: A knowledge and service based paradigm for industry*. New York: Free Press.

Quinn, J. B., Doorley, T. L., & Paquette, P. C. (1990). Beyond products: Services-based strategy. *Harvard Business Review*, *68*(2), 58–67.

Robinson, T., Clarke-Hill, C. M., & Clarkson, R. (2002). Differentiation through service: A perspective from the commodity chemicals sector. *The Service Industries Journal*, *22*(3), 149–166.

Sawhney, M., Balasubramanian, S., & Krishnan, V. V. (2004). Creating growth with services. *MIT Sloan Management Review*, *45*(2), 34–43.

Schmenner, R. W. (2009). Manufacturing, service, and their integration: Some history and theory. *International Journal of Operations and Production Management*, *29*(5), 431–443.

Sharma, A. & Loh, P. (2009). Emerging trends in sourcing of business services. *Business Process Management Journal*, *15*(2), 149–165.

Shi, Y. J. (2003). Internationalisation and evolution of manufacturing systems: Classic process models, new industrial issues, and academic challenges. *Integrated Manufacturing Systems*, *14*(4), 357–368.

Shi, Y. J. & Gregory, M. J. (1998). International manufacturing networks – to develop global competitive capabilities. *Journal of Operations Management*, *16*(2–3), 195–214.

Shostack, L. G. (1982). How to design a service. *European Journal of Marketing*, *16*(1), 49–63.

Shostack, L. G. (1984). Design services that deliver. *Harvard Business Review*, *62*(1), 133–139.

Skinner, W. (1969). Manufacturing – missing link in corporate strategy. *Harvard Business Review*, *47*(3), 136–145.

Skinner, W. (1974). The focused factory. *Harvard Business Review*, *52*(3), 113–121.

Slack, N. (2005). Operations strategy: Will it ever realize its potential? *Gestão and Produção*, *12*(3), 323–332.

Slack, N., Lewis, M., & Bates, H. (2004). The two worlds of operations management research and practice: Can they meet, should they meet? *International Journal of Operations and Production Management*, *24*(4), 372–387.

Spring, M. & Araujo, L. (2009). Service, services and products: Rethinking operations strategy. *International Journal of Operations and Production Management*, *29*(5), 444–467.

Spring, M. & Dalrymple, J. F. (2000). Product customisation and manufacturing strategy. *International Journal of Operations and Production Management*, *20*(4), 441–467.

Toni, A. D., Filippini, R., & Forza, C. (1992). Manufacturing strategy in global markets: An operations management model. *International Journal of Operations and Production Management*, *12*(4), 7–18.

van Biema, M. & Greenwald, B. (1997). Managing our way to higher service-sector productivity. *Harvard Business Review*, *75*(4), 87–95.

Vandermerwe, S. & Rada, J. (1988). Servitization of business: Adding value by adding services. *European Management Journal*, *6*(4), 314–324.

Vos, G. C. J. M. (1991). A production-allocation approach for international manufacturing strategy. *International Journal of Operations and Production Management*, *11*(3), 125–134.

Voss, C. (1992). Applying service concepts in manufacturing. *International Journal of Operations and Production Management*, *12*(4), 93–99.

Wilkinson, A., Dainty, A., & Neely, A. (2009). Changing times and changing timescales: The servitization of manufacturing. *International Journal of Operations and Production Management*, *29*(5): Guest editorial.

Wise, R. & Baumgartner, P. (1999). Go downstream: The new profit imperative in manufacturing. *Harvard Business Review*, *77*(5), 133–141.

Yin, R. K. (1994). *Case study research: Design and methods*. Thousand Oaks, CA: Sage.

Young, S. T., Kwong, K. K., Li, C., & Fok, W. (1992). Global manufacturing strategies and practices: A study of two industries. *International Journal of Operations and Production Management*, *12*(9), 5–17.

Youngdahl, W. E. (1996). An investigation of service-based manufacturing performance relationships. *International Journal of Operations and Production Management*, *16*(8), 29–43.

Youngdahl, W. E. & Loomba, A. P. S. (2000). Service-driven global supply chains. *International Journal of Service Industry Management*, *11*(4), 329–347.

Zehrer, A. (2009). Service experience and service design: Concepts and application in tourism SMEs. *Managing Service Quality*, *19*(3), 332–349.

CHAPTER 3

# Supply Chain Finance: Concept and Modeling

**Miao He, Changrui Ren, Qinhua Wang, and Jin Dong**
*IBM Research – China, Building 19, Zhongguancun Software Park, 8 Dongbeiwang West Road, Haidian District, Beijing 100193, China*

## 3.1 Inefficient Financial Supply Chain

According to the white paper of the Paystream Advisors (2008), the inefficiencies in a supply chain comprise four aspects. The first is *manual and inefficient processes*. When compared with an automatic system, manual- or paper-based transaction processing usually incurs higher processing costs as well as more errors in transaction files. Second, there exist *conflicting interests between the suppliers and buyers*. The buyers, as many as 73% among the 500 top corporations in Europe (Demica, 2007), try to extend the payment terms to improve their working capital, whereas the suppliers desire liquidity as well. Third, *the lack of cash flow visibility* leads to the concern of suppliers who have to maintain excess cash to "hedge" against uncertainties around cash flow. The last inefficiency is due to *high cost of supplier financing*. The traditional supplier financing solutions such as factoring or asset-based lending incur high costs to the supplier, and thus the whole supply chain.

This chapter introduces how supply chain finance (SCF) can address the challenge of conflicting interests between the buyers and suppliers as well as reduce the high costs of supplier financing. Before diving into the SCF solutions, we use the following two examples to illustrate the conflicting interests:

*Example 1*: It is observed that more and more suppliers offer a payment term named "open account," which allows the buyer a credit of 30 to 90 days (David & Stewart, 2007) to gain a competitive advantage. Yet to reflect the opportunity costs, the supplier will usually price higher for compensation.

*Example 2*: The capital shortage of a buyer would result in suboptimal decisions for order quantity, which in turn leads to a suboptimal sales quantity for its suppliers.

Service Science, Management, and Engineering.


DOI: 10.1016/B978-0-12-397037-4.00003-X

It is noted from the two examples that both the parties get hurt explicitly in the costs or revenue and implicitly in the supplier–buyer relationship.

It is common for the *capital-constrained supply chain players*, especially small and medium-sized enterprises (SMEs), to borrow money from the financial institutions to maintain their daily operations. Conventionally, the financial institutions will either negotiate a very high interest rate or even decide not to fund the SMEs due to high expected financial risks.

## 3.2 Introduction to SCF Solutions

The Aberdeen Group (2006) defines SCF as "a combination of Trade Financing provided by a financial institution, a third-party vendor, or a corporation itself, and a technology platform that unites trading partners and financial institutions electronically and provides the financing triggers based on the occurrence of one or several supply chain events."

In an SCF solution, the financial institutions will evaluate the credit risks from the perspective of a supply chain instead of a single player. For instance, expecting minimal chance of a creditworthy buyer's default, the institutions thus expect low probability of the supplier's default. Therefore, they may agree to fund the supplier in this trade with an interest rate much lower than the supplier can get using its own credit.

One way to categorize the rich set of SCF solutions is to identify the critical event triggers. According to the study by Demica (2007), the critical points may include raw material, intermediary production, point of shipment, point of custom clearance, and arrival at vendor-managed inventory hub. We adapt the definition of three critical event triggers in a white paper of the Bank of America, although we are aware of many other definitions (see Fig. 3.1).

These event triggers divide the variety of SCF solutions into the following three categories: preshipment finance, inventory-in-transit finance, and postshipment finance. Table 3.1 summarizes popular SCF solutions.

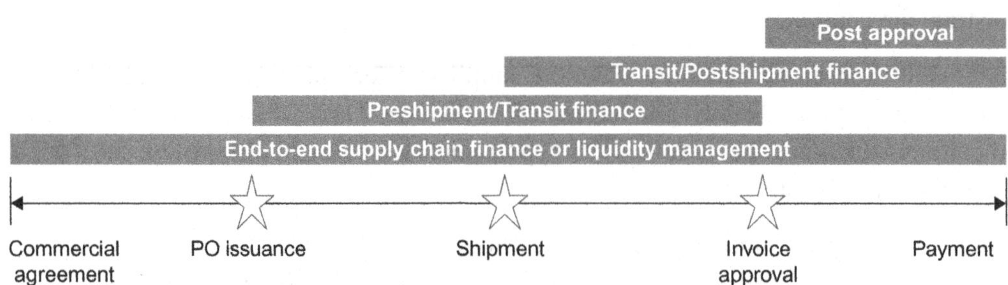


**Figure 3.1: Events-driven supply chain financing model.**
*Source:* White paper by the Bank of America.

**Table 3.1: SCF solutions.**

| Finance Type | Time Interval | SCF Solutions |
|---|---|---|
| Preshipment finance | PO issuance - shipment | • Raw material finance (supplier)<br>• Production finance (supplier) |
| Transit finance | Shipment - Invoice approval | • Inventory-in-transit finance (supplier) |
| Postshipment finance | Invoice approval - payment | • Accounts receivable finance (supplier)<br>• Early payment discount (supplier)<br>• Accounts payable finance (buyer) |

Alternatively, we can divide the SCF solutions based on obligors. If the obligor is on the supply side, then we can call them supplier finance solutions, and buyer finance solutions on the buyer side. The obligors of each financing method are indicated in parentheses in the last column of Table 3.1.

In the subsections below, we review the SCF solutions in detail (Table 3.1).

### 3.2.1 Preshipment Finance

Preshipment finance is applied if the suppliers need money after receiving the purchase order (PO) from the buyer and before the shipment. The payment term of "cash in advance" does not require the preshipment finance. However, the "open account" term, gaining in popularity in recent years, may force the suppliers to ask financial institutions for working capital. The cash obtained via preshipment finance, typically, covers the purchase of the raw materials and the cost of manufacturing. In the SCF solutions, the buyer will leverage its credit for the supplier to negotiate a better interest rate. Usually, participating suppliers and financial institutions are linked directly to a buyer's purchasing system for access to outstanding purchase orders and related data, and the financial institutions assess the suppliers based on their performance track record with the buyer.

### 3.2.2 Transit Finance

According to EZB Limited, the inventory in transit on average ties 26% of the invoice amount on either the suppliers' or the buyers' balance sheet. As the capital tied to the inventory is unproductive, it can be released via SCF solutions. In the supplier finance, the financial institutions fund the supplier when the inventory is on its way but before invoice approval. The inventory-in-transit finance requires visibility of the material flow. Typically, the financial institutions will involve third-party logistics (3PL) providers to closely monitor the in-transit goods so as to mitigate risks. Therefore, financial institutions derived from large logistics providers, such as UPS capital, have advantages in the visibility of the in-transit inventory. Some 3PLs, instead of the buyers, even leverage their own credit for the supplier to earn commission and control the risks by taking on the logistics service at the same time.

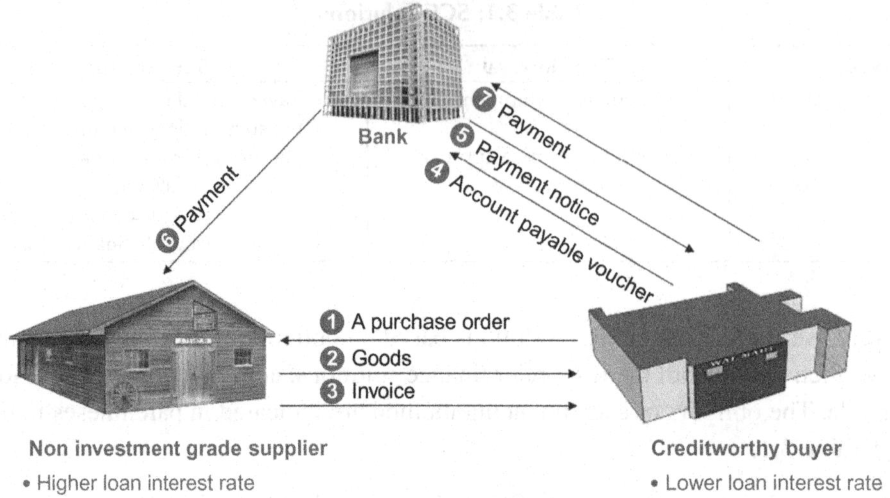


**Figure 3.2: Finance against inventory in transit (to supplier).**

Therefore, 3PLs have developed another profit center through credit leveraging in addition to the traditional logistics service.

Figure 3.2 depicts a simplified process of transit finance. Receiving the PO from a creditworthy buyer, the noninvestment grade supplier ships the goods (either from inventory or from assembly line) to the buyer. The goods will travel about 4 weeks on the ocean to arrive at the buyer. As the buyer and supplier have built a tight relationship in the past, the buyer is willing to use its credit for the supplier to get finance from the bank, which means the bank will pay the supplier immediately after receiving the buyer's account payable voucher. The time is much earlier than the payment maturity date agreed upon between the supplier and the buyer. When the payment comes to maturity, the buyer will issue the payment directly to the bank.

### *3.2.3 Postshipment Finance*

According to the Bank of America, postshipment finance to suppliers includes accounts receivable (AR) finance and early payment discount. Although both SCF solutions are against the AR, we can distinguish them by whether the buyer is directly involved or not.

In AR finance, the suppliers can sell their AR at a discount to a third-party financial institution (the factor) for immediate cash. On invoice maturity, the factor will collect the full payment from the buyer just as in traditional factoring. However, the AR finance in the SCF solution leverages the buyer's credit for the supplier, implying that the factors are willing to buy the AR at a higher rate than in traditional factoring, because they value the linkage between the buyers and suppliers. In return for leveraging the buyer's credit, the suppliers will price lower and extend the terms for the buyer to share the saved financial cost. Thus, the whole supply chain benefits. An electronic platform linking the three parties (the buyer, the supplier, and the financial institutions) can accelerate the processing of the AR finance. We point out that AR finance does not affect the buyer's capital flow, because the buyer is only responsible to pay on time either to its suppliers or to the factors.

With regard to early payment discount, also known as dynamic payable discounting, it allows a buyer and its suppliers to dynamically change their payment terms. In most cases, on a platform established by the buyer, the buyer will notify the amount and the due date of the approved invoices and mark those that are available for early payment. The suppliers can decide a competitive discount to entice the early payment if their ARs are ready (*supplier-initialized early payment discount*). Symmetrically, the buyer also can move first to declare its acceptable discount rate (*buyer-initialized early payment discount*). The early payment discount relieves the capital drought of the suppliers and meanwhile benefits the buyer with interest rate significantly higher than the deposit rate. For example, a 2% discount for early payment of 20 days is commensurate with an annual interest rate as high as 36%, which is a considerable return for short-term investments. Similar to all the other SCF solutions, an electronic platform will enhance the real time decisions for both parties. It is valid for the buyer to borrow from the financial institutions to capture the discount.

It might happen that a capital-constrained buyer needs to finance against its accounts payable (AP) to pay the suppliers on time. In such a case, the financial institutions pay the supplier on invoice maturity, and they allow the buyer to pay them back some time later with extra interest. Traditional AP financing has nothing to do with the supply chain. But in the SCF solutions, the funder will make a decision based on the whole supply chain's robustness, for example, the buyer's relationship with the buyer's buyer. Moreover, it is possible for the supplier to extend the payment term in exchange for the buyer's credit, which is also a method of AP finance.

## 3.3 Mathematical Representations of Supply Chain Finance

Here, we discuss how to model supply chain finance in a mathematical form, so that we can solve the model for optimal (or near optimal) SCF policies. Different modeling frameworks have different positions. For instance, one can model and optimize the policies for the suppliers or buyers, assuming that the behavior or behavioral pattern of the other party is known. Alternatively, some researchers model the game between the suppliers and the buyers.

We will first review existing models in the Operations Research literature, the majority of which focuses on the suppliers, buyers, and the interactions between the two parties. Later, we will describe an innovative modeling framework, proposed by the authors, from the perspective of the funders such as the banks.

### 3.3.1 A Survey

Applying optimization to a firm's financial decision-making process dates back to more than a half-century ago. In 1965, Robichek, Teichroew, and Jones formulate and solve a linear program for a multiperiod problem to minimize the total costs, with satisfying cash demand in each period as a must. The finance officer needs to choose the best one of six alternative financing options in each period, including line of credit, pledging of AR, stretching of AP, term loans, and investments of excess cash. The cash demand in each period is assumed to be known in advance in this chapter. Eppen and Fama (1971) assume uncertain, in contrast to known, requirements on cash inflows and outflows in each period and formulate a stochastic dynamic program to choose the best operational policy, that is, buying or selling "stocks" and "bonds," to maintain a proper cash level. Interested readers can find more research in the literature (e.g., Charnes & Thore, 1966; Kallberg, White, & Ziemba, 1982). Recently, Protopappa-Sieke and Seifert (2010) interrelate the financial decisions with operational decisions in a supply chain and study the trade-offs between most commonly used financial measures and operational measures.

Along with the emergence of SCF solutions, researchers from the Operations Research society are making efforts to introduce optimization techniques into the decision-making process in SCF implementation.

Caldentey and Chen (2010) recognize that suboptimal order quantity resulting from capital constraint hurts the sales revenue of both the retailer (the buyer) and its supplier. Therefore, they highlight the importance of financing a capital-constrained retailer. The retailer is modeled as a newsvendor who may not place the optimal order quantity indicated by the traditional newsvendor solution as the result of limited budget. However, the supplier can negotiate a contract term $(\omega, \alpha)$ to finance the retailer (*internal financing*), where $\omega$ is the wholesale price and $\alpha\,(\alpha \leq 1)$ is the fraction of money paid on ordering. The retailer will place an order optimally against the supplier's offer. After the uncertain customer demand is realized, the supplier collects the minimum of the unpaid ordering cost and the retailer's leftover wealth, which implies a limited liability of the retailer. Alternatively, the retailer can turn to financial institutions (*external financing*). The authors model *the game* between the supplier and the buyer in which the supplier takes the Stackelberg lead. Then, they solve for the Stackelberg equilibrium.

Their numerical results show that for all the budget levels of the retailer, the payoff of the whole supply chain under internal financing is always greater than or equal to that under external financing, which in turn is always greater than or equal to that without any financing. The optimal solutions of the three scenarios (external financing, internal financing, and no financing) converge as the available budget tends to infinity. The authors also measure the payoff of the retailer and supplier separately; internal financing always brings the supplier maximum reward but attains the maximum for the retailer only when its budget $B$ is greater than or equal to some critical budget level $B^*$. If $B < B^*$, the retailer earns more by external financing.

Kouvelis and Zhao (2009) propose a capital-constrained newsvendor problem similar to the study by Caldentey and Chen (2010), but their internal financing is dynamic payable discounting, instead of fractional payment $\alpha$ on ordering (Caldentey & Chen, 2010). Note that dynamic payable discounting is an important postshipment finance option. The essence is that a supplier sells its AP to a third party or its buyers at a discount, while the third party or buyer will settle the payment immediately.

In the study by Kouvelis and Zhao (2009), the game between the retailer and the supplier also reaches a Stackelberg equilibrium in which the supplier takes the Stackelberg lead, that is, the supplier decides the discounting rate and the wholesale price. The banks can provide the external financing with a risk-free interest rate. Numerical results reveal that internal financing outperforms external financing in terms of the payoff of the whole supply chain and the supplier. However, it is not always the case for the retailer.

In the work of Basu and Nair (2008), the authors formulate a stochastic dynamic program to enhance the quantitative decision making for the buyers to capture the dynamic payable discounting "$r_1$/10, net 30," which means a discount of $r_1$ is offered if the buying firm pays in full within 10 days after the departure of the shipment, and otherwise, a full payment within 30 days is required. The reward period of the buying firm can be generated by choosing a combination of the following three interest rates: the equivalent interest rate of early payment discount with one's own capital, denoted as $r_1$; the equivalent interest rate of early payment discount by borrowing a third party's capital, denoted as $r_2$; and the interest rate of depositing the buyer's leftover money into the banks, denoted as $r_3$. It is obvious that $r_1 > r_2 > r_3$. The firm is reluctant to apply its credit line for bank loans, because the bank loan credit is reserved for other daily operations. The firm needs to decide the following two factors: (1) whether to capture the AP discounting, and (2) if yes, how much money of its own should be used to capture the discount, given uncertain amounts of AP and AR in future periods. The objective is to maximize the total expected reward till the end of the horizon. They do not solve the model or show the policy in the capture of dynamic payable discounting.

As our model for financial institutions requires properly modeling of the credit risks of supply chain players, we review some of the popular methods for credit risk evaluation. Crouhy (2006) elaborates and carefully compares the four credit risks models, namely, CreditRisk proposed by Morgan, KMV model by KMV Corporation (1993), CreditRisk+ by Credit Suisse Financial Products, and CreditPortfolioView by Mckinsey. In the CreditRisk model, the transition probabilities of a firm's credit rating are based on the average historic defaults and credit migration. Also, it assumes that all firms in the same rating class have the same transition probabilities. Based on the method proposed by Merton (1974), the KMV method models the dynamics of an individual firm's asset value within a time period $T$. The corporation becomes bankrupt when its asset value is less than its debt. A firm's asset value is lognormally distributed under Merton's model, which is a robust assumption according to KMV's empirical study.

CreditRisk+ uses a Poisson distribution to estimate the probability of $n$ defaults within a period of time, given a large group of obligors. CreditPortfolioView simulates the joint conditional distribution of default and migration probabilities for various rating groups in different industries and for each country, which is conditioned on the value of macroeconomic factors like the unemployment rate, the rate of growth in GDP, the level of long-term interest rates, foreign exchange rates, government expenditures, and the aggregate savings rate. Interested readers may consult Altman and Sauders (1998) and Gordy (2000) for more information.

### 3.3.2 Approximate Dynamic Programming

Despite the abundance of approximate dynamic programming (ADP) textbooks, we will introduce minimum knowledge but enough for the user to understand our SCF modeling in this section. ADP is both a modeling and a solution algorithm design framework, which fits the multiperiod problems with uncertainty, and meanwhile, decisions in the current period will have downstream impact on future periods. We use the following two examples to illustrate such problems.

*Example 1*: In multiperiod inventory management, a retailer decides the order quantity in each period to satisfy the customer demands that are uncertain. Their decision in each period will affect the decisions in future periods. Intuitively, if the retailer orders more than necessary in this period, then they will order less in future periods to keep inventory at an appropriate level.

*Example 2*: In multiperiod fleet scheduling, the scheduler will decide how to allocate the drivers and different types of (loaded and empty) trucks to satisfy the observed demand as well as the uncertain potential demand that will randomly appear across geographical regions. For example, if it is highly possible to have demand raised in region A the next day, then the scheduler could send an empty truck head to region A today to improve the service level and customer satisfaction.

#### 3.3.2.1 Key Modeling Components and Notational Style

According to Powell (2007), the ADP modeling framework includes the following five key components:

**State variables:** These describe what we need to know at a point in time. We usually use $S_t$ to denote the state in period $t$, which can be a vector.
**Decision variables:** These are the variables we control; choosing these variables ("making decisions") represents the central challenge of dynamic programming. It is denoted as $x_t$, representing the decision made in period $t$, which can be a vector.
**Exogenous information process:** These variables describe information that arrives to us exogenously, representing the sources of randomness. In the ADP framework, we use $W_{t+1}$ to describe exogenous information that arrives between periods $t$ and $t + 1$.

**Transition function:** This is the function that describes how the state evolves from one point in time to another. We use $S_{t+1} = S^{M,W}(S_t, x_t, W_{t+1})$ to describe the process. The state in period $t + 1$ will depend on the state in period $t$, decision in period $t$ and exogenous information that becomes known at the beginning of period $t + 1$.

**Objective function:** We are trying to either maximize a contribution function (profits, revenues, rewards, and utility) or minimize a cost function. This function describes how well we are doing at a point in time.

We introduce the concept of contribution function here, which is also an important element in an ADP modeling framework.

**Contribution function:** How much/many we can earn in each period by taking an action. If we measure by how much/many we are to pay, then we can call it the *lost function*. We usually use $R_t$ to denote the reward function and $C_t$ to denote the loss function in period $t + 1$.

We emphasize that state variables can summarize much more than the status of physical resources such as the inventory or number of busy/free employees. Powell (2007) lists the following three types of states:

**The physical state:** This is a snapshot of the status of the physical resources we are managing and their attributes. This might include the amount of water in a reservoir, the price of a stock, or the location of a sensor in a network.

**The information state:** This encompasses the physical state as well as any other information we need to make a decision, compute the transition, or compute the objective function.

**The knowledge state:** If the information state is what we know, the belief state (also known as the knowledge state) captures how well we know it. This concept is largely absent from most dynamic programs, but arises in the setting of partially observable processes (when we cannot observe a portion of the state variable).

### *3.3.2.2 Curses of Dimensionality*

Traditional stochastic dynamic programming such as the Markov decision process (MDP) also addresses the same set of problems as does ADP. However, it is well known that the *curses of dimensionality* significantly restrict the MDP solution algorithm, backward dynamic programming, regarding application to large-sized problems.

The solution algorithm for MDP lies in the Bellman equation. In a reward maximization problem, the Bellman equation is

$$V_t(S_t) = \max_{x_t \in X_t} \left\{ R_t(S_t, x_t) + \underset{w_{t+1} \in W_{t+1}}{E} \left( V_{t+1}(S_{t+1} \mid S_t, x_t, w_{t+1}) \right) \right\}, \tag{3.1}$$

where $V_t(S_t)$ is the value function for state $S_t$ and can be interpreted as the *expected* reward if the system starts evolving from $S_t$ in period $t$ till the end of the planning horizon.

The Bellman equation decomposes the multiperiod calculation into single period efforts. However, it suffers from the following curses of dimensionality:

- Large state space: We need to calculate a $V_t(S_t)$ value for each possible value of $S_t$, $\forall_t$.
- Large outcome space: The exogenous information process introduces randomness into the problem. When computing the expectation of the second term in Eq. (3.1), we need to sum up or integrate over all the possible realizations of stochastic processes.
- Large decision space: We need to enumerate all the decisions to choose the one that maximizes the value function, as indicated by the "max" operator in Eq. (3.1).

According to Powell (2007), ADP combines the flexibility of simulation with the intelligence of optimization, and therefore it addresses the curses of dimensionality encountered in the traditional stochastic dynamic programming very well.

Later, we succinctly discuss the common strategies in the ADP solution algorithm to overcome the curses of dimensionality based on the study by Powell (2007). We refer interested readers to this book for more details and extensions.

#### *3.3.2.3 ADP Solution Algorithm Design*

Solving the stochastic dynamic formulation of a problem results in optimal policies, where a policy is defined to be a rule (or function) that determines a decision, given the available information in state $S_t$.

There are a range of policies that can be grouped under the following four categories (Powell, 2007):

**Myopic policies:** These are the most elementary policies. They optimize costs or rewards in a particular period, but do not explicitly use forecasted information or any direct representation of decisions in the future. However, they may use tunable parameters to produce good behaviors over time.

**Lookahead policies:** These policies make decisions now by explicitly optimizing over *some horizon* by combining some approximation of future information with some approximation of future actions.

**Policy function approximation policies:** These are functions that directly return an action given a state, without resorting to any form of imbedded optimization, and without directly using any forecast of future information.

**Value function approximation policies:** These policies, often referred to as greedy policies, depend on an approximation of the value of being in a future state as a result of a decision made now. The impact of a decision now on the future is captured purely through a value function that depends on the state that results from a decision now.

The ADP framework we followed is based on the value function approximation policies. The Bellman equation in the ADP solution algorithm becomes

$$\hat{v}_t^n(S_t) = \max_{x_t \in X_t}\left\{R_t(S_t,x_t) + \underset{w_{t+1} \in W_{t+1}}{E}\left(\overline{V}_{t+1}^{n-1}(S_{t+1} \mid S_t^n, x_t, w_{t+1})\right)\right\}. \tag{3.2}$$

Note that we use $\overline{V}_t(S_t)$ in Eq. (3.2) to replace $V_t(S_t)$ in Eq. (3.1), where $\overline{V}_t(S_t)$ is an *estimation of the value* in state $S_t$, in contrast to $V_t(S_t)$, which is an exact computation of the value in state $S_t$.

ADP algorithms iteratively update the estimated value with new observations,

$$\overline{V}_t^n(S_t) = (1-\alpha)\overline{V}_t^{n-1}(S_t) + \alpha\hat{v}_t(S_t), \tag{3.3}$$

where $n$ is the iteration count, $\hat{v}_t(S_t)$ is the new observed value as computed from Eq. (3.2), and $\alpha$ is the stepsize or smooth factor, which determines the percentage of contributions of old estimates and the new observations, respectively.

In the ADP solution framework, one need not visit every state and compute a value for each state. Instead, we only compute and update value of the *visited states*. In other words, ADP reduces the number of states that we need to calculate a value for, which partially alleviates the problem of large state space. We call the approximation method a *lookup-table approximation*, reflecting that the resulted approximation is still in the form of a lookup table.

In a more sophiscated algorithm design, we are allowed to build a function that calculates the value of being in a particular state based on some formula. It is the *continuous value function approximation*. For example, we might think that the value of an inventory system is quadratic with the inventory position, reflecting too few items that lead to unsatisfied customers, whereas too many items on hand incur high inventory cost. In such a case, we further reduce the large state space to a small parameter space. In the inventory example, we only need to estimate three parameters (the quadratic coefficient, the linear coefficient, and the constant term).

To tackle the curse of large outcome space, there are basically two methods. One is to sample the outcome space at the second term in Eq. (3.2). In other words, we only compute the expectation over a subset of $W_{t+1}$, denoted as $\Omega_{t+1}$. Then, Eq. (3.2) becomes

$$\hat{v}_t^n(S_t^n) = \max_{x_t \in X_t}\left\{R_t(S_t,x_t) + \underset{w_{t+1} \in \Omega_{t+1}}{E}\left(\overline{V}_{t+1}^{n-1}(S_{t+1}^n \mid S_t^n, x_t, w_{t+1})\right)\right\}. \tag{3.4}$$

The other way is to use the *postdecision state*, $S_t^x$, and the value function for the postdecision state, $V_t(S_t^x)$. $S_t^x$ is the expected state to arrive from state $S_t$, given decision $x_t$, that is,

$S_t^x = S^{M,x}(S_t, x_t)$. To transit to the next period state $S_{t+1}$, we need to add exogenous information to $S_t^x$, that is, $S_{t+1} = S^{M,W}(S_t^x, W_{t+1})$. $V_t^x(S_t^x)$ captures the expected reward of the system starting from postdecision state $S_t^x$. Now Eq. (3.3) becomes

$$\hat{v}_t(S_t^n) = \max_{x_t \in X_t}\{R_t(S_t^n, x_t) + \overline{V}_{t+1}^{x,n-1}(S_t^{x,n} \mid S_t^n, x_t)\}. \tag{3.5}$$

Note that we will use the observed value for predecision state $\hat{v}_t(S_t)$ in period $t$ to update the estimated value for postdecision state $V_{t-1}^x(S_{t-1}^x)$ in period $t-1$. This update is valid because the transition from $S_{t-1}^x$ to $S_t$ depends only on the exogenous information. Therefore, $\hat{v}_t(S_t)$ is a sample of the value of being in state $S_t^n$, while it is also a sample of value of the decision that put us in state $S_{t-1}^{x,n}$.

Equation (3.5) eliminates the computation of the expectation and, therefore, solves the curse of large decision space.

Table 3.2 summarizes a generic algorithm of lookup-table ADP implementation with postdecision state.

After comparing Table 3.2 with traditional dynamic programming, we can observe the following differences:

- Traditional dynamic programming is backward for stochastic problems, whereas ADP is moving forward.
- Traditional dynamic programming requires a value calculation for each state in each period, whereas ADP only updates those states visited within some specified times of iteration $N$.

**Table 3.2: Generic lookup-table ADP algorithm with postdecision state.**

**Step 0**. Initialization:
 Step 0a. Initialize $\overline{V}_t^0, \forall t \in T$.
 Step 0b. Set $n = 1$.
 Step 0c. Initialize $S_0^1$.
**Step 1**. Choose a sample path $\omega^n$.
**Step 2**. Do for $t = 0, 1, 2, \ldots, T$:
 **Step 2a**. Solve:
 $\hat{v}_t(S_t^n) = \max_{x_t \in X_t}\{R_t(S_t^n, x_t) + \overline{V}_{t+1}^{x,n-1}(S_t^{x,n} \mid S_t^n, x_t)\}$
 and let $x_t^n$ be the value of $x_t$ that solves the maximization problem.
 **Step 2b**. If $t > 0$, update $\overline{V}_{t-1}^{n-1}$ using
 $\overline{V}_t^n(S_t) = (1 - \alpha_{n-1})\overline{V}_t^{n-1}(S_t) + \alpha_{n-1}\hat{v}_t(S_t)$.
 **Step 2c**. Find the postdecision state
 $S_t^{x,n} = S^{M,x}(S_t^n, x_t^n)$
 and the next predecision state
 $S_{t+1}^n = S^{M,W}(S_t^{x,n}, W_{t+1}(\omega^n))$.
**Step 3**. Increment $n$. If $n \le N$, go to Step 1.
**Step 4**. Return the value functions $(\overline{V}_t^N)_{t=0}^T$.

- Traditional dynamic programming calculates the expectation across all the possible outcomes, given a probability distribution, whereas ADP only samples the possible outcomes using $\omega^n$.
- Traditional dynamic programming does not allow a postdecision state but ADP does.

#### *3.3.2.4 Cross-Industry Applications*

ADP has proved to be effective in obtaining satisfactory results within short computational time in a variety of problems across industries. Nascimento and Powell (2010) apply ADP to help a fund decide the amount of cash to keep in each period. The objective is to achieve a balance between meeting shareholder redemptions (the more cash the better) and minimizing the cost from lost investment opportunities (the less cash the better). The proposed algorithms can obtain near-optimal results in considerably less time, compared with the exact optimization algorithm. Castanon (1997) applies ADP to dynamically schedule multimode sensor resources. Simao, Day, Geroge, Gifford, Nienow, and Powell (2009) use an ADP framework to simulate over 6000 drivers in a logistics company at a high level of detail and produce accurate estimates of the marginal value of 300 different types of drivers. Bertsimas and Demir (2002) solve the famous multidimensional knapsack problem with both parametric and nonparametric approximations. He, Zhao, and Powell (2010) model and solve a clinical decision problem using the ADP framework. We stress that ADP becomes a sharp weapon, especially when the user has insights into and makes smart use of the problem structure.

### 3.3.3 A Three-Stage Supply Chain Finance Modeling Framework

Here, we are going to describe a modeling framework that measures the implications of decisions with a *financial institution's* income and costs, other than the suppliers and the buyers as in Section 3.3.1. More specifically, we model how the financial institution can make its decisions optimally in a *supplier-financing* problem in which a *noninvestment grade supplier* will leverage the credit of a *creditworthy buyer* to enhance its cash flow.

In a typical SCF solution, the bank negotiates with the obligor and then nails down the loan amount and loan interests. However, the one-shot deal will not be risk free, although the bank comprehensively evaluates the risks from the perspective of a supply chain. Taking the preshipment finance against PO, for example, if the supplier goes bankrupt during production, the supplier will not issue payment to the bank. On the other hand, the creditworthy buyer itself also has the chance to be a defaulter, with a much lower probability than its supplier. The Lehman Brothers situation provided a good lesson in 2008. Moreover, some external factors such as macro economics, local economics, political factors, and regulations will also expose the bank to risks. We do not model the external uncertainties in this exploratory study.

We proposed an innovative business model that divides the "one-time loan" into $T$ stages, and in each stage, the financial institution will fund a fraction of the full loan amount. The multistage supply chain finance will mitigate the financial risks of the financial institution, because it can dynamically adjust its policy based on the observed financial states of the supply chain players, damage of the shipments, etc. The goal is to maximize the expected profit of the financial institution, which equals the income from interests minus the expected unpaid amount as a result of the realized risks.

We characterize multistage supply chain finance as a dynamic and uncertain problem. "Dynamic" points out that multistage decisions are involved, whereas "uncertain" refers to the risks. Hence, we choose to model the problem using a stochastic dynamic program.

Later, we discuss the modeling of each component in detail.

#### *3.3.3.1 State Variable*

To describe the status of the system at stage $t$, we define the notation as below:

$p_{t't}$ = the percentage of the full loan amount permitted by the financial institution to the supplier at stage $t'$, which is before stage $t$, that is, $t' < t$. $\forall t = 1,...,T$;
$p_t = (p_{1t}, \ldots, p_{t-1,t})$, a vector that records the whole history of the percentages of loan that the bank has approved before stage $t$, $\forall t = 1, \ldots, T$;
$V_t^B$ = the asset value of the buying firm at the beginning of stage $t$, $\forall t = 1,\ldots,T$;
$V_t^S$ = the asset value of the supplying firm at the beginning of stage $t$, $\forall t = 1,\ldots,T$;
$I_t$ = the value of the shipments at the beginning of stage $t$, $\forall t = 1,\ldots,T$;
$S_t = (P_t, V_t^B, V_t^S, I_t)$, $\forall t = 1,\ldots,T$.

We include the historic percentages of loans to the state, because it delimits the feasible region of the current decision. It is self-evident that $\sum_{t'=1}^{T} p_{t't} \leq 100\%$, because the financial institution will not borrow more than the full amount applied by the supplier. Note that the financial institution has the option of lending less than the full loan. And the "one-shot deal" is a special case of our problem, where the bank funds the supplier with 100% at a particular stage.

We track the asset values of the buying and supplying firms because our model assumes that the probability of defaults is implied by their asset value. If the supplying firm's asset value drops below the contract value of goods in the manufacturing stage, then it goes bankrupt and cannot fulfill the contract at all. If their asset values drop below the payment after the shipment, then the financial institution can only collect the leftover wealth.

We track the shipment value because this model considers the possible loss during transit as a risk for the financial institution. Loss in transit can be caused by cargo theft, improper handling, etc., which will incur overhead cost and operational cost due to unexpected event handling.

The resulting state modeling, $S_t$, is a four-dimensional state variable.

#### 3.3.3.2 Decision Variable

We define

$x_t$ = the percentage of full amount that the bank decides to lend to the buyer at the beginning of stage $t$ and $\sum_{t'=1}^{t} p_{t't} + x_t \leq 100\%$ as we have discussed.

#### 3.3.3.3 Exogenous Information Process

The exogenous information, $W_{t+1}$, captures all uncertainties that happen during stage $t$ and become known at the beginning of stage $t + 1$. The exogenous processes in this model include the exogenous process of the buyer's credit risk, the supplier's credit risk, and the inventory-in-transit value.

We adapt Merton's model (1974) to design the asset value change over time, which assumes that the asset value follows the dynamics of geometric Brownian motion, that is,

$$\hat{V}_t = V_0 \mathrm{e}^{(\mu - \sigma^2/2)t + \sigma\sqrt{t}Z_t}, \tag{3.6}$$

where $V_0$ is the initial asset value of a firm, $\hat{V}_t$ is the asset value at time $t$, which is a random variable before $t$, $\mu$ is the mean of the instantaneous asset return rate, $\sigma^2$ is the variance of the instantaneous asset return rate, and $Z_t \sim N(0, 1)$.

Note that Merton's model assumes that a firm's asset value can be tracked through its stock price, which implies that we are modeling two listed companies. Other models can be used to characterize the default risk of the buyer and supplier, which will not affect the modeling framework.

For inventory-in-transit risk, we simply use a decreasing linear probability density function to model it, as shown in Fig. 3.3.

The probability distribution coincides with our empirical observations that larger loss occurs with smaller probability. Mathematically, we can write the density function as

$$f(\hat{\delta I}) = \begin{cases} -\dfrac{2}{(I_0)^2}\hat{\delta I} + \dfrac{2}{I_0}, & I_0 \neq 0, \\ 0, & I_0 = 0, \end{cases} \tag{3.7}$$

where $I_0$ is the contract value between the buyer and the supplier, $\hat{\delta I}$ is the random loss amount of the shipment when it arrives at the buyer, and $f(\hat{\delta I})$ is the probability of having loss amount $\hat{\delta I}$. If the supplier fails to send out the shipment due to bankruptcy, then there will be no damage. It is easy to extend the risk model for inventory-in-transit by adding nonlinearity.

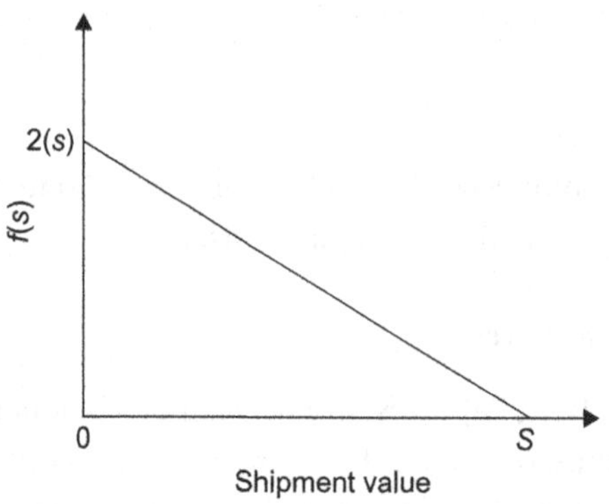


**Figure 3.3: Risk model of the inventory in transit.**

Note that in our model, we let the planning horizon $T = 3$ for two reasons. First, too many stages will cost the bank much labor to track a single business, whereas it might scare off the clients due to the tedious procedures. In addition, a three-stage division inherently matches the supply chain finance phases, where each stage exposes the bank to a different combination of risks.

Consistent with Table 3.1, stage one lasts from the PO issuance to the shipment, stage two covers the time from shipment to invoice approval, and stage three is between invoice approval and payment maturity. The credit risk of the buyer and supplier, modeled in Eq. (3.4), imposes uncertainties on the bank all the time until it collects the payment from the buyer. Stage two will specifically suffer from the inventory-in-transit risk, which will become known at the beginning of stage three.

In summary, our exogenous information process is

$$W_{t+1} = \begin{cases} (\hat{V}^B_{t+1}, \hat{V}^S_{t+1}), & t = 1,3, \\ (\hat{V}^B_{t+1}, \hat{V}^S_{t+1}, \hat{\delta I}), & t = 2. \end{cases} \tag{3.8}$$

#### 3.3.3.4 *Transition Function*

Based on the notational system, we now write the transition function as follows:

$$p_t = (p_{t-1}, x_t), \quad \forall t = 1,2,3, \tag{3.9}$$

$$V^B_t = \hat{V}^B_t = V^B_{t-1}\, e^{(\mu-\sigma^2/2)D_{t-1}+\sigma\sqrt{D_{t-1}}Z_t}, \quad \forall t = 1,2,3, \tag{3.10}$$

$$V^S_t = \hat{V}^S_t = V^S_{t-1}\, e^{(\mu-\sigma^2/2)D_{t-1}+\sigma\sqrt{D_{t-1}}Z_t}, \quad \forall t = 1,2,3, \tag{3.11}$$

$$I_1 = I_0, \tag{3.12}$$

$$I_2 = \begin{cases} I_0, & V_t^S \geq I_0, \\ 0, & \text{otherwise.} \end{cases} \tag{3.13}$$

$$I_3 = I_2 - \hat{\delta} I. \tag{3.14}$$

In Eq. (3.9), we update the vector of loan history $p_t$ by adding $x_t$. From the element level, we will let $p_{t't} = p_{t't-1}$ for $t' < t - 1$ and let $p_{t-1,t} = x_t$. We use $D_t$ to represent the number of days between the beginning of period $t$ and the beginning of period $t + 1$ in Eqs (3.10) and (3.11). The asset value fluctuation of the buyers and suppliers will not be affected by decisions of the financial institution, but will follow the stochastic process under our assumption. Readers should be aware that in many other problems, decisions and exogenous information together decide the state transition.

For the inventory value transition, the shipment value is the contracted value at the beginning of stage one. At the beginning of stage two, the supplier will either ship the contracted goods if it functions normally or ship nothing due to bankruptcy. At the beginning of stage three, we may face some loss due to unexpected events. Note that, based on Eqs (3.7) and (3.14), $I_2 = 0$ will lead to $I_3 = 0$.

#### *3.3.3.5 Reward Function*

The reward for the financial institution is negative in stage one, which represents the loan to the supplier at the beginning of the stage and its risk-free interest with *annual rate* $r_f$ at the end of the stage. The risk-free interest is incurred because, otherwise, the financial institution can invest this money in projects with no risks. Note that we assume that the buyer will apply for exactly the same value as the contract value $I_0$ from the supplier. Therefore, the reward function in stage one becomes

$$R_1 = -I_0 x_1 (1 + r_f D_1 / D_{\text{year}}), \tag{3.15}$$

where $D_{\text{year}}$ is the total number of days per year.

In the second stage, the negative rewards (or, the costs) including the risk-free interest from loan in stage one (the first term of Eq. (3.16)) as well as the money borrowed to the supplier at the beginning of stage two and its interest in the second stage (the second term). We replace $x_1$ with $p_{12}$ in Eq. (3.16), both of which refer to the loan percentage in stage one:

$$R_2 = -I_0 p_{12} (1 + r_f D_1 / D_{\text{year}}) r_f * D_2 / D_{\text{year}} - I_0 x_2 (1 + r_f D_2 / D_{\text{year}}). \tag{3.16}$$

In stage three, the financial institution's deterministic costs, $C_3$, comprise the risk-free interest from loan in stages one and two, the money borrowed to the buyer at the beginning of stage three and its interest, that is, let:

$$C_{31} = -I_0 p_{13}(1 + r_f D_1 / D_{\text{year}})(1 + r_f D_2 / D_{\text{year}}) r_f D_3 / D_{\text{year}},$$

$$C_{32} = -I_0 p_{23}(1 + r_f D_2 / D_{\text{year}}) r_f D_3 / D_{\text{year}},$$

$$C_{33} = -I_0 x_3 (1 + r_f D_3 / D_{year}),$$

$$C_3 = C_{31} + C_{32} + C_{33}.$$

Note that the financial institution can only collect the minimum of the final shipment value and the buyer's asset value at the end of stage three, which indicates that the buyer cannot afford the whole shipment value if its asset value is lower than that. Mathematically, we use $\min\{I_3, V_4^B\}$ to capture this, where $V_4^B$ is the asset value (exogenous information) becoming known at the beginning of stage four, which equals that at the end of stage three. Moreover, the financial institution can also collect the interest from the supplier if the supplier's asset value can afford it. Let $r(r > r_f)$ represent the interest rate negotiated between the supplier and the financial institution. If we replace $r_f$ in $R_1$, $R_2$, and $C_3$ with $r$, we have $R_1'$, $R_2'$, and $C_3'$. Then, the interest amounts to $R_1' + R_2 + C_3' - I_0(p_{13} + p_{23} + x_3)$, where the first three terms indicate the capital plus interest that the financial institution should take back. Deducting the capital from the sum, we get the interest that should be paid by the supplier. At the end of stage three, the financial institution can collect $\min\{R_1' + R_2' + R_3' - I_0(p_{13} + p_{23} + x_3), V_4^S\}$. Another uncertainty for the financial institution originates from the inventory in transit. Suppose there is a loss $(I_2 - I_3)$ during ocean transportation, and assume that the carrier will eventually compensate for the loss. Yet, it will delay the banks' progress in collecting money from the buyer and incur overhead costs. Let us assume that the cost is proportional to the loss with multiplier $c$ due to exception handling. Then, the total reward at stage three becomes

$$R_3 = C_3 + \min\{I_3, V_4^B\} + \min\{R_1' + R_2' + R_3' - I_0(p_{13} + p_{23} + x_3), V_4^S\} + (I_2 - I_3)c. \quad (3.17)$$

### 3.3.3.6 Objective Function

The objective is to maximize the total reward of the bank,

$$\max_{x_t \in X_t} \left\{ \sum_{t=1}^{3} R_t \right\}. \quad (3.18)$$

#### 3.3.3.7 Curses of Dimensionality

If we solve our problem with classic dynamic programming, we will face the curses of dimensionality.

- Large state space: The state variable $p_3$ will have arrangement $A(100, 2) = 9900$ possible values, assuming that the banks only consider percentages as multipliers of 1%. If the shipment value $I_0$ = USD 500,000, then we will have 500 possible values of $I_3$ if we agree that every USD 1000 makes a difference. Assume that we have a supplier with an asset value USD 5,000,000 and a USD 10-billion buyer, and we discretize the buyer asset value at an interval of USD 1000 and that of the supplier at an interval of 10 million, we will need 500 states for $V_t^S$ and 1000 states for $V_t^B$. Multiplying the number of possible values for each dimension of the state variable, the state space will become as large as 2.4 trillion.
- Large outcome space: We have three exogenous processes, that is, the supplier's credit risk, the buyer's credit risk, and the inventory-in-transit risk. If we discretize the outcome space based on the same assumption we have made when evaluating the state space size, we will get a 250 million by 250 million transition matrix.
- Medium decision space: The state space will be an arrangement $A(100, 3)$, which is nearly 1 million.

The curses of dimensionality in this problem call for ADP to come into play, which will save a significant amount of time with satisfactory numerical results. Yet, the algorithmic issues including value function approximation and parameter settings require careful design before successful implementation.

For the value function approximation, an intuitive inference is that the value function of the financial institutions increases with the asset values of the supplier and the buyer before leveling off. When the asset values of the supplier and the buyer are sufficient to pay off the financial institutions, the asset value increase will not bring more income to the financial institutions. On the other hand, the value function decreases with the loss size. We conjecture that a separable additive piecewise linear values function $F(S_t^x)$ will perform well in our problem.

Mathematically, let

$$F(S_t^x) = F^S(V_t^S) + F^B(V_t^B) + F(I_t^x) \tag{3.19}$$

where $F^S(V_t^{S,x})$ is the value function with regard to the supplier's asset value, $F^B(V_t^{B,x})$ is the value function with regard to the buyer's asset value, and $F(I_t^x)$ is a value function with the shipment value.

Each value function is approximated with a piecewise linear function and the slopes and intercepts are iteratively updated with the new value observed. The algorithm framework is identical to Table 3.2, except that we will substitute the lookup-table representation of the value function with the separable additive piecewise linear value function.

We will not address the stepsize rule selection in this chapter, but a rich variety of literature addressing this can be found.

## 3.4 Future Research

In this chapter, we briefly introduce the concept of supply chain finance and some of the most common SCF solutions. Incorporating financial flow management, the supply chain players can actively manage the supply chain from previously ignored perspectives and extract previously invisible values for enterprises in the chain, buyers or suppliers.

We also review the existing quantitative research from different perspectives, which embed optimization techniques. After characterizing the stochastic and dynamic nature of the SCF problems, we introduce an ADP framework which is designed to solve stochastic dynamic problems in a short time but with satisfactory results.

Following that, we introduce the multistage supply chain finance concept and model it as a stochastic dynamic program. Our model maps an important problem in supply chain finance to its mathematical representation in a neat but realistic way. We consider and trade off three types of risks against the financial institutions' expected income. This research not only adds a new solution to the current SCF solution set but also potentially supports financial institutions' quantitative decision-making process. Given the model complexity, we attempt to apply ADP to solve our problem.

Going forward to extend our model, we suggest incorporating the external risks such as microeconomic risk, political risk, etc. Our modeling framework is flexible enough to embrace these factors by adding more exogenous processes and model the asset values as the buyer and supplier as a function of these external factors, respectively.

Another extension to the existing game models between the retailer and the supplier (Caldentey & Chen, 2010; Kouvelis & Zhao, 2009) is to move on from the newsvendor problem setting, which assumes that the leftover newspapers (or similar perishable goods) will have no value or very limited value at the end of a period and will be discarded. We believe that it is possible to extend the game between the supplier and buyer to be multiperiod, which allows the leftover goods at an end of a period to satisfy future demands. This extension fits into the reality that most of the supplier–buyer relationship is long term, and there will be game between them during their periodic cooperation.

It will be a lot more interesting if we can model multiple suppliers, multiple buyers, and multiple financial institutions that are aware of the competition they are facing, and each player is making decisions aiming to maximize their own benefits considering other players' response. In such a relation network, the modeling will be extremely complex, and algorithm design would be a substantial challenge.

## References


Aberdeen Group (2006). Get Ahead with Supply Chain Finance: How to Leverage New Solutions for End-to-End Financial Improvement. July 21.

Altman, E. & Sauders, A. (1998). Credit risk measurement: Developments over the last 20 years. *Journal of Banking & Finance, 21*(11–12), 1721–1742.

Bank of America (n.d). *Merging supply chains, emerging supply chain financial needs,* pp. 1–8.

Basu, P. & Nair, S. (2008). Dynamic payables discounting: A supply chain finance perspective. *Proceedings of the Production and Operations Management Society Annual Meeting,* La Jolla, CA, pp. 1–11.

Bertsimas, D. & Demir, R. (2002). An approximate dynamic programming approach to multidimensional knapsack problem. *Management Science, 48*(4), 550–565.

Caldentey, R. & Chen, X. F. (to appear). The role of financial services in procurement contracts. In P. Kouvelis, O. Boyabatli, L. Dong, & R. Li (Eds.), *Handbook of integrated risk management in global supply chains.* John Wiley.

Castanon, D. A. (1997). Approximate dynamic programming for sensor management. *Proceedings of the 36th Conference on Decision & Control,* 1202–1207.

Charnes, A. & Thore, S. (1966). Planning for liquidity for financial institutions: The chance-constrained methods. *Journal of Finance, 21*(4), 649–674.

Chen, F., Drezner, Z., Ryan, K. J., & Simchi-Levi, D. (2006). Qualifying the bullwhip effect in a simple supply chain: The impact of forecasting, lead times, and information. *Management Science, 46*(3), 436–443.

Crouhy, M., Galai, D., & Mark, R. (2006). A comparative analysis of current credit risk analysis. *Journal of Banking & Finance, 24*(1), 59–117.

David, P. & Stewart, R. (2007). *International logistics: The management of international trade operations* (2nd ed). Beijing: Tsinghua University Press.

Demica (2007). The growing role of supply chain financing in a changing world. Demica Report Series.

Eppen, G. D. & Fama, E. F. (1971). Three asset cash balance and portfolio optimization problems. *Management Science, 17*(5), 311–319.

Eppen, D. G. (1979). Effects of centralization on expected costs in a multilocation newsboy problem. *Management Science, 25*(5), 498–501.

Gordy, M. B. (2000). A comparative anatomy of credit risk models. *Journal of Banking & Finance, 24*(1–2), 119–149.

He, M., Zhao, L., & Powell, W. B. (2010). Optimal control of dosages in controlled ovarian hyperstimulation. *Annals of Operations Research, 178*(1), 223–245.

He, M., Ren, C., Wang, Q., & Dong, J. (2011). An innovative stochastic dynamic model to three-stage supply chain finance. *Proceedings of 2011 International Conference on Service Operations and Informatics (SOLI),* Beijing, pp. 1–6.

Kallberg, J. G., White, R. W., & Ziemba, W. T. (1982). Short term financial planning under uncertainty. *Management Science, 28*(6), 670–682.

KMV Corporation (1993). *Credit monitor review.* San Francisco.

Kouvelis, P. & Zhao, W. H. (2009). *Financing the newsvendor: Supplier vs. bank, optimal rates and alternative schemes* (Working Paper).

Lee, L. H., Padmanabhan, V., & Whang, S. (1997). Information distortion in a supply chain: The bullwhip effect. *Management Science, 34*(4), 546–558.

Lee, L. H. & Whang, S. (2000). Information sharing in a supply chain. *International Journal of Manufacturing Technology and Management, 1*(1), 79–93.

Merton, R. C. (1974). On the pricing of corporate debt: The risk structure of interest rates. *The Journal of Financing, 29*(2), 449–470.
Nascimento, J. & Powell, W. B. (2010). Dynamic programming models and algorithms for the mutual fund cash balance problem. *Management Science, 56*(5), 801–815.
Paystream Advisors (2008). *Dynamic payables discounting & supply chain finance: A buyer's guide to working capital solutions,* pp. 1–18.
Powell, W. B. (2007). The optimizing-simulator: Merging simulating and optimization using approximate dynamic programming. *Proceedings of the 2005 Winter Simulation Conference*, pp. 96–109.
Protopappa-Sieke, M. & Seifert, W. R. (2010). Interrelating operational and financial performance measurements in inventory control. *European Journal of Operational Research, 204*(3), 439–448.
Robichek, A. A., Teichroew, D., & Jones, J. M. (1965). Optimal short-term financial decisions. *Management Science, 12*(1), 1–36.
Sabbaghi, A. & Sabbaghi, N. (2004). Global supply-chain strategy and global competitiveness. *International Business & Economics Research Journal, 3*(7), 63–76.
Simao, H. P., Day, J., George, A. P., Gifford, T., Nienow, J., & Powell, W. B. (2009). An approximate dynamic programming algorithm for largescale fleet management: A case application. *Transportation Science, 43*(2), 178–197.
Vickery, K. S., Jayaram, J., Droge, C., & Calantone, R. (2003). The effects of an integrative supply chain strategy on customer service and financial performance: An analysis of direct versus indirect relationships. *Journal of Operations Management, 21*(5), 523–539.
Zhang, Y. S. & Zipkin, P. (1990). A queueing model to analyze the value of centralized inventory information. *Operations Research, 38*(2), 296–307.

CHAPTER 4

# Designing and Assessing Participatory Public Services for Emerging Markets

**G.R. Gangadharan, Anshu Jain, and Nidhi Rajshree**
*IBM Research — Bangalore, India*

A service system is a system for creating an experience that provides value for both the consumer and provider (Shostack & Kingman-Brundage, 1991). Service systems in the public sector are often created by ad hoc methods, and frequently suffer from defects that impair the systems' capabilities to provide value in a systematic and predictable way (OECD, 2005). An effective service design process can help alleviate these problems by exploiting engineering techniques including modeling, simulation, testing, and prototyping.

Most public enterprises, and especially those of emerging market countries, are organized hierarchically with decision making concentrated at the apex of the organization (Denhardt & Denhardt, 2002). Centralization ensures fairness of treatment for both service recipients and employees, clears chains of accountability, and minimizes the exposure of service delivery to political forces. However, institutions that are organized as hierarchies also find it challenging to address nonroutine problems that demand networked solutions and responsiveness to external public forces (Benington, 2007). In response to the criticisms that centralized hierarchies are stultified and inefficient, some public organizations have moved toward decentralization. More recently, organizational boundaries in the public sector have become porous as service delivery becomes the responsibility of a network of public and private organizations, such as government agencies, private firms, and nonprofit organizations, bound together by contracts and grants, with a common commitment to address relevant services (Bowonder, Gupta, & Singh, 2002).

In developed markets, high-speed Internet and social media facilitate participation of the public and are more targeted in identification of their needs. Early government adapters of these web-based technologies are exploiting them to become more adaptive to the needs of the public (Orlikowski, 2000). In emerging markets, the widespread proliferation of mobile communications technology can be seen as an enabler. However, e-government initiatives in these markets are yet to exploit this in a meaningful way (Bhatia, Bhatnagar, & Tominaga, 2009), although other service providers are making substantial progress.

**Service Science, Management, and Engineering.**


DOI: 10.1016/B978-0-12-397037-4.00004-1

A public service has multiple stakeholders, with each deriving some value. The three key stakeholders are the public, the government leadership, and the operations or delivery organizations. Additional stakeholders in the Indian context are the nongovernment organizations and social groups, including self-help groups and their facilitators, or the network of middlemen, which provides opportunities to manipulate the system to the advantage of their clients (Bertrand, Djankov, Hanna, & Mullainathan, 2007).

In this chapter, we observe the public service design process in a European framework project *Citizens Collaboration and Co-creation in Public Service Delivery* (COCKPIT) and two real-world public service delivery processes in India (one of the biggest emerging economies in the world). Based on these case studies, we identify the factors that affect service delivery in emerging markets and propose an extended framework based on COCKPIT for the emerging markets to enhance service delivery.

This chapter is organized as follows: Section 4.1 discusses a participatory service design methodology, namely COCKPIT, used in the European context. Section 4.2 illustrates, using case studies, the fundamental challenges faced by the emerging markets in designing and delivering public services. Section 4.3 presents the details of the design methodology for COCKPIT for emerging markets (CEM). Section 4.4 illustrates the CEM prototype in designing and delivering a public service in an Indian context and discusses the mapping of fundamental challenges in its design to provide better service delivery. Sections 4.5 and 4.6 present the methodologies to assess corruption and transparency, respectively, as these are the primary factors hindering the delivery of public service in emerging markets. Finally, concluding remarks are in Section 4.7.

## 4.1 COCKPIT: A Participatory Service Design Methodology in a European Context

COCKPIT adapts a highly synergistic approach toward the definition of a new governance model for the next-generation public service delivery decision-making process by combining the research areas of citizens' opinion mining, Web 2.0, and service science in the context of the public sector. COCKPIT views the next generation of public services as complex service systems, that is, as an assembly of "business" services, which are provided by people, technology, and organizations. COCKPIT proposes a public service delivery decision-making process that has the following responsibilities:

- It engages citizens in the decision-making process by innovatively collecting and appropriately feeding their opinions and wishes.
- It empowers citizens in making substantiated and responsible contributions in the decision-making process by disclosing vital information about the cost and operation of public services in a user-friendly way.

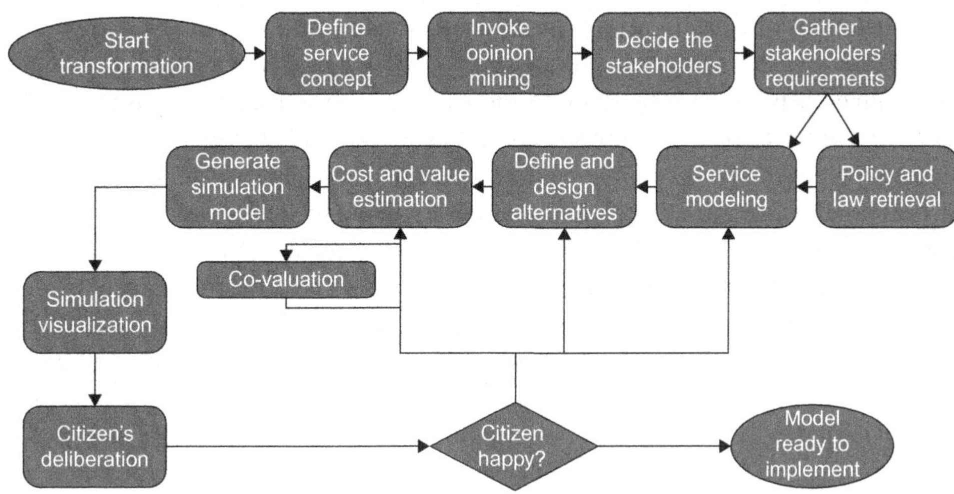


**Figure 4.1: COCKPIT methodology.**

- It incorporates advanced public service optimization, simulation, and visualization tools into the decision-making process.

This methodological approach of COCKPIT on public service delivery is illustrated in Fig. 4.1. Citizens' interactions with Web 2.0 social media are monitored for their opinions and needs on public services in a specified domain. These requirements are then fed into the initial service design process. And then, government decision makers start modeling the service using the public service engineering tool and feed the high-level requirements for the service outcomes. Citizens' opinions and wishes on the selected public services are now available for decision makers to take into consideration. During specification of the service, the existing policy and legal frameworks relating to the corresponding services are consulted using existing text retrieval technology on policy and legal digital libraries. Experts in service delivery feed their constraints related to resources and infrastructure requirements.

Decision makers perform adjustments in service designs, using simulation and visualization, which will reflect in the budgetary and operational constraints. Once decision makers have arrived at a specification for the delivery of the public service, citizens will be presented with a visual simulation of the service in a deliberative platform. The informed judgment of the citizens on the simulated operation and related costs of the selected public services are reported and returned to the decision makers for further consideration and final decisions.

COCKPIT projects a win–win situation between service stakeholders, already established in the commercial world, in the realm of Government and Public Administration services. The fundamental idea of COCKPIT is that Web 2.0 and social media constitute an emerging de facto mass collaboration and cooperation platform between citizens themselves and between citizens and public administrations.

The next generations of citizens, accustomed to collaboration and service co-creation from an early age, are expected to challenge the assumption that public services are provided from producers with credentials (public administrations) to passive consumers (citizens). For this, COCKPIT's governance model empowers and engages citizens in the public service delivery decision-making process by incorporating their opinions in an improved and rather innovative way. Instead of guessing citizens' opinions and wishes about the public services or conducting expensive user surveys and other conventional citizens' feedback surveys, citizens' opinions and wishes are automatically gathered in an efficient, secure, and trusted way from their volunteered interactions in the social media. Subsequently, the gathered opinions and wishes about public services are fed at appropriate states of the service delivery policy and decision-making process. In this way, public service co-creation is realized in the sense that decision makers, who until now could only make assumptions about citizens' and other stakeholders' opinions or use expensively generated feedback, are able to gain access to the public's opinions and wishes in an efficient way.

Empowering and engaging citizens in the conceptualization and design of public services by innovatively extracting and feeding their opinions and wishes in the decision-making process can only accentuate positive attributes toward government and public administration, such as confidence, balance, fairness, and effectiveness. However, the fundamental dilemma is: How should a public administration try to collaborate with Web 2.0 communities of citizens in co-creating public services while not compromising on its responsibility to serve the citizenry as a whole and to implement better services, more efficiently? COCKPIT answers this with a new idea of openness associated with transparency, accountability, and empowerment of citizens in the service delivery decision-making process.

## 4.2 Challenges of Service Provision in Emerging Markets

The COCKPIT method was devised in the European Union context, with assumptions about high maturity of the service design processes, high awareness of citizens about their rights, and high penetration of social media technologies. These assumptions do not apply to the emerging markets. Moreover, given the developing nature of the economy and society, the motivations and needs of citizens are substantially different and more basic in nature. Thus, it is important to identify the key motivators and drivers that impact the delivery of public services in the emerging market context. We analyze the following two real-world public service delivery processes in India. We have used a combination of quantitative and qualitative research techniques for our study. Our main sources of quantitative data were surveys of the official staff involved in case studies and the beneficiaries of case studies. We conducted the survey during the months of September and October 2010 (see Appendix).

### 4.2.1 Case Study: The National Rural Employment Guarantee Scheme

The Mahatma Gandhi National Rural Employment Guarantee Scheme[1] (NREGS), which was launched in 2005 by the Indian Government, focuses on people in rural areas who do not have a regular source of income and aims at providing guaranteed employment for 100 days at certain minimum wage levels to eligible citizens who seek employment.

The NREGS is administered by the central government at three levels: village administration level (*gram panchayat*), block level (*mandal*), and state level.

A survey was conducted using a questionnaire for the beneficiaries of the scheme (see Appendix). The questionnaire was designed to assess few factors such as the citizens' level of awareness of the scheme, grievance redressal mechanisms, and incidences of corruption. We have also studied the official processes of NREGS at the *mandal* (block) level.

Some of the key observations that we noted during the survey are given as follows:

A vast majority of the beneficiaries of any government scheme in rural areas are illiterate and lack the basic understanding of public services offered and their rights to gain benefits from the services. Hence, it is difficult to gain any useful feedback from the beneficiaries regarding the design and delivery of the public services.

The government plays a dual role as a service provider and a designer in all such cases; thus, there is a basic conflict of interest. In this context, and given the extremely low levels of capabilities, it is questionable whether the government would find it desirable or feasible to use the citizens' (beneficiaries') feedback.

Because the NREGS is the key flagship scheme of the government, covering a large number of people, the supply side has been strengthened considerably. This is visible in the infrastructure created for the delivery of this service, in terms of the administration and process machinery. Information and communication technology (ICT) has been widely used, including biometric smart cards for payment systems, mobile phones with embedded software to measure the completed work, and robust accounting systems at various levels. In contrast, it stands out that while strengthening the supply side, the demand side has been ignored to a great extent. The intended beneficiaries suffer due to their low education level and insufficient level of dissemination of information.

The village administration (*gram panchayat*) along with the *mandal* office is responsible for service delivery, awareness creation, payments, grievance handling, and training/skill development. Since this structure is being followed across the country, it lends a considerable amount of power to the functionaries at the very end of the delivery channel, abstracting

[1]http://nrega.nic.in/netnrega/home.aspx.

transparency in the service delivery process. Institutional audit and review mechanisms have not been followed in the delivery channels. Payment to the beneficiaries is made in cash. Each beneficiary has a bank account, with a biometric smart card which an agent of the bank swipes on a machine and carries out the transaction. In spite of such a detailed payment procedure, most of the beneficiaries are unaware that they have a bank account. This unawareness leads to potential deviations and corruption in the delivery system.

### 4.2.2 Case Study: The Targeted Public Distribution System

A scheme for distributing essential commodities such as food grains and kerosene at much cheaper prices to citizens of India is known as a Public Distribution System (PDS).[2] The Targeted Public Distribution System (TPDS) is a food subsidy scheme in India that focuses on the people below the poverty line (BPL). TPDS is operated jointly by the central and state governments, with the former being responsible for procurement, storage and bulk allocation of food grains. The state governments are responsible for distributing food grains to consumers through a network of fair price shops, including identifying BPL families, issuing BPL cards, and supervising and monitoring the functioning of the fair price shops. The state governments are also responsible for moving food grains from district headquarters to the PDS shops, which requires storage at the sub-district level (Saxena, 2009).

Based on our analysis on TPDS, this procedure for selection of BPL beneficiaries is opaque and bureaucratic. Furthermore, the process does not involve *gram sabhas*. There exist many ghost ration cards under names of dead or nonexistent citizens. The food grains are black marketed and this deficiency is covered by distributing bad-quality grains. There are hardly any grievance redressal mechanisms in place.

### 4.2.3 Existence of Corruption and Lack of Transparency

The key factors that affect the public service delivery in emerging markets include some obvious issues of pricing and costing constraints, lack of IT infrastructure, and diversity in the citizen population. Some other implications of the context are the absence of necessary information sources, lack of transparency in public service decision making and operations, and also many instances of corruption, nepotism, and ignorance, which significantly hamper the participation process (Gangadharan, Jain, Rajshree, Hartman, & Agrahari, 2011; OECD, 2005). Among these implications, based on our observations of the case study and the literature survey, we have identified the existence of corruption and lack of transparency as two significant challenges in service provisioning in emerging markets.

[2]http://www.fcamin.nic.in/dfpd/EventListing.asp?Section=PDS&id_pk=1&ParentID=0.

Corruption is a major issue related to delivery of human-aided public services. Systemic deficiencies, such as lack of transparency, deficit of information, weak accounting practices, and obsolete policies, provide a favorable environment to breed different forms of corruption. Corruption leads to lack of trust of citizens in the public services.

An ideal participatory public service design approach should ensure that the citizens are completely convinced about the value that is delivered to them by making visible all the costs and considerations involved in the operations of services. Information dissemination is also important in increasing trust, reducing corruption, and bypassing existing power structures. Menon (2006) argues that "the roots of corruption in India lie in outdated administrative procedures and rules ... the systems are nontransparent in order to maintain a command and control style of administration and maintaining poor levels of information available to the citizens about the rules of the game."

Lack of awareness of citizens and lack of dissemination of information regarding services serve as fundamental reasons for existence of corruption and lack of transparency in service provision.

In NREGS, there is dissatisfaction among the beneficiaries in the daily wages paid to them for the work delivered. Also there have been many instances of muster rolls being fudged with entries corresponding to ghost job cards. The *gram sabhas* are also involved in procuring funds for projects that have never happened or bloated funds for existing projects.

In TPDS, there is dissatisfaction among the beneficiaries on the quality of essential commodities distributed to them. The fair price shop (point of sale under PDS) owners manipulate the muster rolls to show bloated quantities of the items distributed. Many undeserving citizens who are above the poverty line get false BPL cards and get benefits from the service. The discussion paper from United Nations (Armstrong, 2005) discusses the revival in awareness and openness about issues related to integrity, transparency, and accountability in public administration with a perspective from the emerging markets. However, the role of technology in enabling the above issues is still not as significant as it should be.

## 4.3 COCKPIT for Emerging Markets (CEM)

COCKPIT for emerging markets (CEM) adapts the existing COCKPIT methodology to cater for the challenges faced in public service delivery in emerging markets. The CEM methodology is illustrated in Fig. 4.2.

To address the lack of awareness of public services by the citizens, CEM proposes the use of simpler information dissemination media such as posters, mailings, and awareness campaigns that are run with the help of self-help groups and nongovernmental organizations (NGOs). Furthermore, elicitation and dissemination of information can be supported by developing simple interfaces, such as spoken, graphical, and textual interfaces. We also leverage

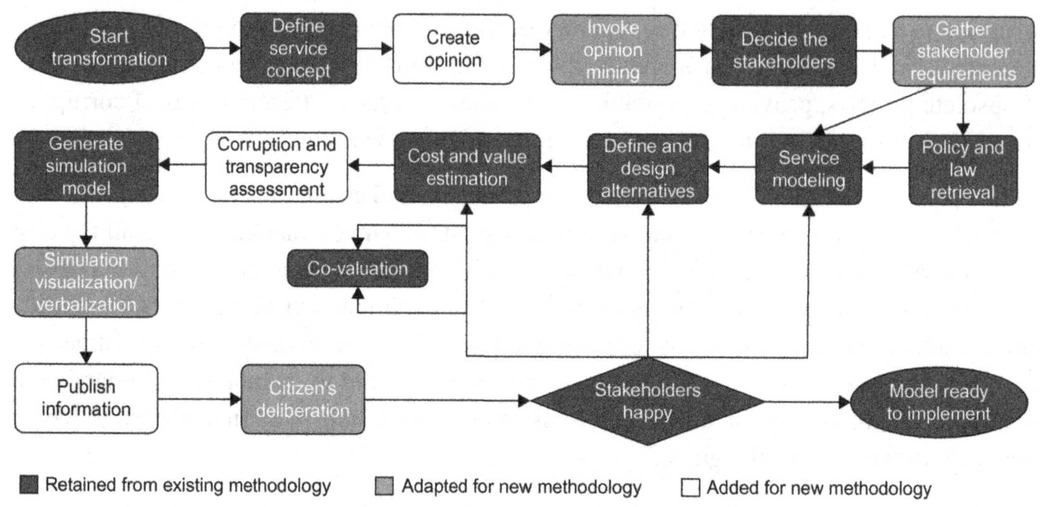


**Figure 4.2: COCKPIT methodology for emerging markets.**

the popularity and pervasiveness of mobile phone handsets for communication in CEM. Some efforts are already under way using the Spoken Web technology (Agarwal, Kumar, Nanavati, & Rajput, 2009).

Citizens' opinions and needs on public services can be gathered by simple interactions, supported by facilitators including NGOs and self-help groups. Technologies such as Optical Mark Recognition sheets can be used for gathering opinions from citizens due to their low literacy rate and lack of interaction with Web 2.0 and social media. The collected information can then be digitized and fed into the initial service design process.

After identifying the stakeholders and their requirements, the existing policy and legal framework relating to the corresponding service are consulted using existing text retrieval technology on policy and legal digital libraries.

The government decision makers then model the service by inputting the high-level requirements for the service outcomes. Citizens' opinions and wishes on the selected public services are available to the decision makers to be taken into consideration. Our CEM service design tool provides a number of service components with built-in transparency features and governance options for guarding against corruption. Any interaction between different participants is modeled as a separate entity with attributes describing its transparency and potential for corrupt subversion. The CEM contains explicit measures of transparency and potential for corruption as significant indicators of quality of service. These measures are aggregated over the tasks in the service delivery process, and indices of transparency and vulnerability to corruption are given as parameters of each alternative service design. The CEM incorporates feature modeling capabilities to provide a variety of service offerings, for example, the same service at different price points with varying service levels. The resulting service designs are simulated and visualized in order for the decision makers to make adjustments that will reflect budgetary and operational constraints.

Once decision makers have arrived at a specification for the delivery of the public service, citizens will be presented with a visual or verbal simulation of the service in a deliberative platform. We propose that mobile phones and video broadcasts to NGOs and self-help groups could be used as the medium to communicate the verbal or visual simulation of these services and elicit a public response. This response could then be communicated back to the decision makers as in the first round of requirements gathering.

## 4.4 CEM Prototype and Implementation

In this section, we give an illustration of the CEM methodology using NREGS. Our CEM implementation of NREGS does not represent the views of the service providers and/or associated third-party sources in the NREGS under the Government of India.

In the current implementation, the laborers are required to give their personal details and a fingerprint to the local officials to receive a job card. This job card is, in effect, a bank account, but most of the recipients are unaware of this and do not have any experience with any financial services. When the work is performed, an official measures the amount of work done, transfers this information to the central office, and credits a payment to the laborers' bank accounts. Subsequently, the officials visit the village, validate the identity of the laborer using a fingerprint, and transfer cash based on the credit available in the bank account.

There are numerous opportunities for exploitation inherent in this system, whereby all the power lies in the hands of the officials, and the service recipients are ignorant of their rights and the processes to express grievances. In fact, Shankar, Gaiha, and Jha (2010) point out that the chief beneficiaries of the NREGS program are those connected to the authorities and with higher levels of education and affluence.

The initial step in the redesign of such a service is to enlist local independent facilitators with knowledge of the local language. These facilitators are equipped with a list of questions and a set of instructional materials in the local language that details the rights of the citizens and explains the processes in a simple and clear manner. The questionnaire focuses on the experiences of the laborers with the scheme and prompts them for details of where the system is vulnerable to exploitation and where their grievances lie.

The CEM is then used to describe the service in terms of its three major processes: issue of the job card, performance of work and its measurement, and transfer of funds. The tasks in each process are assessed for transparency and vulnerability. The tool offers alternatives with higher degrees of transparency from a library of components with standard methods for improvement – for example, adding the transmission of SMS to the recipients at various stages of the process that are not currently visible to them, similar to the work volume assessment.

Each touchpoint is evaluated, and additional checks and balances are proposed for vulnerable points of interaction. In the case of funds transfer, the amount of credit in the recipients' bank accounts is sent concurrently to the recipient and to a central auditing function.

The tool also proposes the use of standard patterns for service, such as a separate and more informative process for first-time users of the system, and the generation and dissemination of instructional materials in the form of posters and checklists.

The tool then generates a number of alternative service delivery options and evaluates each of them according to the criteria of cost, time, transparency, and vulnerability. It also generates a simulation model that is parameterized, so that alternatives can be explored by skilled designers. A simplified explanation of the service alternatives is prepared for the facilitators to take to the villagers in a second round of consultation.

During the second and subsequent rounds of consultation, the service alternatives are evaluated by the villagers in a simplified manner, asking only for pairwise comparisons of different features. The results of these comparisons are fed back into a third-party tool that uses the analytic hierarchy process to rank the features and inform the service designer about the value perceived by the consumers.

## 4.5 Assessing Corruption in Public Services in Emerging Markets

There exist various tools to measure and assess corruption in public services. The most commonly used tools are surveys and institutional diagnostics. Public opinion surveys combined with expert opinion and public official surveys are used to measure the perceived corruption in a public service. Qualitative and quantitative institutional diagnostics, such as audits and, compliance with anticorruption measures, are used to assess the strengths and weaknesses of a government institution (Zöllner & Teichmann, 2007).

Using the traditional design patterns approach in software engineering (Gamma, Helm, Johnson, & Vlissides, 1994), we define recurring patterns of corruption by a framework of five dimensions:

1. **What:** defines the problem or the situation
2. **Use when:** defines the constraints under which the solution must be applied
3. **Why:** defines the rationale
4. **How:** defines the solution
5. **Example:** indicates the instances where the solution has worked.

Table 4.1 lists a summary of the primary patterns we have described using the above framework.

**Table 4.1: Key patterns of corruption.**

| Name | Pattern | | | | |
|---|---|---|---|---|---|
| | **What** | **Use when** | **Why** | **How** | **Example** |
| Target Identification Irregularities | Fuzzy eligibility criteria for identifying public service beneficiaries. Wrong recipients get the benefits of the service | Targeted section eligible for service, administered targeting, criteria for targeting weak | Errors of inclusion, errors of exclusion | Accurate metrics, self-targeting, decentralized targeting, channels for appeal | Combination of administered and self-targeting in Tunisia: milk subsidies shifted to reconstituted milk packaged in cheaper cartons less attractive to wealthy consumers (Iqbal, 2006) |
| Consolidation of Power | Too many discretionary powers to service managers and delivery personnel in determining critical outcomes | Critical outcomes can be flipped based on subjective assessment | The importance of a favorable outcome is highly valued by the beneficiary | Transparency in decision making, supported by concrete facts, downward accountability, Social Audits | Hong Kong Independent Commission Against Corruption (ICAC), used supervision, accountability and publicity to prevent corruption (Scott, Carstairs, & Roots, 1998) |
| Conflict of Interest | Beneficiary of service related to the decision makers | Human discretion applied by decision makers can provide benefits to undeserving people | Nepotism and, favoritism often cited as impediments to rightful decisions | Establish accountability chain, regular audits and verification, signed declarations when making such decisions | Recommendations to curb nepotism and other forms of corruption in European Union (Li, 2000), such as increased accountability, transparency, independent EU prosecutors body, strengthening fraud investigation |

We illustrate our approach to detect corruption patterns through an example of the Target Identification Irregularities Pattern as applied to the TPDS. In India, TPDS services are provided to citizens falling below the poverty line, which ensures the availability of essential food grains to the poor. Many states in India suffer from unclear criteria for inclusion to BPL. Moreover, the procedures used to screen the families for determining their eligibility are not well defined. As a result, many poor households are in the Above Poverty Line (APL) category and are denied access to subsidized food grains under TPDS. On the contrary, many of those with political patronage have found their place in the BPL category.

We adapt the following methods to increase the accuracy of eligibility criteria for identification of the poor:

- *Accurate measurement of targets identification criteria*: Targets identification criteria should be accurately measurable. In the case of TPDS, generally the indicator is poverty where measurement of wealth is difficult to achieve. More appropriate indicators could be health (nutritional status) or education.
- *Administered targeting/self-targeting with limited privileges*: Service can be designed by combining both administered and self-targeting. Administered targeting selects the target population based on some standard or condition that the citizen must meet. On the other hand, self-targeting services involve some conditions or characteristics that discourage some citizens from accessing the service. In the case of PDS service, the targeted population may be identified by distributing ration cards based on some administered criteria of poverty; however, these cards may be self-targeting if they are valid only for purchase of inferior commodities (Iqbal, 2006).
- *Dynamic measurement of criteria*: Targeting based on criterion that is prone to change needs indicators that look at causal links rather than outcomes. In the case of PDS, the indicators of poverty are static measures of living standards such as family income or expenditure, which can change over time. Better indicators for a family earning a living from agriculture could be sufficient rain for a healthy harvest (Jenkins & Goetz, 2002).
- *Withdrawal of service*: Withdrawal of benefits for those who no longer meet the criteria can be tapered over time rather than suddenly cut off. In the case of PDS, this can be done by continued supply of essential food commodities to those BPLs who have reached the poverty line.
- *Decentralized targeting with local context*: This makes selected indicators more relevant to the local situation. In the case of PDS, geographic indicators such as rain can help better identification. In desert areas, assets such as land might not be appropriate indicators as these would be mostly covered by sand dunes and will not be productive.
- *Channels for appeal*: This is a means of tackling many forms of corruption. Surprisingly, many beneficiaries or victims of corruption are hardly aware of the channels that exist.

Similarly, other patterns of corruption could be matched with the existing public services to assign them indices of vulnerability to corruption.

## 4.6 Transparency Assessment of Public Services in Emerging Markets

As described in Section 4.2, one of the key requirements from modern day governments in public service design is transparency and openness of the service to citizens. Although transparency, visibility, and openness are certainly used as an abstract objective or goal during the design of services, there is no easy way to represent them in a formal service design and to quantify, measure, and analyze transparency of service design objectively. Existing service design tools leave transparency to the designer's discretion. They do not have inbuilt methods or tools to design transparency in the system or enforce them on designers. There is an increasing need for a framework that helps service designers define transparency of their design. To assess transparency, an obvious prerequisite was to model the concepts relating to transparency in a service design. Transparency is essential at any citizen-facing interface in a service. Such interfaces are popularly described as touchpoints. The concept of touchpoints has been known for a few decades now, since the formal analysis of service encounters (Bitner, 1990). To allow transparency assessment, we conceived a method and framework that can allow users to model transparency touchpoints in business processes and service design. This model can then be enhanced and leveraged with intelligence in the system related to various analytics and computations on top of the model.

Based on the inputs of transparency touchpoints, the proposed system computes and outputs a transparency score for the service, which objectively quantifies the levels of transparency, openness, and visibility of the service. This input can then be used to decide the optimal design between multiple design alternatives and to compare related services with respect to transparency. Our method allows modeling of transparency beyond the traditional line of visibility in a service blueprint model and allows the designers to expose even relevant information and notifications from the backstage processes to increase the visibility. We use various attributes of these touchpoints, such as distance to customer, desired level of transparency, and total number of processes of the system, to calculate a transparency metric.

A Transparency Score is computed by aggregating the transparency over all the touchpoints and normalizing the aggregate score over all the tasks in the process. Although the absolute value of transparency score may not have immediate relevance and meaning, the transparency score can become a scale to compare the transparency of different service designs and an assistant to the decision-making process in selection of the ideal design.

$$\text{Transparency Metric} = \frac{\sum_{i=\text{TP}} f(L_i)g(D_i)}{h(n)}$$

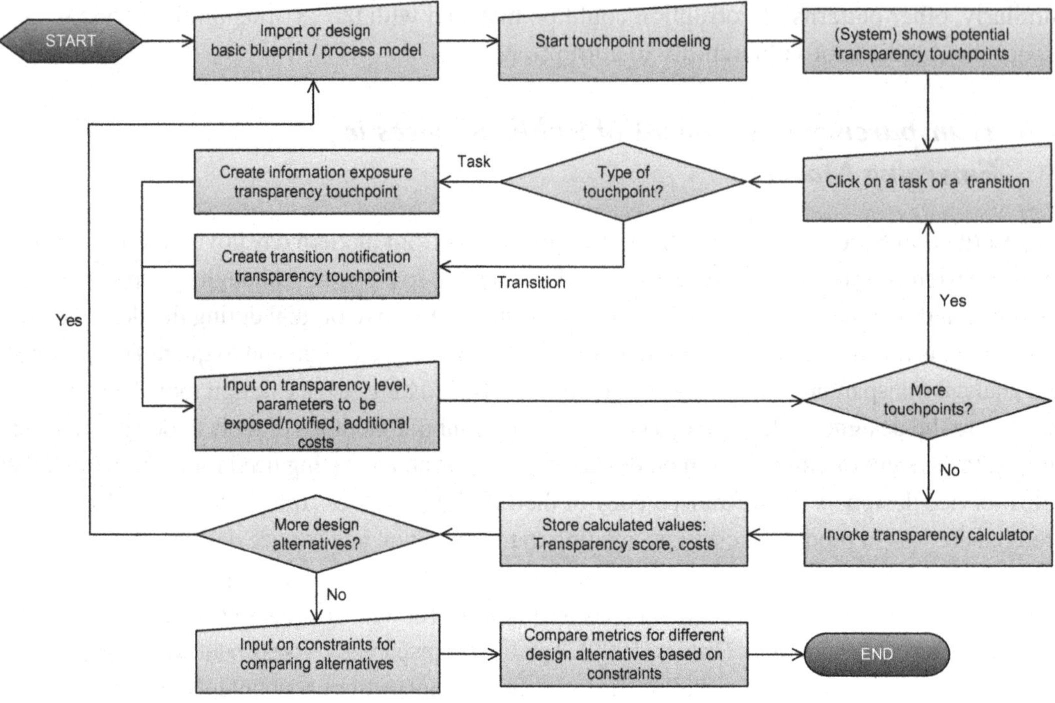


**Figure 4.3: Flow of activities for performing transparency assessment.**

where TP is a transparency touchpoint in the system, $L_i$ the transparency level associated with touchpoint ($TP_i$), $D_i$ the distance from the line of interaction to a transparency touchpoint ($TP_i$), and $n$ the total number of tasks in the system or the total number of tasks in the system that are below the line of interaction.

Cost associated with transparency touchpoints is given by

$$\text{Costs} = \sum_{i=\text{TP}} c_i$$

where $c_i$ is the cost associated with a transparency touchpoint ($TP_i$).

The cost component is important when evaluating the cost benefit of alternative designs in the case of different transparencies. Figure 4.3 shows the basic flow of a system for performing a transparency assessment of a service design.

## 4.7 Concluding Remarks

Modernized public sector agencies share some of the objectives of private sector agencies such as efficiency, goal accomplishment, maximum utilization of employees and resources, and minimization of operational costs. However, they have to address other equally important goals such as maximizing transparency, minimizing the incidences of corruption, creating value for citizens, and thereby increasing the complexity of public service delivery manifold. In this chapter, we have described a methodology for participatory service design in emerging markets. The key issues we addressed are related to adapting the methodology to a service clientele that may be illiterate and lacking in knowledge of their rights. We have primarily focused on the issues of lack of transparency and existence of corruption, including methods for assessing these qualities and providing alternatives to current practices. We have used real-world case studies from an emerging market country to motivate our methodology.

## Acknowledgments

This work is partially supported by the European Union, within the research project COCKPIT (Citizens Collaboration and Co-creation in Public Sector Service provision [Call: FP7-ICT-2009-7.3 Grant agreement no.: 248222]). We also express our gratitude to Alan Hartman, Krishna Kummamuru, and Arun Sharma.

## References

Agarwal, S. K., Kumar, A., Nanavati, A., & Rajput, N. (2009). *Content creation and dissemination by-and-for users in rural areas*. Paper presented at the IEEE/ACM International Conference on Information and Communication Technologies and Development (ICTD), Doha, Qatar, April 17–19.

Armstrong, E. (2005). Integrity, transparency and accountability in public administration: Recent trends, regional and international developments and emerging issues. Economic and Social Affairs report, United Nations.

Benington, J. (2007). *From private choice to public value*? Retrieved from http://www.cihm.leeds.ac.uk/document_downloads/John_BeningtonPrivate_Choice_to_Public_Value.pdf

Bertrand, M., Djankov, S., Hanna, R., & Mullainathan, S. (2007). Obtaining a driving license in India: An experimental approach to studying corruption. *Quarterly Journal of Economics, 122*, 1639–1676.

Bhatia, D., Bhatnagar, C., & Tominaga, J. (2009). How do manual and e-government services compare? Experiences from India. *Information and communications for development*, World Bank Report.

Bitner, M. J. (1990). Evaluating service encounters: The effects of physical surroundings and employee responses. *Journal of Marketing, 54*(April), 69–82.

Bowonder, B., Gupta, V., & Singh, A. (2002). *Developing a Rural Market ehub: The case study of e-Choupal Experience of ITC*. Indian Planning Commission Report.

Denhardt, J. V. & Denhardt, R. B. (2002). *The new public service: Serving, not steering*. Armonk, NY: M.E. Sharpe Publishers.

Gamma, E., Helm, R., Johnson, R., & Vlissides, J. (1994). *Design patterns: Elements of reusable object-oriented software*. Boston, MA: Addison Wesley.

Gangadharan, G. R., Jain, A., Rajshree, N., Hartman, A., & Agrahari, A. (2011). *Participatory service design for emerging market*. Paper presented at the Proceedings of the IEEE International Conference on Service Operations and Logistics, and Informatics, SOLI, Beijing, July 10–12.

Iqbal, F. (2006). *Sustaining gains in poverty reduction and human development in the Middle East and North America*. Washington, DC: The International Bank for Reconstruction and Development, World Bank.

Jenkins, R. & Goetz, A. M. (2002). Civil society engagement and India's Public Distribution System: Lessons from the Rationing Kruti Samiti in Mumbai. In *Workshop of the World Development Report (WDR)*, Oxford, UK, 4–5 November.

Li, K. (2000). *Recommendations for the curbing of corruption, cronyism, nepotism, & fraud in the European Commission*, 24 B.C. Int'l & Comp. L. Rev. 161.

Menon, V. (2006). Anti-corruption in India: Issues and strategies. In V. K. Chand (Ed.), *Book: Reinventing public service delivery in India*. New Delhi: Sage publications.

OECD (2005). *Modernising government: The way forward.* Paris, France: OECD Publishing.

Orlikowski, W. J. (2000). Using technology and constituting structures: A practice lens for studying technology in organizations. *Organization Science*, 11, 404–428.

Saxena, N. C. (2009). *Modernisation of public distribution system in India through computerisation.* Retrieved from www.scccommssioners.org.

Scott, T., Carstairs, A., & Roots, D. (1998). Corruption prevention: The Hong Kong Approach. *Asian Journal of Public Administration, 10*(1), 110–119.

Shankar, S., Gaiha, R., & Jha, R. (2010). *Information and corruption: The national rural employment guarantee scheme in India* (ASARC Working Papers). Australian National University.

Shostack, G. L. & Kingman-Brundage, J. (1991). How to design service. In: C. Congram & M. Freidman (Eds.), *The AMA handbook for the service industries* (pp. 243–261). New York: AMACOM.

Zöllner, C. & Teichmann, I. (2007). *Mapping of corruption and governance measurement tools in Sub-Saharan Africa.* Transparency International Policy and Research Department, UNDP.

## Appendix

## Questionnaire for NREGS Beneficiaries

1. Are you currently a member of the NREGS?
   a. Yes
   b. No
2. If No, why not?
   a. Not aware
   b. Not eligible
   c. Application in progress
3. If Yes, how did you learn about the NREGS?
   a. TV
   b. Radio
   c. Newspaper
   d. Pamphlet
   e. Personal visit by officials
4. How did you submit your application for registration for the scheme?
   a. Written
   b. Oral
5. Did you pay any fee/charges to register for the scheme?
6. Are you aware of any verification procedure conducted before your job card was issued?
7. After registering for the scheme, within how many days did you receive your job card?
8. In case of non-receipt of job card, are you aware of whom to contact for grievance redress?
9. Are you aware of the application for job procedure after receiving job card?
10. Did you receive a receipt once your application was processed?
11. How far is the place where your work has been allocated?
12. If your workplace is beyond 5 km, do you receive any additional wages?
13. What is your current daily wage rate?

14. Do men and women receive the same wage rate?
15. What is the frequency at which you receive the wages?
16. Do you usually receive timely payments?
17. Are you provided the following facilities at your work place?
    a. Creche
    b. Drinking water
    c. Shed
18. How do you receive your wages?
    a. Direct cash payment
    b. Bank account
    c. Post office
19. Did you pay any charges for opening your bank account?
20. Is the registration to *gram panchayat* open throughout the year?
21. Would you be interested in giving feedback to the government for improvement in the scheme?
22. What would be an appropriate incentive?
    a. Money
    b. Clothes/utensils/food
23. If yes, would you be interested in SMS- or recorded IVR-type polls or feedback?
24. How can the government reach you to solicit your feedback regarding the scheme?
    a. Post cards
    b. Telephone/mobile
    c. Village sabha/open forum
25. Are you also involved in
    a. Other government services
    b. Micro-finance services
    c. NGO-delivered services
26. What is the nature of the work allocated to you?
27. Do you have any influence in choosing the type of work allocated to you?
28. Are you aware of any corruption or unfairness in the system?
29. What parts of the Employment Guarantee service do you find difficult/painful?
30. What improvements would you like to see in the current scheme?
31. Can you compare the scheme with any other government service/scheme? How would you rate the scheme?
32. Do you understand how much the scheme costs the government to deliver? Are you interested in this question?

CHAPTER 5

# Recommendation Algorithms for Implicit Information

Xinxin Bai*, Jinlong Wu†, Haifeng Wang*, Meng Zhang*, Jun Zhang*, Yuhui Fu*, Xiaoguang Rui*, Wenjun Yin*, and Jin Dong*

*IBM Research – China, Beijing, China

†IBM Research – China, Beijing, China and Peking University, Beijing, China

## 5.1 Introduction

With the rapid development of data storage technology and further product or service (called *item* uniformly in this chapter) diversification, many types of enterprises, such as e-commerce and retail enterprises, have been accumulating more and more transaction data from their customers, who have subsequently had to choose between too many different items. To help customers find out what they need, recommender systems have been developed to offer personalized suggestion for them.

For customers, recommender systems can help them find items which they are interested in. For enterprises, recommender systems can improve the loyalty of their customers by enhancing the user experience and further convert more browsers to consumers. Hence, many companies are making a great deal of effort to establish and improve their own commercial recommender systems. One of the most successful companies is *Amazon*, which has spent over 10 years in developing its recommender system. In return, Amazon added about another 20% sales from recommendations in 2002 (Linden, Smith, & York, 2003; Paterek, 2007). Another famous example is *Netflix*, an online movie rental company. Netflix established the Netflix Prize competition (Netflix, 2009) to improve the accuracy of its movie recommender system—*Cinematch*—by 10%.

A recommender system first achieves user preferences by analyzing their histories of usage behaviors then uses the resulting models to produce personalized recommendations for the users. For example, an online store can learn customer preferences according to product ratings and product purchase history, i.e. browse, buy, add to the cart, etc. Product ratings are usually called *explicit ratings*, while product purchase histories are usually called *implicit ratings*) (Cho & Kim, 2004; Hu, Koren, & Volinsky, 2008). In light of the




DOI: 10.1016/B978-0-12-397037-4.00005-3

different ways to produce recommendations, recommender systems are usually classified into three types (Albadvi & Shahbazi, 2009; Cho & Kim, 2004):

- *Content-based filtering (CBF)*: CBF methods provide recommendations based on features of users and items. Usually, features of an item such as price and type are given in advance. Features of one user are created according to his (her) consuming items. Recommendations to one user are those items whose features best match those of the target user. CBF has some major drawbacks: (1) CBF depends on preprocessing features of items, which are usually not easy to obtain and (2) CBF usually recommends to one user items that are similar to ones he (she) has already consumed. Hence, other types of items which the user is not familiar with are usually not presented to him (her).
- *Collaborative filtering (CF)*: CF methods do not require any explicit feature of items or users. They only rely on past behavior of users, which might cause explicit or implicit ratings of users to items. CF first analyzes interactions of user–item pairs on the basis of known ratings and then uses the interactions to produce recommendations. Its fundamental philosophy is that two users probably continue liking similar items if they liked similar ones (user-based CF) or one user likes items which are similar with ones he (she) liked (item-based CF). CF does not rely on explicit preprocessing features. Hence, it is domain free and can circumvent difficulties faced by CBF. But at the same time, CF introduces some new difficulties: (1) Cold start: CF can rarely produce effective recommendations to a new user, since there are few ratings available for the new user. Similar difficulty exists for a new item. (2) Scalability: CF systems may produce recommendations to millions of users from tens of thousands of items. The conventional CF methods need to calculate the similarity of each pair of users or items and to store all of them in the main memory of computers, which is challenging.
- *Hybrid of CBF and CF:* Hybrid filtering methods combine CBF and CF methods to overcome the drawbacks of both CBF and CF and benefit from their advantages. To combine CBF and CF the following three methods are suggested (Albadvi & Shahbazi, 2009): (1) Weighted combination: At first, CBF and CF methods are applied separately, then their resulting predictions are combined to achieve the final predictions. (2) Mixed combination: At first, CBF and CF methods are applied to obtain recommendations separately, then the two lists of recommendations are combined together to achieve the final list of recommendations. (3) Sequential combination: Initially user features are formed by CBF methods. When there are enough ratings for the users, CBF methods can be replaced by CF methods to obtain the final recommendations.

CF systems may depend on user preferences for items from different sources. The most precise method is to allow users to rate items according to their interests. For instance, Netflix encourages users to express their preferences to watched movies by using different numbers of stars (1, 2, 3, 4, or 5). But in many practical applications, explicit ratings of items are not available. Internet users often browse news from many websites quickly. It is impossible to ask them to rate each news item they watched. Supermarkets store plenty of transaction data, but usually obtain little direct information about whether one user likes one product or not. When explicit preferences are not available, CF systems use implicit information to infer user preferences (Oard & Kim, 1998). For supermarkets, implicit information may include the purchase history. For e-commerce websites, implicit information may include what products users bought, added to the cart, and browsed.

So far, most of the published work focuses on exploring how to use explicit ratings, especially after Netflix instigated the Netflix Prize competition (Netflix, 2009) in October 2006. However, as we stated before, only implicit information is available to construct CF systems in some practical circumstances. Implicit information does not supply user preferences as precisely as explicit information does. It does, however, reduce users' burden since explicit ratings to items are not necessary. Some works have been published investigating implicit information, such as Hu et al. (2008) and Oard and Kim (1998).

Compared with explicit information, implicit information exhibits distinct characteristics (Hu et al., 2008). Values of explicit information usually represent the level of preferences, but values of implicit information usually represent frequencies of user actions. For instance, ratings 1 and 5, respectively, represent "totally dislike" and "quite like" in the Netflix Prize data set. However, in our supermarket data set, ratings represent how many times an item is bought by a user. More detailed distinctions between explicit and implicit information can be found in the study by Hu et al. (2008).

In this chapter, we first explore some methods to analyze implicit information. The motivation of the exploration is to establish an effective recommender system for a supermarket to produce personalized recommendations for its customers. The implicit information is how many times a customer bought a product in 2 months. Apart from the direct purchase history, several features for each user are extracted by experts according to the purchase history of the user. We then propose a hybrid model to integrate all known information and then extend the method suggested by Hu et al. (2008) to solve the hybrid model efficiently.

This chapter is organized as follows. In Section 5.2, we introduce the notation, recommendation framework, and quality metric used throughout the chapter. In Section 5.3, we briefly review the conventional neighborhood-based algorithm and then propose a more efficient similarity measure and recommendation strategy. Some extensions to MF models are introduced to analyze implicit information in Section 5.4. In Section 5.5, we propose a hybrid model to combine implicit and feature information and then develop an efficient and well-scalable algorithm. In Section 5.6, we first introduce the supermarket data set and the preprocessing step we use, and then show the results of our new models applied to the new data set and compare them with the results of old models. Finally, we conclude in Section 5.7. This chapter is an extension of the conference version (Bai et al., 2011).

## 5.2 Preliminaries

In this section, we will start by introducing the notation that we will use throughout this chapter. Then, we take a brief look at the recommendation framework and discuss the quality metrics that are used to evaluate the prediction results.

### 5.2.1 Notation

Let $U$ and $I$ be the numbers of users and items, respectively. Denote the associated value of user $u$ and item $i$ as $r_{ui}$. For data sets with explicit information, $r_{ui}$ represents the rating of user $u$ to item $i$. In the Netflix Prize data set, $r_{ui} \in \{1, 2, 3, 4, 5\}$. For data sets with implicit information, $r_{ui}$ usually reveals observations for user actions (Hu et al., 2008). In the supermarket data set that we use in our experiments, $r_{ui}$ represents the number of times user $u$ purchased item $i$. Although in this chapter we mainly focus on analyses of implicit information, we simply call $r_{ui}$ the rating of user $u$ to item $i$ consistently. Algorithms for explicit information usually consider ungiven ratings of user–item pairs as missing data and ignore them when recommendation models are constructed. This approach does not work for implicit information since there is no negative information in this situation (Hu et al., 2008). Hence, here we set $r_{ui}$ to be 0 if no action happened between user $u$ and item $i$, as suggested in the study by Hu et al. (2008). In our supermarket data set, $r_{ui} = 0$ means that user $u$ did not purchase item $i$.

Denote the training set by $\mathcal{P}$. Without confusion, we also use $\mathcal{P}$ to express the set of all user–item pairs with nonzero ratings on the training set, namely, $\mathcal{P} = \{(u,i)|r_{ui} > 0\}$. Denote $\{i|(u,i) \in \mathcal{P}\}$ by $\mathcal{P}_u$ and $\{u|(u,i) \in \mathcal{P}\}$ by $\mathcal{P}^i$. We usually use bold letters to express tensors, e.g., bold lower case letters for vectors and bold capitals for matrices. For example, $\mathbf{R} = (r_{ui}) \in \mathbb{R}^{U \times I}$ represents the rating matrix. The hyperparameters of models are expressed by Greek letters, such as $\alpha$, $\beta$, and so forth.

### 5.2.2 Recommendation Framework and Quality Metric

We always construct our models in the training set, and evaluate their effectiveness in the test set. After models are constructed, predictions for all items are acquired from the models, and then items with the highest predictions are recommended to users.

To evaluate the recommendation quality, two metrics—*recall* and *precision*—are widely used. Recall is defined as the ratio of the number of items that appear in both the recommendation list and the test set to the number of items in the test set. Precision is defined as the ratio of the number of items that appear in both the recommendation list and the test set to the size of the recommendation list. However, recommendation lists with larger sizes yield higher recalls and lower precisions compared with lists with smaller sizes. Hence, a metric called $F1$, which combines recall and precision, is widely used to balance both of them and produce more reasonable quality evaluation (Albadvi & Shahbazi, 2009; Cho & Kim, 2004). $F1$ is defined as follows:

$$F1 = \frac{2 \times \text{recall} \times \text{precision}}{\text{recall} + \text{precision}}. \tag{5.1}$$

The size of the recommendation list can be found by cross-validation to obtain the highest $F1$. But in practical applications, the size of the recommendation list is usually restricted by practical situations. For instance, a supermarket usually requires that one sheet should cover the whole recommendation list. Furthermore, according to our experiences and experiments, when the size of the recommendation list is fixed, an increase in precision causes an increase in $F1$. So, we simply use precision to evaluate the recommendation quality and fix the size of the recommendation list to 10, since the purpose of the chapter is to compare different models for recommendations but not to construct a practical recommender system. Note that in this chapter, the *recommendation precision* means the average value of the precisions of all users.

## 5.3 Neighborhood Models

Neighborhood (also called $k$ nearest neighbors or kNN for short) models have been used to obtain recommendations since the inception of recommender systems (Albadvi & Shahbazi, 2009; Breese, Heckerman, & Kadie, 1998; Cho & Kim, 2004), and are still popular nowadays because they are easy to understand and realize. KNN approaches obtain the predictive value of one user–item pair by using the weighted average of ratings of its neighbors. They consist of three steps:

1. Similarity formation: This step finishes the calculation of similarity between users (user-based kNN) or items (item-based kNN). Some widely used similarity measures include *Pearson correlation* and *Cosine*. For example, cosine similarity between users $u$ and $v$ is calculated as follows:

$$S^{c}_{uv} = \frac{\sum_{i \in \mathcal{P}_{uv}} r_{ui} r_{vi}}{\sqrt{\sum_{i \in \mathcal{P}_{uv}} r^2_{ui} \sum_{i \in \mathcal{P}_{uv}} r^2_{vi}}}, \tag{5.2}$$

   where $\mathcal{P}_{uv}$ is defined as the set of items which were rated by users $u$ and $v$, namely, $\mathcal{P}_{uv} = \mathcal{P}_u \cap \mathcal{P}_v$.
2. Neighbor selection: After the similarity matrix is computed, this step chooses neighbors for the target user or item. For user-based kNN models, users with the highest similarity with the target user form his (her) neighborhood. Similarly, item-based kNN models form the target item's neighborhood. This technique is usually called *best-n-neighbors* (Albadvi & Shahbazi, 2009; Cho & Kim, 2004). Usually, kNN models share the same neighbor selection strategy.
3. Prediction generation: The final step forms the predictive ratings of the target user for all items, and return items with the highest ratings as the recommended items for the target

user. One popular and effective strategy to obtain the prediction for an item is to use the frequency with which the item was purchased by the neighbors of the target user, that is,

$$\hat{r}_{ui} = \sum_{v \in N_u} \delta_{vi}, \tag{5.3}$$

where

$$\delta_{vi} = \begin{cases} 1, & \text{if } r_{vi} > 0 \\ 0, & \text{if } r_{vi} = 0 \end{cases} \tag{5.4}$$

and $N_u$ is the neighborhood of user $u$ selected in the previous step.

In many practical situations, some items of little importance are purchased frequently, but some others of greater importance are seldom purchased. In our supermarket data set, plastic shopping bags were purchased with high frequency. But these transactions tell little about the interests of users. Hence, it is reasonable to consider the importance of items when similarity is calculated. We suggest the use of the following *weighted cosine* similarity measure:

$$S_{uv}^{\text{wc}} = \frac{\sum_{i \in \mathcal{P}_{uv}} c_i r_{ui} r_{vi}}{\sqrt{\sum_{i \in \mathcal{P}_{uv}} c_i r_{ui}^2 \sum_{i \in \mathcal{P}_{uv}} c_i r_{vi}^2}}, \tag{5.5}$$

where $c_i$ is the weight associated with item $i$. In our experiments, we take

$$c_i = \log\left(\frac{U}{|\mathcal{P}^i| + 1}\right), \tag{5.6}$$

which is motivated by the popular term *frequency-inverse document frequency (tf-idf)* weight in text mining. The choice of $c_i$ in Eq. (5.6) produces larger weights for items that were purchased less frequently. It works well for our supermarket data set.

We also use the shrinkage technique proposed for sparse explicit ratings in the study by Bell, Koren, and Volinsky (2007), since typically a user purchases a small proportion of products in the supermarket data set. Hence, finally, we calculate the similarity between users $u$ and $v$ as follows:

$$S_{uv}^{\text{swc}} = \frac{|\mathcal{P}_{uv}|}{|\mathcal{P}_{uv}| + \gamma} \cdot S_{uv}^{\text{wc}}, \tag{5.7}$$

where $\gamma$ is the shrinkage parameter that can be determined by cross-validation.

The prediction strategy in Eq. (5.3) totally ignores rating values, which is the number of times users purchased products in the supermarket data set. But if a user frequently purchased a product, it seems reasonable to guess that the product attracts more interest from the user than products purchased only once. Here, we suggest a new strategy to consider the above factor:

$$\hat{r}_{ui} = \sum_{v \in N_u} \delta_{vi} \log(r_{vi} + 1). \tag{5.8}$$

More details on predictions using kNN models can be found in Section 5.6.

## 5.4 Matrix Factorization Models

Because of the computation complexity advantage, matrix factorization models have received much attention for sparse collaborative filtering (CF) problems with explicit ratings. In this section, we will mainly discuss the extensions to matrix factorization models and the usage for implicit collaborative filtering problems that will make use of all unknown ratings of user–item pairs.

### 5.4.1 Matrix Factorization for Explicit Ratings

Recently, matrix factorization (MF) models have attracted much attention because of their linear scale and effectiveness for sparse collaborative filtering (CF) problems with explicit ratings. One typical example is their impressive results for the Netflix Prize problem (Koren, Bell, & Volinsky, 2009; Paterek, 2007; Takács, Pilászy, Németh, & Tikk, 2008; Wu, 2009). MF models assume that each user's preferences can be described by a small number of factors, and each item's characteristics can also be expressed by the same number of factors. One user likes one item if the user factor vector matches well with the item factor vector. On the contrary, one user dislikes one item if their factor vectors do not match well. The user and item factors are achieved by solving the following minimization problem:

$$\min_{\mathbf{X},\mathbf{Y}} \sum_{(u,i)\in\mathcal{P}} [(r_{ui} - \mathbf{x}_u^T\mathbf{y}_i)^2 + \lambda(\| \mathbf{x}_u \|_2^2 + \| \mathbf{y}_i \|_2^2)], \tag{5.9}$$

where $\mathbf{x}_u$ and $\mathbf{y}_i$ are user $u$'s and item $i$'s factor vectors of length $K$, and $\lambda$ is the regularization parameter. $\mathbf{X} \in \mathbb{R}^{U\times K}$ is defined as $(\mathbf{x}_1,\ldots,\mathbf{x}_U)^T$, and $\mathbf{Y} \in \mathbb{R}^{I\times K}$ is defined similarly. *Stochastic gradient descent* (Paterek, 2007) or *alternate least squares* (Bell & Koren, 2007) are usually used to solve the above unconstrained optimization problem.

### 5.4.2 Weighted Matrix Factorization for Implicit Ratings

As we stated in Section 5.2.1, MF for explicit ratings (5.9) only uses given ratings of user–item pairs and ignores all unknown ratings when it is constructed. So, the computational complexity of MF only depends linearly on the amount of given ratings but not the product of the numbers of users and items.

However, we cannot just ignore unknown ratings for implicit CF problems. Unlike explicit ratings, implicit ratings provide no negative ratings (Hu et al., 2008). The unknown ratings in implicit CF problems stand for negative ratings in some sense. That is why unknown ratings are replaced with rating 0 in implicit CF problems. They should be considered when models are constructed. On the other hand, nonzero ratings deliver more information about users and items than zero ratings. So, the models should present different weights for them. As for kNN models, we think that items purchased frequently by a user tell more about the user's interests than items purchased infrequently. But at the same time, we account for the importance of an item itself by the frequency the item was purchased by different users, as suggested in Eq. (5.6). We use variable $c_{ui}$ to represent the weight of rating $r_{ui}$. According to the previous discussion, our final choice for $c_{ui}$ is

$$c_{ui} = 1 + \alpha_1 \log\left(1 + \frac{r_{ui}}{\alpha_2}\right) + \alpha_3 \delta_{ui} \log\left(\frac{U}{|\mathcal{P}^i| + 1}\right), \tag{5.10}$$

where $\delta_{ui}$ is defined in Eq. (5.4), and $\alpha_k$ ($k = 1, 2, 3$) are the hyperparameters, which will be determined using the *Nelder–Mead Simplex* method (Singer & Nelder, 2009) in our subsequent experiments. Note that $c_{ui} = 1$ if $r_{ui} = 0$, which is important when designing scalable algorithms for our new models.

Another distinct aspect between explicit and implicit ratings is that ratings of explicit information represent the levels of preferences, but ratings of implicit information usually represent observations of user actions, which show indirect preferences for items. In our supermarket data set, rating $r_{ui}$ is the number of times user $u$ purchased product $i$. Hence, the magnitude of $r_{ui}$ can represent the trend of user $u$'s preference for product $i$ more or less. We introduce a new variable $p_{ui}$ to represent the preference of user $u$ for item $i$. Our choice for $p_{ui}$ is

$$p_{ui} = \beta_1 \log\left(1 + \frac{r_{ui}}{\beta_2}\right), \tag{5.11}$$

where $\beta_1$ and $\beta_2$ are the hyperparameters, which can be determined by cross-validation or the Nelder–Mead simplex method. In our experiments, we simply take $\beta_1 = \beta_2 = 1$. Note that $p_{ui} = 0$ if $r_{ui} = 0$.

Integrating the comprehensive considerations above, we revise the original MF model in Eq. (5.9) to obtain our new objective function:

$$F(\mathbf{X}, \mathbf{Y}) = \sum_{u=1}^{U} \sum_{i=1}^{I} [c_{ui} (p_{ui} - \mathbf{x}_u^T \mathbf{y}_i)^2 + \lambda_1 \| \mathbf{x}_u \|_2^2 + \lambda_2 \| \mathbf{y}_i \|_2^2], \tag{5.12}$$

where $\lambda_1$ and $\lambda_2$ are the regularization parameters. We call the new model *Weighted Matrix Factorization* (*WMF*) since it is a variant of the original MF model in Eq. (5.9). The same model framework but with different types of $c_{ui}$ and $p_{ui}$ is also proposed in the study by Hu et al. (2008) to analyze implicit information.

The objective function of WMF (5.12) is a sum of $UI$ terms; minimizing it by stochastic gradient descent method will lead to computational complexity of at least $O(UI)$, which is unfeasible for large data sets. Fortunately, the alternate least squares method can be used here to achieve computational complexity which is linear to $|\mathcal{P}|$ if we notice that $c_{ui} = 1$ and $p_{ui} = 0$ if $r_{ui} = 0$ (Hu et al., 2008).

The alternate least squares method first fixes item factor matrix **Y** and updates user factor matrix **X**, then fixes **X** and updates **Y**. The whole loop is repeated until the stopping criterion is satisfied. Most standard stopping criteria can be used here. Since the stopping criterion is not intrinsic, we simply stop the iteration if the recommendation precision in the test set does not increase. The stopping criterion is also used by all the other models in the chapter. Later, we introduce the update of **X** in detail.

When updating the factor vector of user $u$, we differentiate $F$ with respect to $\mathbf{x}_u$ and set the derivative to 0 and obtain

$$\mathbf{x}_u = (\mathbf{Y}^T\mathbf{C}_u\mathbf{Y} + \lambda_1 I\mathbf{I}_d)^{-1}(\mathbf{Y}^T\mathbf{C}_u\mathbf{p}_u), \tag{5.13}$$

where $\mathbf{C}_u = \text{diag}(c_{u1}, \ldots, c_{uI})$, $\mathbf{p}_u = (p_{u1}, \ldots, p_{uI})^T$, and $\mathbf{I}_d$ is the unit matrix.

Note that $\mathbf{Y}^T\mathbf{C}_u\mathbf{Y}$ can be divided into two parts: $\mathbf{Y}^T\mathbf{Y}$ and $\mathbf{Y}^T(\mathbf{C}_u - \mathbf{I}_d)\mathbf{Y}$ (Hu et al., 2008). The first part $\mathbf{Y}^T\mathbf{Y}$ does not depend on $u$ and can be precomputed. The second part $\mathbf{Y}^T(\mathbf{C}_u - \mathbf{I}_d)\mathbf{Y}$ can be achieved in $O(K^2|\mathcal{P}_u|)$ since $c_{ui} = 1$ when $i \notin \mathcal{P}_u$. $\mathbf{Y}^T\mathbf{C}_u\mathbf{p}_u$ can be computed in $O(K|\mathcal{P}_u|)$ since $p_{ui} = 0$ when $i \notin \mathcal{P}_u$. Hence, $\mathbf{x}_u$ can be updated in time $O(K^2|\mathcal{P}_u| + K^3)$ except for the precomputation of $\mathbf{Y}^T\mathbf{Y}$, and the whole user factor matrix **X** can be updated in time $O(K^2|\mathcal{P}| + K^3U)$, which is linear to the size of $\mathcal{P}$.

The update of the item factor matrix **Y** can be derived symmetrically. The new $\mathbf{y}_i$ can be obtained as follows:

$$\mathbf{y}_i = (\mathbf{X}^T\mathbf{C}^i\mathbf{X} + \lambda_2 U\mathbf{I}_d)^{-1}(\mathbf{X}^T\mathbf{C}^i\mathbf{p}^i), \tag{5.14}$$

where $\mathbf{C}^i = \text{diag}(c_{1i}, \ldots, c_{Ui})$, $\mathbf{p}^i = (p_{1i}, \ldots, p_{Ui})^T$. The update of **Y** can be finished in time $O(K^2|\mathcal{P}| + K^3I)$. Hence, the computational complexity of all the updates in one loop using the alternate least squares method is $O(K^2|\mathcal{P}| + K^3(U + I))$.

After the user and item factor matrices **X** and **Y** are found, WMF first produces the predictive values for all the user–item pairs as

$$\hat{p}_{ui} = \mathbf{x}_u^T\mathbf{y}_i \quad (u = 1, \ldots, U; i = 1, \ldots, I), \tag{5.15}$$

then $N$ items with the highest predictive values associated with a user are recommended to the user.

### 5.4.3 Weighted Matrix Factorization with Biases

MF models usually benefit much from user and item biases in explicit CF problems (Paterek, 2007). Here, we add biases to our WMF model and obtain the new objective function:

$$\begin{aligned} G(\mathbf{X}, \mathbf{Y}, \mathbf{b}, \mathbf{d}) = \sum_{u=1}^{U}\sum_{i=1}^{I}[&c_{ui}(p_{ui} - b_u - d_i - \mathbf{x}_u^T\mathbf{y}_i)^2 \\ &+ \lambda_1 \| \mathbf{x}_u \|_2^2 + \lambda_2 \| \mathbf{y}_i \|_2^2 + \lambda_3 b_u^2 + \lambda_4 d_i^2], \end{aligned} \tag{5.16}$$

where $\mathbf{b} = (b_1, \ldots, b_U)^T$, $\mathbf{d} = (d_1, \ldots, d_I)^T$, and $\lambda_j$ $(j = 1,2,3,4)$ are the regularization hyper-parameters. We refer to the new model as *BWMF*.

As the alternate least squares method can be used to solve WMF, it can also be used to solve BWMF efficiently. We can obtain the new $\mathbf{x}_u$ by

$$\mathbf{x}_u = (\mathbf{Y}^T\mathbf{C}_u\mathbf{Y} + \lambda_1 I\mathbf{I}_d)^{-1}(\mathbf{Y}^T\mathbf{C}_u(\mathbf{p}_u - b_u\mathbf{e} - \mathbf{d})), \tag{5.17}$$

where $\mathbf{e}$ is a vector with all the elements equal to 1.

We can use the same method to calculate $(\mathbf{Y}^T\mathbf{C}_u\mathbf{Y} + \lambda_1 I\mathbf{I}_d)$ and $\mathbf{Y}^T\mathbf{C}_u\mathbf{p}_u$. The term $\mathbf{Y}^T\mathbf{C}_u(b_u\mathbf{e} + \mathbf{d})$ can be divided into two parts: $b_u\mathbf{Y}^T\mathbf{e} + \mathbf{Y}^T\mathbf{d}$ and $\mathbf{Y}^T(\mathbf{C}_u - \mathbf{I}_d)(b_u\mathbf{e} + \mathbf{d})$. Both $\mathbf{Y}^T\mathbf{e}$ and $\mathbf{Y}^T\mathbf{d}$ are independent of $u$ and can be precomputed. $\mathbf{Y}^T(\mathbf{C}_u - \mathbf{I}_d)(b_u\mathbf{e} + \mathbf{d})$ can be computed in time $O(K|\mathcal{P}_u|)$ because $c_{ui} - 1 = 0$ when $i \notin \mathcal{P}_u$. Hence, $\mathbf{X}$ can still be updated in time $O(K^2|\mathcal{P}| + K^3U)$.

Symmetrically, we can obtain the new $\mathbf{y}_i$ as

$$\mathbf{y}_i = (\mathbf{X}^T\mathbf{C}^i\mathbf{X} + \lambda_2 U\mathbf{I}_d)^{-1}(\mathbf{X}^T\mathbf{C}^i(\mathbf{p}^i - \mathbf{b} - d_i\mathbf{e})). \tag{5.18}$$

Similarly, we can update $\mathbf{Y}$ in time $O(K^2|\mathcal{P}| + K^3I)$.

The update of the user bias $b_u$ can be obtained from

$$b_u = \frac{\mathbf{e}^T\mathbf{C}_u(\mathbf{p}_u - \mathbf{d} - \mathbf{Y}\mathbf{x}_u)}{\mathbf{e}^T\mathbf{C}_u\mathbf{e} + \lambda_3 I}. \tag{5.19}$$

The denominator can be reexpressed as $\mathbf{e}^T(\mathbf{C}_u - \mathbf{I}_d)\mathbf{e} + (1 + \lambda_3)I$, so it can be computed in $O(|\mathcal{P}_u|)$. The first part of the numerator $\mathbf{e}^T\mathbf{C}_u\mathbf{p}_u$ can be computed in $O(|\mathcal{P}_u|)$. The second part, $\mathbf{e}^T\mathbf{C}_u(\mathbf{d} + \mathbf{Y}\mathbf{x}_u)$, can be further divided into two parts: $\mathbf{e}^T(\mathbf{C}_u - \mathbf{I}_d)(\mathbf{d} + \mathbf{Y}\mathbf{x}_u)$ and $\mathbf{e}^T(\mathbf{d} + \mathbf{Y}\mathbf{x}_u)$. $\mathbf{e}^T(\mathbf{C}_u - \mathbf{I}_d)(\mathbf{d} + \mathbf{Y}\mathbf{x}_u)$ can be computed in time $O(K|\mathcal{P}_u|)$, and $\mathbf{e}^T(\mathbf{d} + \mathbf{Y}\mathbf{x}_u)$ can be computed in time $O(K)$ if $\mathbf{e}^T\mathbf{d}$ and $\mathbf{e}^T\mathbf{Y}$ are precomputed. Hence, the complexity of updating all $\mathbf{b}_u$ values [$b_u$ values] is $O(K|\mathcal{P}|)$ when we precompute $\mathbf{e}^T\mathbf{d}$ and $\mathbf{e}^T\mathbf{Y}$.

Symmetrically, the update of the item bias $d_i$ can be obtained from

$$d_i = \frac{\mathbf{e}^T\mathbf{C}^i(\mathbf{p}^i - \mathbf{b} - \mathbf{X}\mathbf{y}_i)}{\mathbf{e}^T\mathbf{C}^i\mathbf{e} + \lambda_4 U}. \tag{5.20}$$

Similarly, the complexity of updating all $\mathbf{d}_i$ values [$d_i$ values] is $O(K|\mathcal{P}|)$ when we precompute $\mathbf{e}^T\mathbf{b}$ and $\mathbf{e}^T\mathbf{X}$.

Hence, the complexity of all the updates in one loop of BWMF is $O(K^2|\mathcal{P}| + K^3(U + I))$ if we precompute $\mathbf{e}^T\mathbf{b}$, $\mathbf{e}^T\mathbf{d}$, $\mathbf{X}^T\mathbf{e}$, $\mathbf{Y}^T\mathbf{e}$, $\mathbf{X}^T\mathbf{b}$, $\mathbf{Y}^T\mathbf{d}$, $\mathbf{X}^T\mathbf{X}$, and $\mathbf{Y}^T\mathbf{Y}$ before beginning the corresponding updates.

After the model parameters are found, BWMF first produces the predictive values for all the user–item pairs by

$$\hat{p}_{ui} = b_u + d_i + \mathbf{x}_u^T\mathbf{y}_i \quad (u = 1, \ldots, U; i = 1, \ldots, I), \tag{5.21}$$

then $N$ items with the highest predictive values associated with a user are recommended to the user.

## 5.5 A Hybrid Model of Implicit and Feature Information

As introduced in Section 5.1, CF methods encounter the cold start difficulty for a new user (or a new item) because few ratings associated with the user (or the item) are collected in the start-up phase. But in many practical applications, we can get round this difficulty by using CBF methods if features of users or items are available. Naturally, WMF provides a unified framework to combine implicit ratings and feature information to generate recommendations efficiently even in the start-up phase. Later, we introduce the method of incorporating user features into BWMF. The incorporation of item features can be obtained symmetrically.

Denote the feature vector of user $u$ by $\mathbf{f}_u = (f_{u1}, \ldots, f_{uL})^T$, where $L$ is the number of user features. $\mathbf{F} \in \mathbb{R}^{U \times L}$ is defined as $(\mathbf{f}_1, \ldots, \mathbf{f}_U)^T$. We expand BWMF to incorporate the user features and obtain the new objective function:

$$H(\mathbf{X},\mathbf{Y},\mathbf{W},\mathbf{b},\mathbf{d}) = \sum_{u=1}^{U}\sum_{i=1}^{I}[c_{ui}(p_{ui} - b_u - d_i - \mathbf{x}_u^T\mathbf{y}_i - \mathbf{f}_u^T\mathbf{w}_i)^2 + \lambda_1\|\mathbf{x}_u\|_2^2 + \lambda_2\|\mathbf{y}_i\|_2^2 + \lambda_3 b_u^2 + \lambda_4 d_i^2 + \lambda_5\|\mathbf{w}_i\|_2^2], \tag{5.22}$$

where $\mathbf{w}_i$ is the regression coefficient vector associated with item $i$ and $\lambda_j$ ($j = 1,2,3,4,5$) are the regularization hyperparameters. $\mathbf{W} \in \mathbb{R}^{I \times L}$ is defined as $(\mathbf{w}_1, \ldots, \mathbf{w}_I)^T$. We refer to the new model as *HBWMF*.

Analogously, the alternate least squares method can be used to solve HBWMF efficiently if we use the previous schemes: precomputation and dividing $c_{ui}$ into two parts. The complexity of all the updates in one loop of HBWMF is $O((K+L)^2|\mathcal{P}| + K^3(U+I) + L^2U + L^3I)$. The detailed updates of the model parameters are shown in Algorithm 1.

**Algorithm 1 (HBWMF).** Choose the number of factors $K$ and the regularization hyperparameters $\lambda_j$ ($j = 1,2,3,4,5$). Initialize the item factor matrix $\mathbf{Y}$, the item coefficient matrix $\mathbf{W}$, and the bias vectors $\mathbf{b}$ and $\mathbf{d}$, e.g., sampling the values from a uniform distribution or simply setting all of them (except $\mathbf{Y}$) to 0. Perform the following steps iteratively:

1. *Precompute* $\mathbf{e}^T\mathbf{d}$, $\mathbf{Y}^T\mathbf{e}$, $\mathbf{Y}^T\mathbf{d}$, $\mathbf{W}^T\mathbf{e}$, $\mathbf{Y}^T\mathbf{W}$ *and* $\mathbf{Y}^T\mathbf{Y}$.
2. *For user* $u$ ($u = 1,\ldots,U$), *do:*
   a. *update user factor vector* $\mathbf{x}_u$ *using:*

$$\mathbf{x}_u = (\mathbf{Y}^T\mathbf{C}_u\mathbf{Y} + \lambda_1 I\mathbf{I}_d)^{-1} \cdot (\mathbf{Y}^T\mathbf{C}_u(\mathbf{p}_u - b_u\mathbf{e} - \mathbf{d} - \mathbf{W}\mathbf{f}_u)); \tag{5.23}$$

   b. *update user bias* $b_u$ *using:*

$$b_u = \frac{\mathbf{e}^T\mathbf{C}_u(\mathbf{p}_u - \mathbf{d} - \mathbf{Y}\mathbf{x}_u - \mathbf{W}\mathbf{f}_u)}{\mathbf{e}^T\mathbf{C}_u\mathbf{e} + \lambda_3 I}. \tag{5.24}$$

3. *Precompute* $\mathbf{e}^T\mathbf{b}$, $\mathbf{X}^T\mathbf{e}$, $\mathbf{X}^T\mathbf{b}$, $\mathbf{F}^T\mathbf{e}$, $\mathbf{F}^T\mathbf{b}$, $\mathbf{X}^T\mathbf{F}$, $\mathbf{X}^T\mathbf{X}$ *and* $\mathbf{F}^T\mathbf{F}$.
4. *For item i* ($i = 1,\ldots,I$), *do:*
   a. *update item factor vector* $\mathbf{y}_i$ *using:*

$$\mathbf{y}_i = (\mathbf{X}^T\mathbf{C}^i\mathbf{X} + \lambda_2 U\mathbf{I}_d)^{-1} \cdot (\mathbf{X}^T\mathbf{C}^i(\mathbf{p}^i - \mathbf{b} - d_i\mathbf{e} - \mathbf{F}\mathbf{w}_i)); \tag{5.25}$$

   b. *update item coefficient vector* $\mathbf{w}_i$ *using:*

$$\mathbf{w}_i = (\mathbf{F}^T\mathbf{C}^i\mathbf{F} + \lambda_5 U\mathbf{I}_d)^{-1} \cdot (\mathbf{F}^T\mathbf{C}^i(\mathbf{p}^i - \mathbf{b} - d_i\mathbf{e} - \mathbf{X}\mathbf{y}_i)); \tag{5.26}$$

   c. *update item bias* $d_i$ *using:*

$$d_i = \frac{\mathbf{e}^T\mathbf{C}^i(\mathbf{p}^i - \mathbf{b} - \mathbf{X}\mathbf{y}_i - \mathbf{F}\mathbf{w}_i)}{\mathbf{e}^T\mathbf{C}^i\mathbf{e} + \lambda_4 U}. \tag{5.27}$$

5. *Check whether some given stopping criterion is satisfied by using the updated parameters. If one does, then stop; if not continue the previous steps.*

After the model parameters are achieved, HBWMF first produces the predictive values for all the user–item pairs using

$$\hat{p}_{ui} = b_u + d_i + \mathbf{x}_u^T\mathbf{y}_i + \mathbf{f}_u^T\mathbf{w}_i, \quad (u = 1,\ldots,U; i = 1,\ldots,I) \tag{5.28}$$

then $N$ items with the highest predictive values associated with a user are recommended to the user.

## 5.6 Experimental Study

In this section, we will first introduce the supermarket data set we use and then apply our new models to the data set. Experiment results show that our new models can make significant improvements and achieve better performance than previous models.

### 5.6.1 Data Description

All of our experiments are based on a data set from a supermarket. The data set comprises 13,025 users' 2-month purchase data in the supermarket, which contains 9850 different items. Each value ($r_{ui}$) in the data set represents the number of times a user purchased an item. We remove the data of users who purchased less than 10 different items and users who purchased more than 100 different items. Users who purchased less than 10 different items provide too little information, so it is unfeasible to generate precise recommendations for them. Users who purchased more than 100 different items in 2 months are usually small traders. It is not the purpose of this study to provide recommendations for this type of consumer.

After the above cleaning of the data set, the number of users decreases to 8638 and the number of user–item pairs with nonzero values decreases from 274,925 to 244,967. Then, the whole data set is split into two data sets randomly, one of which is used as the training set and the other as the test set. To eliminate the impact of random split on the final recommendation precisions of models, we randomly split the whole data set five times. Without specific declaration, the recommendation precisions of models in this section represent the average precisions of the corresponding models by randomly splitting the whole data set five times. For example, the training set comprises 131,364 nonzero values of user–item pairs, and the test set comprises 130,856 nonzero values of user–item pairs[1] in our first random split.

Apart from the direct purchase data set above, 12 features for each user are extracted by experts according to the purchase history of the user. Typically, one user has one or two nonzero feature values. These user features are used to construct the hybrid model and further improve the recommendation accuracy.

### 5.6.2 Results

#### 5.6.2.1 Results of the kNN Models

First, we implement the kNN models with different similarity measures [(5.2) and (5.7)] and rating strategies [(5.3) and (5.8)]. When the conventional cosine measure (5.2) is used to compute similarity between two users and the most frequently purchased product strategy (5.3) is used to produce ratings and then recommendations, kNN finds an average recommendation precision of 0.2328 in our five random splits of the whole data set if the number of neighbors is chosen to be 150. The numbers of neighbors for all the kNN models are chosen to obtain the best recommendation accuracies, respectively. The average precision increases to 0.2343 with the same number of neighbors, if we use the weighted cosine similarity (5.7) and the log-rating strategy (5.8), where the shrinkage hyperparameter $\gamma = 1$. Our new similarity measure and rating strategy improve the recommendation accuracy.

For comparison, we also use user features to achieve recommendations. The feature variables are first normalized to the same range [0, 1] and then used to calculate the similarity between different users. We use the cosine measure with the shrinkage technique (and the shrinkage parameter $\gamma = 2$) because of the sparsity of the feature values. The final average recommendation precision is 0.2518 (the number of neighbors is still 150), which is much lower than 0.2343.

To improve the accuracy of our new kNN model [(3-5)+(3-8)], we linearly combine its similarity and the similarity calculated from the features. The combining proportion is determined by *grid search* from 1.0 to 0.0 with a step size of 0.05 to achieve the best recommendation accuracy. Finally, we obtain the optimal combining proportion of 0.65, and the average recommendation precision increases from 0.2343 to 0.2376 with the same number of neighbors. We refer to the model as *hybrid kNN*.

[1] Note that $131,364 + 130,856 > 244,967$ because some original values larger than 1 are divided into two positive integer values, which are put into the training and test sets, respectively.

**Table 5.1: Average recommendation precisions of the kNN models in five random splits of the whole data set.**

| Simirity Measure | Rating Strategy | Shrinkage Parameter | Precision |
|---|---|---|---|
| $S_f^{sc}$ | (3-8) | $\gamma = 2$ | 0.2518 |
| $S^c$ | (3-3) | - | 0.2328 |
| $S^{wc}$ | (3-3) | - | 0.2337 |
| $S^{swc}$ | (3-8) | $\gamma = 1$ | 0.2343 |
| $0.35S_f^{sc} + 0.65S^{swc}$ | (3-8) | - | 0.2376 |

Different kNN models use different similarity measures and rating strategies to obtain recommendations. The number of neighbors is chosen to be 150, which produces the best recommendation accuracies for all the kNN models. The superscripts "s", "c", and "w" stand for "shrunk", "cosine", and "weighted", respectively. The subscript "f" indicates the corresponding similarity calculated based on the user features.

Table 5.1 shows the average recommendation precisions of the above kNN models and their parameter settings in detail.

#### *5.6.2.2 Results of Matrix Factorization Models*

As we stated before, the same model framework (5.12) is also proposed by Hu et al. (2008), but with different choices for $c_{ui}$:

$$c_{ui} = 1 + \alpha_1 \log\left(1 + \frac{r_{ui}}{\alpha_2}\right),$$

where $\alpha_1$ and $\alpha_2$ are the hyperparameters, which will be determined using the Nelder–Mead simplex method in our experiments, and $p_{ui}$:

$$p_{ui} = \delta_{ui}.$$

We refer to their model as *HKV*, named after the initial capitals of the surnames of the authors (Hu et al., 2008).

To compare HKV with our new MF models, we first obtain their hyperparameters by applying the Nelder–Mead simplex method to the resulting data sets from our first random split, respectively, with the numbers of factors $K$ fixed to 30. The maximum numbers of iterations for the Nelder–Mead simplex method are both limited to 30. Then, these obtained hyperparameters are used in all the data sets from the five random splits, because, computationally, it is too expensive to run the Nelder–Mead simplex method in each random split. The resulting values of the hyperparameters for the models are listed in Table 5.2.

The final recommendation results of HKV and WMF are shown in Fig. 5.1 when the number of factors $K$ is 30 and the data sets from the first random split are used. From Fig. 5.1, HKV obtains the highest precision of 0.2352 when the number of iterations is 27 and WMF obtains the highest precision of 0.2428 when the number of iterations is 17. WMF produces significantly better recommendation accuracy in the supermarket data set.

**Table 5.2: Values of the hyperparameters in different models, which are obtained by using Nelder-Mead simplex method and the data sets from our first random split.**

| Model | $\alpha_1, \alpha_2, \alpha_3$ | $\lambda_1, \lambda_2, \lambda_3, \lambda_4, \lambda_5$ |
|---|---|---|
| HKV | 2.3251, 1.0307, - | 0.0044, 0.0081, -, -, - |
| WMF | 0.1957, 0.9098, 1.4464 | 0.0077, 0.0075, -, -, - |
| BWMF | 0.0610, 0.9101, 1.5434 | 0.0084, 0.0074, 0.0060, 0.0124, - |
| HBWMF | 0.1247, 0.8602, 1.5719 | 0.0085, 0.0076, 0.0077, 0.0208, 0.0020 |

The numbers of the model factors $K$ are all fixed to 30, and the maximum numbers of iterations for the Nelder-Mead simplex method are all set to 30 when they are trained.

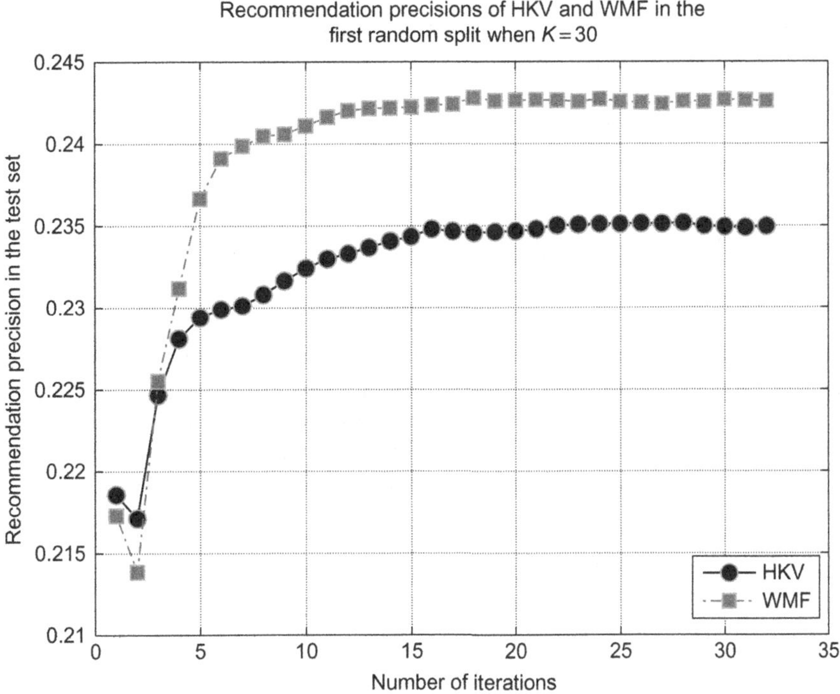


**Figure 5.1: Recommendation precisions of HKV and WMF when the number of factors $K$ is 30 and the data sets from the first random split are used. Their hyperparameters $\alpha_k$ and $\lambda_l$, which can be found in Table 5.2, are chosen by using Nelder-Mead simplex method to obtain the best recommendation accuracies, respectively. The horizontal axis represents the number of iterations, and the vertical axis represents the recommendation precision within 10 recommended products in the test set.**

Average recommendation precisions of HKV and our matrix factorization (MF) models with different numbers of factors in the five random splits are shown in Fig. 5.2. The values of the hyperparameters used in each of the four MF models can be found in Table 5.2. All of our new MF models produce better average recommendation accuracies when $K \geq 10$ than the

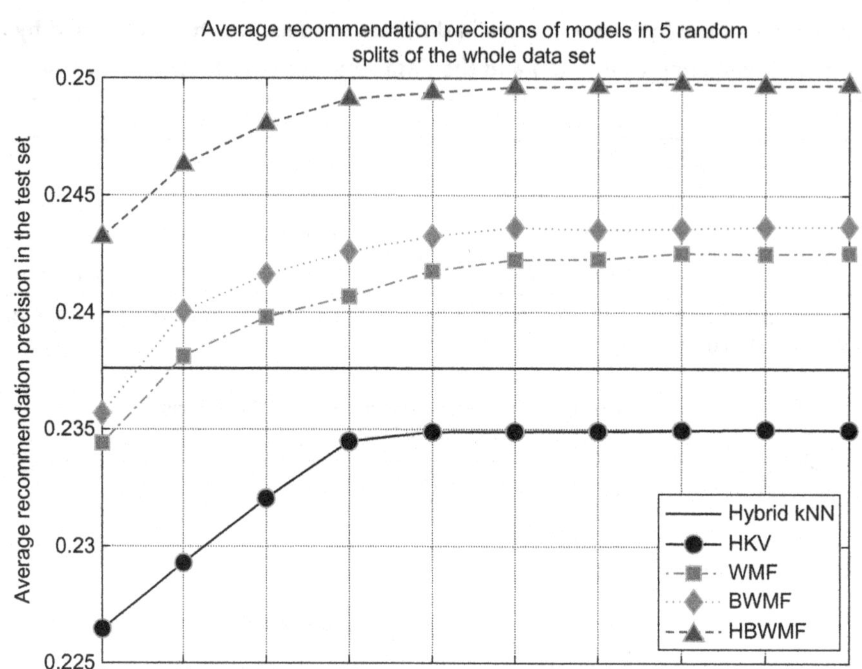


**Figure 5.2: Average recommendation precisions of various models with different numbers of factors $K$ in five random splits of the whole data set. The hyperparameters of each model (except hybrid kNN) are determined by applying Nelder–Mead simplex method to the data sets from the first random split, respectively, when $K$ is fixed to 30. The obtained hyperparameters are then used for all $K$'s.**

hybrid kNN model does. But HKV produces worse accuracies than the hybrid kNN model, even when the number of factors $K$ in HKV is taken as 50. BWMF exhibits slight superiority over WMF for all the values of $K$. But its superiority seems to vanish gradually when $K$ increases. When compared with the other models, HBWMF produces a dramatic improvement from user features. The average recommendation precision increases to 0.2497 from 0.2437 of BWMF when $K = 50$.

## 5.7 Conclusions

In this chapter, we explore some methods that use implicit information. We suggest a new weighted similarity measure that considers different importance of items. A new rating strategy is also proposed to further improve the recommendation accuracy of conventional neighborhood models. When compared with neighborhood models, matrix factorization (MF) models usually generate better predictions for explicit collaborative filtering (CF) problems. We introduce some extensions to original MF models to analyze implicit information. These new

MF models produce significantly better recommendation accuracies than the neighborhood models above in a supermarket data set.

We also propose a hybrid MF model that integrates user or item feature information. We hope that the hybrid model can fix the cold start problem, which is intrinsic in pure CF models. But its effectiveness needs to be verified further. The hybrid model improves the recommendation accuracy in our experiments.

In future, we will explore other methods to solve the new MF models, although the alternate least squares method can be used to produce efficient algorithms for them. Methods that efficiently solve original MF models for explicit ratings are our first candidates, such as variational expectation maximization (VEM) (Lim & Teh, 2007) or Markov chain Monte Carlo (MCMC) (Salakhutdinov & Mnih, 2008). We will also explore the probabilistic interpretations of the new MF models for implicit information.

On the other hand, it is worth exploring more efficient approaches to combine implicit and feature information. One reasonable choice is to use kernel regression instead of linear regression in HBWMF.

Finally, we will also try to consider more practical factors when a practical recommender system is constructed. One reasonable consideration is the repetition of user actions. For example, should we recommend to a user an item which he/she has already purchased? We may introduce a value for each item to represent the probability that the item will be purchased again. Actually, we applied the idea to our new kNN model and obtained a slight improvement of the recommendation accuracy. Another consideration is to produce recommendation lists with different sizes for different users.

## Acknowledgments

The authors thank their colleague for extracting the experimental data used in this chapter from the original database.

## References

Albadvi, A. & Shahbazi, M. (2009). A hybrid recommendation technique based on product category attributes. *Expert Systems with Applications, 36*(9), 11480–11488. doi:10.1016/j.eswa.2009.03.046.

Bai, X., Wu, J., Haifeng, W., Zhang, J., Yin, W., & Dong, J. (2011). *Recommendation algorithms for implicit information*. Paper presented at the proceedings of IEEE International Conference on Service Operations and Logistics, and Informatics (SOLI'2011), Beijing, July.

Bell, R. & Koren, Y. (2007). Scalable collaborative filtering with jointly derived neighborhood interpolation weights. Paper presented at the Proceedings of IEEE International Conference on Data Mining (ICDM'07), Omaha, NE, October.

Bell, R. M., Koren, Y., & Volinsky, C. (2007). Modeling relationships at multiple scales to improve accuracy of large recommender systems. In P. Berkhin, R. Caruana, & X. Wu (Eds.), *Proceedings of the KDD*, San Jose, CA, August (pp. 95–104). ACM.

Breese, J. S., Heckerman, D., & Kadie, C. (1998). *Empirical analysis of predictive algorithms for collaborative filtering.* Paper presented at the Proceedings of the 14th Conference on Uncertainty in Intelligence, Madison, WI, July (pp. 43–52).

Cho, Y. H. & Kim, K. (2004). Application of web usage mining and product taxonomy to collaborative recommendations in e-commerce. *Expert Systems with Applications, 26*, 233–246.

Hu, Y., Koren, Y., & Volinsky, C. (2008). *Collaborative filtering for implicit feedback datasets.* Paper presented at the IEEE International Conference on Data Mining, 263–272. Retrieved from http://doi.ieeecomputersociety.org/10.1109/ICDM.2008.22

Koren, Y., Bell, R. M., & Volinsky, C. (2009). Matrix factorization techniques for recommender systems. *IEEE Computer, 42*(8), 30–37.

Lim, Y. J. & Teh, Y. W. (2007). Variational Bayesian approach to movie rating prediction. In *Proceedings of KDD cup and workshop*, San Jose, CA, August.

Linden, G., Smith, B., & York, J. (2003). Amazon.com recommendations: Item-to-item collaborative filtering. *IEEE Internet Computing, 7*(1), 22–30.

Netflix (2009). Netflix prize. Retrieved from http://www.netflixprize.com

Oard, D. & Kim, J. (1998). Implicit feedback for recommender systems. In *Proceedings of the AAAI workshop on recommender systems*, Madison, WI, July (pp. 81–83).

Paterek, A. (2007). Improving regularized singular value decomposition for collaborative filtering. In *Proceedings of the KDD-cup and workshop*, San Jose, CA, August. ACM Press.

Salakhutdinov, R. & Mnih, A. (2008). *Bayesian probabilistic matrix factorization using Markov chain Monte Carlo.* Paper presented at the Proceedings of the 25th International Conference on Machine Learning *(ICML'08)*, Helsinki, July (pp. 880–887).

Singer, S. & Nelder, J. (2009). Nelder–Mead algorithm. Retrieved from http://www.scholarpedia.org/article/Nelder-Mead_algorithm

Takács, G., Pilászy, I., Németh, B., & Tikk, D. (2008). *Investigation of various matrix factorization methods for large recommender systems.* Paper presented at the Proceedings of the 2nd Netflix-KDDWorkshop, Las Vegas, NV.

Wu, J. (2009). Binomial matrix factorization for discrete collaborative filtering. In W. Wang, H. Kargupta, S. Ranka, P. S. Yu, & X. Wu (Eds.), *International Conference on Data Mining ICDM'09* (pp. 1046–1051). IEEE Computer Society.

CHAPTER 6

# Online Strategies for Optimizing Medical Supply in Disaster Scenarios

**Dennis Güttinger*, Eicke Godehardt†, and Andreas Zinnen‡**

*Goethe University, Frankfurt/Main

†SAP AG, SAP Research

‡University of Luxembourg

## 6.1 Introduction

In recent years, the number of natural disasters and major incidents at mass events has increased. Examples are the big tsunami in Japan in March 2011 and the Love Parade tragedy in Duisburg (Germany) in 2010. Such catastrophes usually engender a large number of casualties that have to be supplied medically as proficiently and quickly as possible. One of the challenges is that in practice casualties are not registered all at once, but usually arrive in small groups. Furthermore, in general, the exact number of casualties is unknown in advance. This conjuncture requires application of reasonable online strategies for assignment of casualty groups to available ambulances and hospitals. Additionally, the time window for making a decision about where to assign a casualty is very small during the chaotic period immediately after a major incident.

However, efficient online strategies rarely exist or they assume explicit arrangements with surrounding medical institutions about the number of casualties taken in.

One of the main reasons is that the problem at hand is very complex, and efficient algorithms that compute an optimal assignment of arriving casualty groups to available ambulances and physicians are unlikely to exist. More precisely, in our formulation, the problem is related to a *coupled multiple machine earliness tardiness problem* with variable costs; see Akyol and Bayhan (2008). Thus, efficient algorithms can only provide approximately optimal solutions.

In this chapter, we compare three different strategies for online assignment of casualties to available ambulances and physicians in hospitals: the D'Hondt assignment strategy, which is used in today's practice, a greedy strategy that takes expected journey lengths into account, and an adjusted version of simulated annealing. Since the present assignment problem is NP-complete, the D'Hondt strategy and our greedy approximation usually will lead to a suboptimal solution.

Service Science, Management, and Engineering.


DOI: 10.1016/B978-0-12-397037-4.00006-5

We will therefore investigate additional benefits when applying simulated annealing iterations subsequent to greedy, which allows us to find solutions as close as possible to the optimal solution.

Comparison of the three strategies will be based on a relaxed version of our assignment problem, which provides a lower bound of the optimal solution and can be efficiently computed.

For our evaluations, we have generated random data samples within reasonable intervals and distribution. We estimated the interval centers from an empirical data set.

Our tests show that greedy clearly outperforms the D'Hondt assignment strategy, but is outperformed by the results yielded by simulated annealing with the greedy result as initialization. The additional computational effort required by simulated annealing compared to D'Hondt and greedy is just acceptable so that our adapted online version of simulated annealing in combination with a greedy initialization is the preferred tool for solving our online assignment problem.

The remainder of this chapter is structured as follows: in Section 6.2, we will give a review of related work. Section 6.3 contains an introduction of penalty functions and triage groups as well as information about the data sets used in this chapter. In Section 6.4, we will formalize our optimization problem and discuss the aforementioned algorithms in detail. In Section 6.5, the experimental results will be discussed, and in the last section, we summarize our results and give an outlook on future prospects.

## 6.2 Related Work

Public security has attracted much interest in recent years. Lin, Batta, and Rogerson (2009) use genetic algorithms and a vehicle assignment heuristic to solve a logistics model for delivering critical items in a disaster relief operation. The public funded research projects SOGRO (SOGRO, 2009) (instant rescue at major accidents with heavy casualties) and SoKNOS (SoKNOS, 2009) (Service-Oriented Architectures Supporting Networks of Public Security) address the optimized care of the injured in disasters. SOGRO focuses on techniques to shorten the first chaotic phase immediately after a major incident and before assignment of casualties to transport vehicles and hospitals, whereas SoKNOS mainly deals with the adaptation of service-oriented architectures, semantics, and human-adapted workplaces in public security. Information processing is also a major topic within SoKNOS, but especially with a focus on visualization.

Several works deal with the special case of optimization and scheduling techniques in emergency and disaster scenarios. Most of them are special cases of network flow problems, and none of them can be transformed into (coupled) Job Shop Scheduling problems like our problem (Barbarosoglu, Ozdamar, & Cevik, 2002; Brown & Vassiliou, 2006).

The Job Shop Scheduling problem (JSS) was first presented by Graham (1966). For the special case of two machines and an arbitrary number of jobs, this problem can be solved efficiently with Johnson's algorithm (Johnson, 1994). Garey (1976) proved that JSS is NP-complete for more than two machines. For this general case, many different approximation techniques for JSS have been suggested. These can roughly be divided into priority dispatch rules, bottleneck-based heuristics, artificial intelligence, local search methods, and meta-heuristics; see Jain and Meeran (1999).

An adjusted version of simulated annealing to solve the same assignment problem, but with the assumption that the total set of casualties is known beforehand, has been applied by Güttinger, Zinnen, and Godehardt (2011). They compare our approach empirically with classic simulated annealing and a greedy strategy. When comparing the adjusted version of simulated annealing and the classic simulated annealing algorithm with respect to their convergence behavior towards the optimum, we can see that the adjusted version converges faster. Furthermore, the risk of getting stuck in a local optimum because of a wrong choice of input parameter is smaller for the adjusted version.

We use the adjusted simulated annealing in this work to test whether a significant improvement can be achieved by applying it for our online assignment problem. The crucial difference between Güttinger et al. (2011) and this work is that we want to provide a tool that can be used *during* the chaotic phase instantly after a major incident. In this setting, the total number of casualties is unknown, and the decision about assignment of a casualty has to be made instantaneously with incomplete information. However, in the study by Güttinger et al. (2011), the introduced techniques are used as a tool for planning security measures in advance of a mass event (e.g., concerts, soccer match) and, thus, *before* a (possible) incident and with complete information. Online strategies have been developed for various combinatorial optimization problems. Grötschel, Krumke, and Rambau (2001a) give a survey of some of the underlying models. They also discuss online optimization of real-world transportation systems and focus on those that arise in production and manufacturing processes, in particular in company internal logistics in Grötschel, Krumke, and Rambau (2001b).

Different online scheduling problems are the topic of Hamidzadeh, Lau, and Lilja (2000), Friese and Rambau (2006), and Bent and Hentenryck (2004). Hamidzadeh et al. (2000) introduce a self-adjusting dynamic scheduling class of algorithms to schedule-independent tasks on the processors of a multiprocessor system. Friese and Rambau (2006) develop and experimentally compare the policies for the control of a system of multiple elevators with capacity one in a transport environment with multiple floors. The authors show by using dynamic column generation that there exists a re-optimization policy implementation that runs in real time. A novel approach that tries to approximate a decision regretted in the context of an online stochastic optimization problem is introduced by Bent and Hentenryck (2004). The authors apply their regret algorithm to the problem of online packet scheduling in networks as well as on the online multiple vehicle routing problems with time windows. Vehicle routing is also the topic of the study by Bagula, Botha, and Krzesinski (2004), who deal with online traffic engineering and present a new routing scheme referred to as least interference optimization (LIO) where the online routing process uses the current bandwidth availability and the traffic flow distribution to achieve traffic engineering in IP networks. A natural online version of linear optimization, which can be applied to several combinatorial problems such as max-cut, variants of clustering, and the classic online binary search tree problem, is considered by Kalai and Vempal (2002).

An approximate dynamic programming approach for making ambulance redeployment decisions in an emergency medical service system is used in the study by Restrepo, Henderson, and Topaloglu (2010). The authors deal with the question of where idle ambulances should be redeployed to maximize the number of calls reached within a given delay threshold.

These authors considered the problem of finding a static deployment of ambulances in an emergency medical service system (Restrepo, Henderson, & Topaloglu, 2009). In this context, the goal is to allocate a given number of ambulances among a set of bases to minimize the fraction of calls that are not reached within a time standard. We are not aware of any approaches similar to ours that improve care of the injured in disasters through a mapping to online job-shop scheduling problems. Therefore, we have developed our own optimization framework to solve this problem, which will be introduced in detail in the next sections.

## 6.3 Experimental Setup

We describe the data used for our simulation studies later. The subsequent explanations that follow refer to Güttinger et al. (2011). The main difference between the data setup used in their study and this work is that we now assume casualties arrive in small groups at different points in time and not as a whole at one specific point in time. We furthermore draw actual journey length and medical treatment duration randomly around expected values and do not take them as deterministic. Before describing the data setup that we used, we first want to introduce and explain the concept of penalty functions.

### 6.3.1 Triage Groups and Penalty Functions

If a high number of casualties have to be medicated at once, the injured are divided into different *triage categories* to ensure an adequate supply. Patients with a higher priority are dealt with first. These groups differ only slightly from country to country. Windle, Mackway-Johnes, and Marsden (2006) give a detailed overview about the differences. For the purpose of this work, we use the definition by Crespin and Neff (2000):

- *T1: Urgent vital thread*, for example, respiratory insufficiency with asphyxia, massive (external) bleeding, heavy shock, serious thorax or abdomen trauma, face and/or respiratory system burns, tension pneumothorax
  ⇨ Immediate treatment necessary

- *T2: Seriously injured*, for example, craniocerebral injury, serious eye and/or face injury, unstable luxation, exposed fracture/joint
  ⇨ Postponed treatment urgency, supervision
- *T3: Slightly injured*, for example, uncomplicated fractures, repositionable luxation, contusion, distortion, graze
  ⇨ Posterior treatment (ambulant, if applicable)
- *T4: Without any viability chance, dying*, for example, open craniocerebral injury with cerebral mass discharge, severest and extended burns
  ⇨ Supervising, death awaiting treatment, terminal care

In general, casualties are assigned a priority according to their triage category and are treated more or less quickly: casualties assigned to T1 are dealt with first, then T2, and T3.

Often, a longer waiting time will affect patients' state of health because the severity of many injuries depends on the therapy time. A tension pneumothorax, for example, is nonhazardous if treated immediately. But without treatment after a short period, a tension pneumothorax can become life-threatening. Note that the concept of triage categories as described above fixes a priority level for each casualty, not taking into account the waiting time. Therefore, assignment switches are not allowed and often lead to a falsified prioritization.

To improve the process, our optimization method considers *penalty functions* over time for all injuries describing the corresponding course of injury. The penalty functions $f_y$ assign

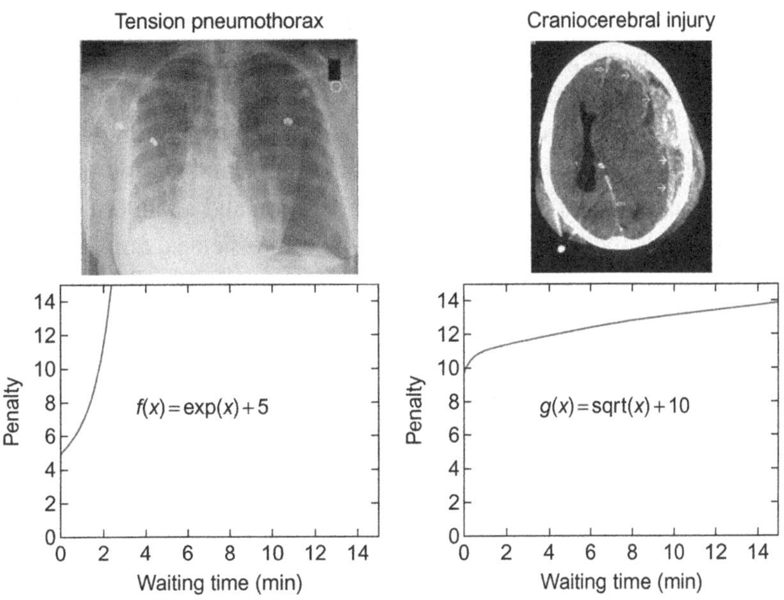


**Figure 6.1: Example pictures and functions illustrating the course of disease over time.**

a penalty value to each injury $y \in Y$ (set of injuries) depending on the waiting time $t$ until medication. Figure 6.1 illustrates the course of injury for a tension pneumothorax (left figure) and a craniocerebral injury (right figure). Obviously, these two conditions have a different initial priority and course. A long waiting time thus has a different impact. In the case of pneumothorax, the penalty increases exponentially with time. For a craniocerebral injury the waiting time also influences the level of injury, but has nothing like the effect as for a tension pneumothorax.

Overall, we define 23 different sample functions for injuries such as luxation, craniocerebral injury, bleeding, or serious dermal burn. They differ in both their initial priority ($t = 0$) and the positive functional course (e.g., $f(x) = \ln(x + 1)$, $f(x) = x^2$, $f(x) = e_x - 1$, $f(x) = \sqrt{x}$). Clearly, a longer waiting time causes a higher penalty value for all the functions. For later evaluation, we assign these 23 functions to the triage groups, as defined earlier, considering the value for $t = 0$ similar to the initial priorities. Of 23 functions, 11 are assigned to T1, 6 to T2, and the remaining 6 to T3. We do not consider T4 since those subjects do not have an impact on the overall result.

We want to point out that the introduction of such functions is useful and feasible after consultation with experts from the city council of Frankfurt, Germany. The chosen functions are, however, given as examples to show the power of our approach and need to be adapted by experts in realistic future scenarios.

### 6.3.2 Data

This section introduces the data sets that will be used to show the feasibility of our methods for optimization of the emergency supply after a mass casualty incident (cf. Sections 6.4 and 6.5). The factors affecting the optimization problem are manifold. The following values have to be specified for the algorithm: number of casualties and corresponding injuries with their expected length of treatment, arrival time of each casualty, actual duration for transport and medication for each casualty (depends on expected duration and a random deviation in each case), number of hospitals including number and type of physicians, number of vehicles, and a matrix specifying the expected transportation time between the location of the catastrophe and hospitals. Since we design our algorithm to work in arbitrary settings, the data sets will consider empirical data as well as artificial data that are randomly drawn. For testing, we evaluate a virtual disaster during a football game in Frankfurt, Germany. The data sets are discussed in detail in the following sections.

#### 6.3.2.1 Empirical Data

For testing, we use the arena disaster scenario at the soccer World Cup 2006 as described by Der Magistrat der Stadt Frankfurt am Main (2006). The number of casualties is fixed at 1000. The distribution between different triage groups (T1–T4) has been assumed to be 20%, 40%, 20%, 20% (see Table 6.1). This leads 400 casualties to be assigned to the triage group T2 and

**Table 6.1: Distribution of 1000 casualties in triage groups.**

| 100% (1000 casualties) | | | |
|---|---|---|---|
| T1<br>20%<br>(200 casualties) | T2<br>40%<br>(400 casualties) | T3<br>20%<br>(200 casualties) | T4<br>20%<br>(200 casualties) |

**Table 6.2: Interval bounds for randomly generated input variables in the training set.**

| Variable | Left Bound | Right Bound |
|---|---|---|
| Size of casualty groups | 1 | 10 |
| Number of casualties | 200 | 1600 |
| Expected medication duration | 5 minutes | 3 hours |
| Initial priority | 10 | 1000 |
| Number of hospitals | 10 | 30 |
| Number of beds | 100 | 1000 |
| Number of physicians | 50 | 500 |
| Number of vehicles | 5 | 25 |
| Vehicle capacity | 5 | 25 |
| Expected journey length to hospitals | 5 minutes | 30 minutes |

200 casualties to each of the remaining groups. Altogether 800 casualties (ignoring T4) with a significant chance of viability have to be treated and considered in the optimization problem. In Section 6.1, we explored the affiliation between the 23 injuries and the triage groups.

We are thus able to draw injuries for the casualties for each group randomly, leading to 800 casualties distributed to T1–T3 as specified by Der Magistrat der Stadt Frankfurt am Main (2006).

To obtain realistic values for the number of hospitals including the number and type of physicians and the number of available vehicles, we have collected data from 20 quality reports from different hospitals (German Department of Justice, 1998) around Frankfurt. Finally, we estimate the transportation time matrix between the arena and the hospitals using a route planning tool. The medication duration for the randomly generated casualty test set is chosen in the same range as for the training set (between 5 minutes and 3 hours; see Section 6.3.2.2 for details).

#### *6.3.2.2 Random Data Samples*

For performance evaluation and comparison, we have generated 1000 instances of random data sets as follows. For each of the aforementioned factors (e.g., number of casualties, injuries), we randomly draw a value from a uniform distribution within reasonable intervals (see Table 6.2). As we take the empirical data set from Frankfurt (see Section 6.3.2.1) as a representative, the corresponding values serve as interval centers for the underlying uniform distribution that is used for drawing the random data samples.

Because the set of instances is generated randomly and has a representative value, the method should generalize many problems in this domain. A consideration of all combinations over the factors is not realistic. The number of degrees of freedom is certainly too high.

If a mass casualty occurs, for example, due to a stampede at a mass event, injured persons that need medical treatment are registered subsequently in groups by medical field help. Thus, decisions about further treatment of casualties (for example, to which hospital a casualty should be moved) have to be made separately for each group not knowing at this point in time if and how many casualties are still to come.

For testing the algorithms used in this work, we therefore model the arriving casualty groups as a Poisson process with variable jump size (which corresponds to the size of a casualty group). The jump size is drawn randomly within a reasonable interval; see first row in Table 6.2.

Given a (fixed) instance (set of casualties, hospitals, vehicles, etc.) drawn from the intervals in Table 6.2, the actual journey length to hospitals and the actual medical treatment duration of casualties are considered to be normally distributed around their expected values with standard deviations. The assumption of randomly biased medical treatment durations and journey lengths is reasonable, since in practice usually there also exist exogenous environment variables that influence these quantities such as, for example, current traffic situation or possible complications during surgery. Consequently, these actual values are unknown when an assignment decision for a casualty has to be taken. They are found *after* a casualty has arrived at the destination hospital or has already been medically treated, respectively. The optimization procedure is therefore done under uncertainty and based on the corresponding expected values. See the next sections for details.

## 6.4 Algorithms for Optimization

In this section, we first define formally the optimization problem (cf. Section 6.1). Then, we describe three different algorithms that we apply to address the problem (cf. Section 6.2).

### 6.4.1 Optimization Problem and Criteria

Figure 6.2 illustrates the previously described optimization problem including relationships between the entities involved. Intuitively, our aim is to find an optimal assignment between casualties and transport vehicles and physicians such that all casualties can be medicated as proficiently and quickly as possible depending on their individual injuries. Since there exist temporal dependencies related to the arrival time of casualties, this optimization is executed iteratively not on the whole set of casualties but on small groups. These groups arrive at intermittent points in time $t_i \in [0, T]$, $i = 1, 2, \ldots, N$ with $T = t_N$ at the disaster area. In the following, we formally introduce the other factors involved. Furthermore, we define the

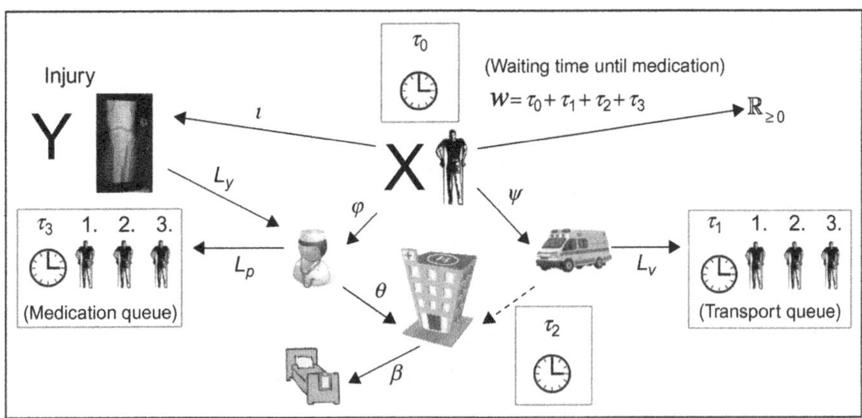


**Figure 6.2: Interrelations between the entities involved in emergency supply for mass events. Core entities are casualties *X* with injury *Y* mapped to a vehicle and a doctor, who in turn belongs to a specific hospital. The optimization is performed subject to waiting time *w*, doctor qualifications and the hospital capacity $\beta$.**

optimization function, which calculates the total level of injury over all patients. Clearly, the goal is to minimize this function.

Let $V$, $P$, and $H$ be a set of vehicles, physicians, and hospitals, respectively. Furthermore, let $M$ be a set of assignment tuples $(\varphi, \psi)$, where $\varphi$ maps all casualties $x \in X$ to a physician $p \in P$ and $\psi$ maps all casualties to vehicle $v \in V$, respectively. Since the complete set of casualties $X$ is unknown beforehand, we define time-dependent subsets $X(t) \subset X$ for each point in time $t \in [0, T]$, where $X(t)$ defines the set of casualties that have already arrived at the disaster area at time $t$. Obviously, $X(t_i) \subset X(t_j)$ for $0 \leq t_i < t_j \geq T$ and $X = X(t)$, since we defined $T$ as the particular point in time when the last group of casualties arrives at the disaster area.

Analogously we define sets $M(t)$ for each point in time $t \in [0, T]$, where $M(t)$ contains all assignment tuples $(\phi_t, \psi_t)$ that map all casualties $x \in X(t)$ to a physician $p \in P$ and to a vehicle $v \in V$, respectively. Again, we have $M = M(t)$.

We define the following band of optimization functions (one function for each $t \in [0,T]$):

$$F(t) = \sum_{x \in X(t)} f_x(w(x)) \tag{6.1}$$

with

$$w: X \rightarrow \mathbb{R}_{\geq 0},\ w(x) = \tau_0(x) + \tau_1(x) + \tau_2(x) + \tau_3(x), \quad \forall x \in X.$$

The function first considers the total waiting time $w(x)$ until medication for casualty $x$. Subsequently, this waiting time serves as a parameter for the penalty function $f_x$ describing the course of injury that casualty $x \in X$ is suffering from. The overall level of injury is the sum over the single level of injury of all the casualties $f_x(w(x))$. The waiting time itself in turn is the sum of the following:

- $\tau_0 : X \rightarrow \mathbb{R}_{\geq 0}$ : Arrival time at registration point
- $\tau_1 : X \rightarrow \mathbb{R}_{\geq 0}$ : Waiting time until transport
- $\tau_2 : X \rightarrow \mathbb{R}_{\geq 0}$ : Actual transport duration for each casualty
- $\tau_3 : X \rightarrow \mathbb{R}_{\geq 0}$ : Medication waiting time in hospitals

  The medication waiting time in hospitals for a casualty is determined using the actual medical treatment duration of all the casualties that are assigned to the same physician and treated earlier. We therefore additionally define

- $\tau_4 : X \rightarrow \mathbb{R}_{\geq 0}$: Actual medical treatment duration for each casualty.

The objective is to find an optimal mapping of casualties to physicians and vehicles for each point in time $t \in [0, T]$. Note that $(\phi_t, \psi_t) \in M(t)$ only assigns already arrived casualties to physicians and vehicles, but does not define the order of treatment. Therefore, we additionally define the functions $(L_p(t))_{p \in P}$, $(L_v(t))_{v \in V}$ as

$$L_p(t): \phi_t^{-1}(p) \rightarrow \{1, 2, \ldots, |\phi_t^{-1}(p)|\},$$
$$L_v(t): \psi_t^{-1}(p) \rightarrow \{1, 2, \ldots, |\psi_t^{-1}(v)|\}.$$

The functions define for all physicians $p$ and vehicles $v$ at all points in time $t \in [0, T]$ a sequence of treatment for patients that are either assigned to a specific doctor $(\phi_t^{-1}(p))$ or vehicle $(\psi_t^{-1}(p))$.

The mappings $(\phi_t, \psi_t)$ and orderings $(L_p(t))_{p \in P}$ and $(L_v(t))_{v \in V}$ should be chosen such that they minimize $F(t)$ for each $t \in [0, T]$ and concurrently satisfy the following two conditions:

- C1: The total number of medicated casualties at time $t$, $\lambda(h, t)$, in a hospital does not exceed the number of available beds $\beta(h)$ for all hospitals $h \in H$.
- C2: A casualty $x$ is assigned to a qualified physician $\varphi_t(x)$: $\phi_t(x) \in L_{\iota(x)}\ \forall x \in X(t), \forall t \in [0, T]$ with classification function $\iota : X \rightarrow Y$ mapping a casualty to an injury and $\pi : y \rightarrow L_y \subset \mathcal{P}(P)$ assigning all injuries $y \in Y$ to the set of qualified doctors.

Note that the ordering sequences $(L_p(t))_{p \in P}$ and $(L_v(t))_{v \in V}$ are bijective, since each casualty has exactly one position in the assigned transport and medication queue. In the worst case, all casualties arrive in one group at the same point in time at the registration point, which means $X = X(t_0)$. Furthermore, each physician $p \in P$ can medicate each type of injury, and the number of available beds is greater than the total number of casualties in each hospital. In this case, there are $|P|^n \cdot |V|^n$ different combinations of feasible maps $\varphi$ and $\psi$ and at most $n! \cdot n!$ possible bijective orderings for each $(\phi, \psi) \in M = M(t_0)$, if $n \in \mathbb{N}$

is the total number of casualties. Thus, altogether we obtain $O(n^n)$ different combinations. In addition, the optimization problem is nonlinear since the objective function is nonlinear:

Obviously, Eq.(6.1) can contain nonlinear penalty functions $f_x$, and thus, using linear optimization algorithms such as Simplex or interior point for solving our optimization problem is impossible. Therefore, we have decided to use a greedy strategy and an adjusted version of simulated annealing as optimization techniques, which are completely independent of the objective function characteristics. We compare them to a D'Hondt assignment strategy, which is currently used in practice. We introduce all the mentioned algorithms in detail in Section 6.4.2.

### 6.4.2 Optimization Algorithms

In this section, we introduce the algorithms used in this work for our simulations: namely, these are the D'Hondt assignment strategy, which is currently used in practice to solve the online assignment problem, a greedy strategy that we have developed, and an adjusted version of simulated annealing from Güttinger et al. (2011) that is used to iteratively improve the initial solution.

Although the D'Hondt strategy and the greedy algorithm that we have designed differ in the way casualties are assigned to vehicles and hospitals, the strategy how to compute a (final) assignment of casualties to physicians (in hospitals) is the same for both techniques. The reason is that this step (assignment of casualties to physicians) is done after casualties have already been delivered to specific hospitals. In other words, the hospital where a casualty is delivered has to be fixed before medical treatment of this casualty can be started. It is therefore independent of the decision in which order a physician in a hospital will treat the casualties that are assigned to this hospital medically. This follow-up optimization algorithm in hospitals is described in Section 6.4.2.1 before the actual assignment strategies are introduced.

#### 6.4.2.1 Follow-up Optimization in Hospitals

It seems reasonable to assume that processing of casualties within hospitals is also done with respect to priorities. This means that each time a physician is available he or she has to decide which casualty of all currently waiting casualties should be treated next such that the sum of penalties for all currently waiting casualties is minimized. If we interpret penalties as costs, this decision problem is equivalent to generalization of the *single machine earliness tardiness problem*, which is known to be NP-complete; see Ow and Morton (1989).

In practice, an available physician would choose a waiting casualty out of all waiting casualties such that the sum of penalties of all other waiting casualties *after* having medicated this specific casualty is minimized. Any other more complex strategies going beyond this consideration can be assumed as impracticable not least because of the urgent circumstances.

This leads to a follow-up optimization procedure in hospitals or, mathematically, the (final) computation of $L_p(t)$ for all $p \in P$ and each $t \in [0, T]$. This procedure is independent of the applied online assignment strategy and, therefore, executed subsequent to the D'Hondt and greedy algorithm introduced later. It consists of the following steps:

1. For each hospital $h \in H$ and each time $t \in [0,T]$:
   a. For each physician $p \in P$ that is idle at time $t$ and that can medicate at least one casualty $x \in X$ currently waiting for medical treatment in hospital $h$:
      - Assign this casualty $x$ of all waiting casualties to an available and qualified physician $p$ (that is, $\phi(x) = p$), such that total level of injury over all other waiting casualties is minimized at the point in time when $x$ is medicated (based on expected medical treatment duration).

#### 6.4.2.2 D'Hondt

The city council of Frankfurt am Main (Germany) has arranged with each regional hospital in negotiations how many casualties can be taken in by this hospital if a mass casualty incident occurs. To ensure a balanced assignment of casualties to hospitals regarding the negotiated accommodation numbers they use the D'Hondt method for online assignment.

The D'Hondt method is the highest averages method and is used for allocating seats in party-list proportional representation. The method is named after the Belgian mathematician Victor D'Hondt. This system, which is used by many legislatures all over the world, slightly favors large parties and coalitions over scattered small parties; see Pukelsheim (2007).

Furthermore, when loading ambulances with casualties, usually an attempt is made to assign all casualties with the same destination hospital to one ambulance (of course, this is only relevant for ambulances with load capacity greater than 1). Practically, this means that each time a casualty is assigned to a vehicle that was empty before, the queue of waiting casualties is scanned for other casualties with the same destination hospital.

Translated to our online assignment problem, the corresponding algorithm is as follows:

1. For each arriving casualty group:
   a. Assign arriving group to hospitals with respect to D'Hondt strategy.
2. For each point in time $t \in [0,T]$:
   a. For each vehicle $v \in V$ that is idle at time $t$:
      i. Transport this casualty $x \in X$ by vehicle $v$ as the next one who is waiting for transport at time $t$ and suffers highest current level of injury $f_x(w(x))$ of all casualties currently waiting for transport.
      ii. Search for waiting casualties that are assigned to the same hospital as casualty $x$ from the previous step and also assign them to the available vehicle $v$ (only if the vehicle has remaining capacity and if there exist such casualties).
3. Compute assignment from casualties $x \in X$ to physicians $p \in P$ (see above).

The assignment of casualties to physicians is done using the "follow-up optimization in hospitals" procedure described in Section 6.4.2.1.

#### 6.4.2.3 Greedy

In this section, we introduce a new greedy online assignment strategy. This greedy strategy processes currently waiting casualties in priority order (that means casualties that suffer more seriouus injuries are dealt with first) and computes locally optimal hospital and vehicle assignments. In contrast to the D'Hondt strategy, greedy considers expected journey lengths to hospitals. Furthermore, when loading ambulances with capacity greater than one, greedy does not attempt to assign all casualties in the same ambulance to the same hospital, but always assigns a casualty to the next available vehicle independent of destination hospitals of other passengers within the same cart load.

Our greedy algorithm consists of the following steps:

1. For each point in time $t \in [0, T]$:
   a. For each vehicle $v \in V$ that is idle at time $t$:
      i. While vehicle $v$ has remaining capacity *and* there are casualties waiting for transport at time $t$:
         1. Transport this casualty $x \in X$ by vehicle $v$ as the next one who is waiting for transport at time $t$ and who has the most serious injury $f_x(w(x))$ of all the casualties currently waiting for transport.
         2. Determine the destination hospital for $x$:
            - Choose a qualified physician $p \in P$ and feasible medication queue position for $x$, such that the estimated value as regards extent of injury (based on expected transport and medication duration of all casualties already registered) is minimized. Set the location of $p$ as the destination hospital for $x$.
2. Compute assignment from casualties $x \in X$ to physicians $p \in P$ (see above).

Remarks:

- Note that it is possible that a vehicle starts delivering casualties although it is not yet full, since the number of casualties still to come in the future is unknown. The same applies to the D'Hondt strategy.
- If a new casualty arrives at their destination hospital, then he or she can be placed in the queue at every position that belongs to another casualty who is either waiting for medication or has not yet arrived at the hospital. Queuing positions of casualties that are currently under medical treatment or have already been medicated are not feasible.
- The assignment to a physician in step B is only tentative because final assignment to physicians is computed subsequently in step 1.a.i.2 with the procedure "follow-up optimization in hospitals"; see above.

#### 6.4.2.4 Simulated Annealing

For our simulations, we use a workload-adapted version of simulated annealing from Güttinger et al. (2011), which uses transition probabilities that are adapted to the current workload of vehicles and physicians.

Simulated annealing itself is a common technique for minimization of functions whose exact analytical characteristics are unknown due to high complexity. Each step of the Simulated simulated annealing algorithm replaces the current solution by a random nearby solution, chosen with a probability that depends on both the difference between the corresponding function values and a global parameter *Temp* (called the temperature), which is gradually decreased (degression coefficient) during the process. The dependency is such that the current solution changes almost randomly when *Temp* is large, but decreases as *Temp* tends to zero. The allowance for increases saves the method from becoming stuck at local optima. For a detailed introduction to simulated annealing, refer to Das and Chakrabati (2005).

In our setting, a solution $S$ is an arbitrary assignment of those casualties that are waiting for transport at the disaster area at a specific point in time to vehicles and physicians. A nearby solution is one with the same assignments except that a casualty is assigned to a different vehicle and/or physician or another queuing position. The performance of simulated annealing is carried mainly by the initial temperature, the degression coefficient, and the number of runs for each temperature.

Two factors have a significant impact on an efficient calculation of a good approximation to the global optimum. First, a good initialization is crucial for a fast convergence. Second, sophisticated generation of nearby solutions strongly affects the speed of convergence.

As already described earlier, we consider an online assignment problem where groups of casualties arrive at intermittent intervals. Each time a group of new casualties arrives, a decision about conveyance and destination hospital has to be taken for each casualty of the group. The idea is now to assign newly arriving casualties tentatively to an ambulance and a qualified physician in an *initialization* step. Afterwards, a simulated annealing procedure is triggered that tries to optimize the assignment of all casualties that are currently waiting for evacuation in an *optimization* step. These two steps that are executed at each point in time when a new casualty group arrives are described in detail in Sections 6.4.2.5 and 6.4.2.6.

#### 6.4.2.5 Initialization

Simulated annealing needs an initial assignment from casualties to vehicles and physicians as an input for subsequent optimization iterations. Following Güttinger et al. (2011), where we have shown that using a greedy result as initial assignment for simulated annealing is superior to using any random initialization strategy, we will also use the greedy algorithm from Section 6.2.3 for initialization in this work. Note that finding a general strategy for choosing the input parameters and computing an initial assignment for simulated annealing is an open research question.

### 6.4.2.6 Optimization

The optimization step is triggered at each point in time when a new casualty group has arrived at the disaster area. Following the computation of an initial assignment for recently arrived casualties in the previous step computed with greedy (see Section 6.4.2.5), the (tentative) assignment of all casualties currently waiting for transport at the disaster area (including those of the recently arrived group) is optimized by simulated annealing.

The choice of a casualty that is reassigned to a different vehicle and physician on trial is based on the following observation:

- A reassignation of casualties with more serious injuries will conjecturally have a greater effect on the global optimization function $F$ and thus speed up convergence of simulated annealing. Therefore, we define the probability of reassigning a waiting casualty $x \in X$ in a specific simulated annealing step by the ratio of his or her current expected injury $f_x(w(x))$ and the overall expected level of injuries $F(t)$ for all currently waiting casualties at time $t \in [0, T]$. Remember $w$ is the function for waiting time and $f_x$ is the function for the corresponding course of disease for casualty $x$:

$$P_t(x) = \frac{f_x(w(x))}{f(t)}$$

The probabilities of choosing a specific vehicle and physician where a casualty is being reassigned are defined as follows:

- It seems plausible that a reassignment of a casualty $x \in X$ to a vehicle $v \in V$ with a low workload will increase the global optimization function less than when assigning $x$ to a vehicle that is already used to the full. In order to estimate the workload of a specific vehicle $v$, we sum up the level of injury of all patients $\psi_t^{-1}(v)$ that are assigned to the specific vehicle (negative term of the numerator), but still waiting for transport. By subtracting this value from the overall level of injuries $F(t)$ of all casualties currently waiting for evacuation at time $t \in [0, T]$, we obtain an antiproportional dependency leading to low values if the vehicle is fully loaded. Finally, we normalize by the sum of the corresponding numerators for all vehicles:

$$P_t(v) = \frac{F(t) - \sum_{x \in \psi_t^{-1}(v)} f_x(w(x))}{\sum_{\hat{v} \in V} \left( F(t) - \sum_{x \in \psi_t^{-1}(\hat{v})} f_x(w(x)) \right)}$$

- Similarly, we define probabilities for the assignment of the chosen casualty $x$ to qualified doctors. Once more, an assignment of $x$ to a physician $p$ with low workload will probably improve the optimization. Let $P_x$ be the set of physicians that are qualified to medicate the injury of $x$.

$$D_{x,t} = \sum_{\hat{p} \in P_x} \sum_{\hat{x} \in \varphi_t^{-1}(\hat{p})} f_x(w(\hat{x}))$$

gives the overall level of injury of all casualties $\hat{x}$ that are assigned to qualified doctors, but not yet under medical treatment at time $t \in [0, T]$. The probability of assigning a casualty $x$ to a physician $p$ is calculated as follows:

$$P_t(p|x) = \begin{cases} 0, & \text{if } p \notin P_x, \\ Q(p,t,x), & \text{if } p \in P_x \end{cases}$$

where

$$Q(p,t,x) = \frac{D_{x,t} - \sum_{\hat{x} \in \psi_t^{-1}(p)} f_x(w(\hat{x}))}{\sum_{\hat{p} \in P_x} \left( D_{x,t} - \sum_{\hat{x} \in \psi_t^{-1}(\hat{p})} f_x(w(\hat{x})) \right)}$$

If $p \notin P_x$, an assignment of $x$ to $p$ is not feasible because physician $p$ does not have the necessary qualifications. If $p$ has the necessary qualifications, the probability is high in the case of a low workload for $p$. Once more, the negative term of the numerator specifies the injuries of all casualties that are mapped to the specific physician and have not yet been medically treated. By subtracting this value from the overall level of injury of patients medicated by qualified physicians, the numerator will be higher for doctors with a low workload. The denominator normalizes the specific values by summing up the corresponding values for all qualified physicians $\hat{p} \in P_x$.

Note that for the previous considerations about assignment probabilities at each point in time only those casualties were taken into account that have already arrived at the disaster area, since we assume that the type and final number of casualties still to come in the future are unknown. Furthermore, for computing ambulance and physician assignment probabilities those casualties that have already been transported to the destination hospital or have already been treated medically, respectively, are omitted, since their assignment is irrevocable.

Besides this, we want to point out that in this and in the following paragraph all computations of casualty levels of injury are based on *expected* journey lengths to destination hospitals and medical treatment durations, since these values are drawn randomly *after* evacuation and medical treatment of a casualty has already started (and thus when assignment is already irrevocable).

The previous considerations finally lead to the following algorithm:

- Input: Initial temperature $T_{\text{init}}$, degression coefficient $s \in (0,1)$, iteration coefficient $m$.
  1. For all $i = 1, 2, \ldots, N$ do
     a. Update assignment $A_{\text{best}}$ by assigning newly arrived casualties from *the ith* group to ambulances and qualified physicians

b. Compute value of objective function $F_{best}$ for $A_{best}$.
c. Set $T := T_{init}$.
d. Repeat
   i. For all $j = 1, 2, \ldots, m$ do
      1. Choose casualty $x$ from a set of waiting casualties with probability $P_t(x)$.
      2. Draw vehicle $v \in V$ with probability $P_t(v)$ and physician $p \in P$ with probability $P_t(p|x)$.
      3. Compute value of objective function $F_{new}$ for assignment $A_{new}$ that would result when reassigning casualty $x$ to vehicle $v$ and physician $p$.
      4. Draw uniform random variable $u \in [0,1]$.
      5. If $\exp\left(\frac{F_{best} - F_{new}}{T}\right) > u$ then
         a. Set $A_{best} := A_{new}$ and $F_{best} := F_{new}$.
   ii. $T = T \cdot s$.

   Until $|F_{best} - F_{new}| < \varepsilon$
- Output: Approximatively optimal assignment $A_{best}$.

### 6.4.3 Relaxations

In this subsection, we will define a relaxed version of our optimization problem that will serve as a standard of comparison for all the above optimization strategies. In this relaxed version, we completely abstract from any waiting times for transport and medication. This means that the number of vehicles and the number of physicians of a specific type in a hospital is unlimited. As a consequence, each casualty can be transported immediately by a vehicle and medically treated by a physician as soon as the affected person has arrived at a hospital where the specific type of injury can be treated. It follows that the sequences $(L_p)_{p \in P}$ and $(L_v)_{v \in V}$ in which casualties are processed are irrelevant and it is an optimal strategy to transport each casualty to the nearest hospital where he or she can be treated medically.

The above considerations yield a lower bound for the minimal (optimal) value of our optimization function $F$ with respect to the original problem definition. This lower bound can be computed in $O(|X| \cdot |P|)$, where $|X|$ is the total number of casualties and $|P|$ is the number of physicians.

## 6.5 Experiments and Results

In this section, we will compare the performance of the algorithms introduced so far: the D'Hondt assignment strategy, a greedy algorithm, and a workload-adapted version of simulated annealing, which uses the result of greedy as initial configuration. The performance evaluation is done with respect to a lower bound, which can be efficiently computed for each test instance. For this purpose, all the above-mentioned algorithms were implemented in the C programming language. The empirical and randomly generated data sets used for our experiments were provided in tabular form as character-separated value files which were loaded by programs to compute an optimized assignment of casualties to vehicles and physicians. Programs for generating random data sets were also implemented in C.

Since simulated annealing additionally requires an initial temperature, the number of runs with constant temperature and a temperature degression coefficient as input, we first try to find a combination of these parameters that performs well in all instances in a training step.

### 6.5.1 Training

In order to find adequate parameters for our scenario, we continuously generate different combinations of these three parameters randomly and calculate for each combination and several randomly generated instances an assignment of casualties to vehicles and physicians with simulated annealing. The combination leading to the minimum standardized average level of injury is saved and used later in the test. Note that results for different number of casualties cannot be compared directly. Therefore, for all computations in this and in the following subsection, we normalize with respect to the total number of casualties.

In the study by Güttinger et al. (2011), we have shown that evaluation of 100 parameter combinations is sufficient for problems of the current type. On this basis, the performance on 100 randomly generated instances was compared, and the combination of parameters was chosen where average level of injury of each casualty was the least severe. In doing so, the following parameter combination turned out to show the best performance:

- Initial temperature: 75
- Number of runs with constant temperature: 82
- Temperature degression coefficient: 0.02

These parameters will be used for all further simulated annealing computations.

### 6.5.2 Test

Our tests consist of two parts. First, we want to evaluate the performance of our algorithms on an empirical data set. In the second part, we will show that these results can be generalized for arbitrary data sets that are representative of our problem definition.

#### 6.5.2.1 Empirical Evaluation

We have evaluated the performance of our algorithms on the empirical data set from Frankfurt introduced in Section 6.3. The final value of the average level of injury per casualty is about 30% smaller for simulated annealing than for greedy (129.054 for simulated annealing and 181.24 for greedy). The final value for D'Hondt of 966.24 is many times larger than for the other strategies. The lower bound value for the given empirical assignment

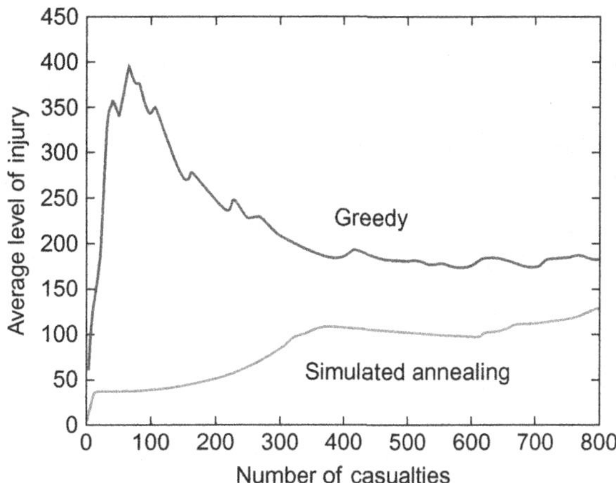


**Figure 6.3 Evolution of average casualty level of injury with respect to the number of casualties that have already been medically treated when applying greedy and simulated annealing on test set.**

problem in its relaxed form equals 45.28. The resulting approximation bound values (quotient of algorithm approximation result and lower bound value for the relaxed problem definition) are 21.33 for D'Hondt, 4.0 for greedy and 2.85 for simulated annealing, respectively.

The underperformance of the D'Hondt strategy can easily be explained by the fact that it does not take journey length to hospitals into account, which in turn has a significant influence on the casualty's severity of injury.

The evolution of average level of injury per casualty for simulated annealing and greedy is plotted in Fig. 6.3 (we have omitted the illustration of D'Hondt, since this strategy does not seem to be able to compete on the test set). On the abscissa, the number of casualties that have already been treated medically is displayed, and the level of average injury these casualties suffer from the ordinate values.

We can see that at the beginning average level of injury for greedy increases rapidly and reaches its peak when having medicated almost 100 casualties. Each of them suffers an injury value of 400 on average at this point in time. Afterwards, the average level of injury decreases and levels off at 180 after having treated 400 casualties medically. The trend of level of injury for simulated annealing is clearly smoother. After a rapid increase of average level of injury for the first 10 casualties, the graph slopes slightly upwards to a level of 100 after having medicated about 400 casualties and stagnates subsequently. Finally, the average level of injury slightly increases again for the last 200 casualties up to a level of 130.

Altogether we can conclude that the assignments computed by simulated annealing are much steadier than that by greedy with respect to the number of medicated casualties. In particular, the casualties medicated first within the greedy strategy obviously suffer more serious injury.

This progress can be connected to the natural course of an incident: the first casualties that are registered can be transported and medically treated immediately, since available capacities are still in good supply. As more casualties are registered, bottlenecks in connection with transport evolve and congestions in hospitals occur. Particularly, in this phase the choice of a suitable order of casualties is crucial. Apparently, in this respect, simulated annealing benefits from its more detailed solution space examination in comparison to greedy when deriving a good casualty transport and medication process order. After a period of time when more casualties have arrived, the situation eases off.

#### *6.5.2.2 Generalization of Empirical Results*

To be able to generalize the empirical test set results from the above, we have generated 1000 additional random data samples (different from those that were used in the training step) as described in Section 6.3. We applied each of the algorithms on every data set and computed the average level of injury with respect to the number of instances and casualties. Furthermore, we have computed the quotient of average casualty level of injury derived by a technique and the lower bound for average casualty injury resulting from the relaxed problem definition. The results are shown in Table 6.3.

The second column contains the average level of injury of each casualty. As already detected for the empirical test set, the results show a clear outperformance of greedy in comparison to the D'Hondt assignment strategy. The final results for both strategies were available after about 1 second of computation time in each instance (all computations in this work were executed on an AMD Opteron Processor with 2.60 GHz and 8 GB of RAM).

We can furthermore see that the average improvement for simulated annealing in comparison to greedy is smaller for the general case than for our empirical test set. However, the average level of injury suffered by each casualty can still be leveled down by about 10% when applying workload-adapted simulated annealing iterations subsequent to greedy.

The average improvement of 10% for the general case when applying simulated annealing subsequent to greedy is also much smaller than in Güttinger et al. (2011) where this strategy brought 99% improvement compared to greedy results. This fact can be explained by the smaller size of the solution space in our setting compared to Güttinger et al. (2011), which can be seen as a special case of our problem definition (all casualties are registered at the same time).

**Table 6.3: Optimization results.**

| Technique | Average $F$ | Lower Bound Quotient |
|---|---|---|
| D'Hondt | 8673.57 | 17.15 |
| Greedy | 1332.68 | 2.66168 |
| Simulated annealing (with greedy initialization) | 1203.13 | 2.39815 |

Due to temporal dependencies, the number of feasible assignments is less than that in the work of Güttinger et al. (2011), and thus the greedy result is "less suboptimal" in this work.

Since additional computational effort in the execution of simulated annealing is negligible (about 1 second for each arriving casualty group on average), it is nevertheless the strategy of choice as a 10% improvement can have a vital effect in the context of a mass casualty incident.

## 6.6 Conclusions

In this final section, we want to summarize our results and give an outlook on further interesting research questions regarding this topic.

### 6.6.1 Summary

In this work, we compare several online approaches for optimizing the emergency supply after a major incident. For a given set of physicians, hospitals, and transport vehicles, the algorithms introduced in this work compute an assignment of casualties arriving in groups that suffer from specific types of injuries to available transport and medical capacities. Our objective is to minimize total injury to all casualties, where the level of injury of a casualty is measured by the value of a specific penalty function. For each type of injury, we define such a penalty function that depends on waiting time until medical treatment and describes the current state of health of a casualty.

In our simulation studies, we show that a simple greedy strategy clearly outperforms an assignment based on the D'Hondt algorithm, which is, to the best of our knowledge, the only technique currently used in practice to deal with the problem at hand. We have furthermore analyzed the additional benefit when applying a workload-adapted version of simulated annealing subsequent to each greedy iteration. In doing so, it appeared that the average level of injury for each casualty (and so also total level of injury) yielded by the greedy strategy can be additionally reduced by about 10% with an acceptable computational expenditure.

Thus, altogether we can conclude that the combination of a greedy strategy and a subsequent application of a workload-adapted version of simulated annealing works well for the given online assignment problem.

### 6.6.2 Future Prospects

Despite the promising results of our adjusted simulated annealing algorithm introduced in this work, there is still room for improvements. One issue for future research will be to try out other optimization techniques like genetic approaches to see if they can provide better results. Furthermore, the influence of the chosen penalty functions on the results should be studied with respect to their number and type.

## References


Akyol, D. & Bayhan, G. (2008). Multi-machine earliness and tardiness scheduling problem: an interconnected neural network approach. *The International Journal of Advanced Manufacturing Technology*, *37*(5), 576–588.

Bagula, A., Botha, M., & Krzesinski, A. (2004). Online traffic engineering: the least interference optimization algorithm. In *IEEE International Conference on Communications, Paris.*

Barbarosoglu, G., Özdamar, L., & Cevik, A. (2002). An interactive approach for hierarchical analysis of helicopter logistics in disaster relief operations. *European Journal of Operational Research, 140, 118–133.*

Bent, R. & Hentenryck, P. V. (2004). Regrets only! Online stochastic optimization under time constraints. In *Proceedings of the 19th AAAI.*

Brown, G. & Vassiliou, A. (2006). Optimizing disaster relief: real-time operational and tactical decision support. *Naval Research Logistics, 40.*

Crespin, U. & Neff, G. (2000). *Handbuch der Sichtung*. Edewecht,Germany: Stumpf & Kossendey-Verlag.

Das, A. & Chakrabati, B. K. (2005). *Quantum annealing and related optimization methods* (Vol 679). Heidelberg, Germany: Springer Verlag.

Der Magistrat der Stadt Frankfurt am Main (2006). Medizinisches Schutzkonzept im Rahmen der Fußballweltmeisterschaft, pp. 1-29.

Friese, P. & Rambau, J. (2006). Online-optimization of multi-elevator transport systems with reoptimization algorithms based on set-partitioning models. *Discrete Applied Mathematics*, *154*(13), 1908–1931.

Garey, M. R. (1976). The complexity of flowshop and jobshop scheduling. *Mathematics of Operations Research*, *1*, 117–129.

German Department of Justice (1998). The municipal authorities of Frankfurt/Main (Germany) (2006). Graham, R. (1966). Bounds for certain multiprocessing anomalies. *Bell System Technical Journal, 54*, 1563–1581.

Grötschel, M., Krumke, S., & Rambau, J. (2001a). Combinatorial online optimization in real time. Technical report, Konrad-Zuse-Zentrum für Informationstechnik Berlin.

Grötschel, M., Krumke, S., & Rambau, J. (2001b). Online optimization of complex transportation systems. Technical report, Konrad-Zuse-Zentrum für Informationstechnik Berlin.

Güttinger, D., Zinnen, A., & Godehardt, E. (2011). Optimizing emergency supply for mass events. In *Proceedings of the fourth international ICST conference on simulation tools and techniques, Barcelona.*

Hamidzadeh, B., Lau, Y., & Lilja, D. (2000). Dynamic task scheduling using online optimization. *IEEE Transactions on Parallel and Distributed Systems*, *11*(11), 1151–1163.

Jain, A. S. & Meeran, S. (1999). A state-of-the-art review of job-shop scheduling techniques. *European Journal of Operations Research, 113*, 390–434.

Johnson, S. M. (1994). Optimal two- and three-stage production schedules with set-up times included. *Naval Research Logistics Quarterly*, *1*, 61–68.

Kalai, A. & Vempala, S. (2002). Geometric algorithms for online optimization. *Journal of Computer and System Sciences, 64*, 26–40.

Lin, Y. H., Batta, R., & Rogerson, P. (2009). *A logistics model for delivery of critical items in a disaster relief operation: heuristic approaches.* Working paper.

Ow, P. & Morton, T. (1989). The single machine early/tardy problem. *Management Science*, *35*(2), 177–192.

Pukelsheim, F. (2007). Seat bias formulas in proportional representation systems. In *Proceedings of the fourth ECPR general conference.*

Restrepo, M., Henderson, S., & Topaloglu, H. (2009). Erlang Loss Models for the static deployment of ambulances. *Health Care Management Science*, *12*(1), 67–79.

Restrepo, M., Henderson, S., & Topaloglu, H. (2010). Approximate dynamic programming for ambulance redeployment. *INFORMS Journal on Computing*, *22*, 266–281.

SOGRO (2009). Instant rescue at a big accident with a number of casualties. Retrieved from http://www.sogro.de/

SoKNOS (2009). Service-Oriented Architectures Supporting Networks of Public Security. Retrieved from http://www.soknos.de/

Windle, J., Mackway-Jones, K., & Marsden, J. (2006). *Emergency triage*. Cambridge, England: Blackwell Publishers.

CHAPTER 7

# Evaluating Traffic Signal Control Systems Based on Artificial Transportation Systems

**Fenghua Zhu, Zhenjiang Li, and Yisheng Lv**

*State Key Laboratory of Management and Control for Complex Systems, Institute of Automation, Chinese Academy of Sciences, Beijing 100190, China*

## 7.1 Introduction

Traffic Signal Control Systems (TSCS) have played an important role in traffic management throughout the world, especially in big cities. However, there is still no effective method to evaluate their performance and reliability. The main difficulty of the evaluation lies in the reproduction of the transportation environment in the laboratory, as transportation systems are too huge and too complex to be described by traditional simulation methods.

Currently, the evaluations of TSCS are mainly carried out through field operational tests (FoTs). Moore, Jayakrishnan, McNally, and MacCarley (1999) evaluated SCOOT in Anaheim through FoTs from 1994 to 1998. The evaluation was carried out in 12 scenes, by comparing the traffic flows before and after installation of SCOOT. Besides evaluating in-peak and off-peak periods, the experiments before and after NHL hockey games were also conducted to assess SCOOT's capacity to adapt to a sudden change in traffic flow. The best-known evaluation of TSCS was the FoTs of Traffic-responsive Urban Control (TUC) performed by Kosmatopoulos et al. (2006) in Europe. The evaluation was carried out in three traffic networks with quite different traffic and control infrastructure characteristics: Chania, Greece (23 intersections); Southampton, the United Kingdom (53 intersections); and Munich, Germany (25 intersections), where TUC was compared with the respective resident Traffic Signal Control System, TASS, SCOOT, and BALANCE. The main limitation of FoTs is their very high cost, as they involve an ex-post evaluation method that cannot be carried out until the system is deployed in the field. Furthermore, the evaluation result is closely connected with the environment and it is not really applicable elsewhere.

Traffic simulation was considered as a significant innovation when it was introduced in transportation R&D (Mirchandani & Head, 2001; Tate & Bell, 2000; Wilson, Millar, & Tudge, 2006; Zhou, 2008). Theoretically, traffic simulation software can be used in evaluation applications much more widely than FoTs. However, current traffic simulation

Service Science, Management, and Engineering.


DOI: 10.1016/B978-0-12-397037-4.00007-7

software still plays a very limited role in evaluation and the results of simulations are much less important and less reliable than those of FoTs. There are two main reasons for this. One is that most traffic simulations place too much emphasis on the travel process and too little on how the travel demands are generated (Bhat & Koppelman, 1999; Davidson et al., 2007; Kitamura & Fujii, 1998). For example, most traffic simulation software uses OD data as input. It is not only very costly to collect OD data but also very difficult, if not impossible, to transfer OD data to individual travel plans, which are needed in microsimulations. The second reason is that most traffic simulations limit their perspectives in the traffic subsystem itself and build their models using top-down methods. It is well known that transportation is an open giant complex system involving nearly all aspects of our society. Not only is the number of contained elements increasing rapidly, but the connections between the transportation system and the urban environment are also getting closer and closer. All these make the top-down reductionist method of traditional simulation very ineffective.

Wang proposed the concept of Artificial Transportation Systems (ATS) to develop holistic artificial traffic systems from the bottom up (Wang & Tang, 2004a, 2004b; Wang & Liu, 2008; Zhang, Wang, Zhu, Zhao, & Tang, 2008). The main idea of ATS is to obtain a deep insight into vehicle movement and traffic evolution by extracting basic rules of individual vehicle and local traffic behavior and by analyzing the complex phenomena emerging from the interactions among individuals and, furthermore, to develop an artificial system to "replicate" a real traffic system in the laboratory. ATS promote traditional traffic simulation to a higher level and wider perspective and have shown promise in resolving the modeling, experimentation, evaluation, and decision-making problems in transportation. Regarding the evaluation of TSCS, ATS not only enable us to construct authentic evaluation environments but also provide feasible ways to generate reasonable traffic demand in various scenarios.

Here, the emphasis is to introduce our work and to carry out a case study evaluating TSCS based on ATS. The rest of the chapter is organized as follows: Section 7.2 gives a brief introduction of ATS. Section 7.3 describes the main ideas of the proposed evaluation method. Section 7.4 gives more details about the evaluation method, where a specific evaluation platform is built on ATS. Section 7.5 introduces the evaluated system, the GreenPass system. In Section 7.6, experiments are carried out to verify the effectiveness of this method, and Section 7.7 concludes and discusses future possibilities for this methodology.

## 7.2 Developing an Artificial Transportation System from the Bottom Up

Based on the concepts and methods of artificial societies and complex systems, ATS differ from other computer traffic simulation programs in three main aspects. First, similar to computer games, they can generate artificial traffic behaviors and other related data using only population statistics and behavioral models. This is useful for testing and validation in many transportation applications, especially at the level of logical correctness (Wang, 2010). Second, they provide

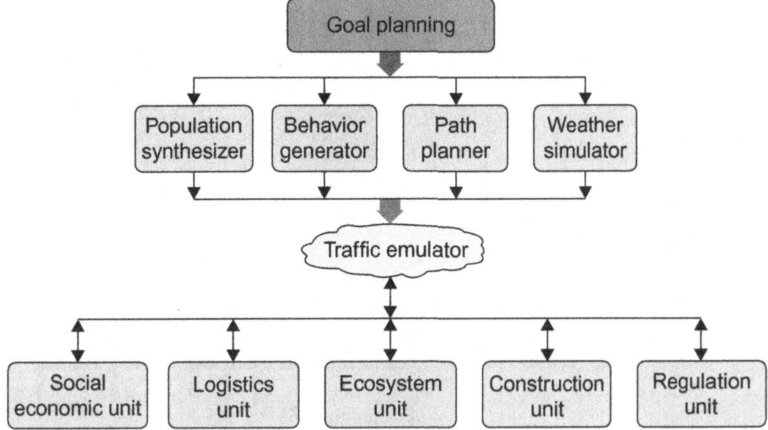


**Figure 7.1: The structure of ATS.**

a hierarchical environment for integrating and exchanging information for traffic modeling and analysis at different resolution levels, from microscopic, to mesoscopic, to macroscopic. Third, they offer a platform or a "living traffic lab" for computational experiments for transportation analysis and synthesis (Wang & Lansing, 2004). The structure of ATS is shown in Fig. 7.1.

Besides these aspects, ATS are based on agent programming and object-oriented techniques for social and behavioral modeling. Recipes and architectures developed for conventional traffic simulations, especially those developed for Transims, DynaSmart, and DynaMIT, are also useful for the construction of ATS. In the following, we will briefly introduce how to generate traffic demand based on individuals' activities. For more details about ATS, readers are referred to Wang and Tang (2004a, 2004b), Wang and Liu (2008), Zhang et al. (2008), and Wang (2010).

Because individuals' behaviors take the place of the OD matrix as input data in ATS, the first step in building ATS is to generate a reasonable population for the specified area. We implement a separate module, namely the Artificial Population Module (APM), for this task and, as mentioned before, model each person as one agent in this module. APM provides mechanisms to assign attributes to an agent as well as how these attributes change over time. In the design process of APM, a number of theories and models in sociology and anthropology are adapted. For example, the population structures in APM are divided into three types (Fig. 7.2) namely increasing, decreasing, and static, which are also widely used by sociologists in classifying populations.

In reality, travel is an induced activity, that is, people travel to perform activities in different places. Therefore, the key step in developing an artificial transportation system is establishing a complete all-day activity plan for each person. We first classify the population into five classes, such as child, student, commuter, unemployed, and senior, and then construct the framework of a complete all-day activity plan for each class. For example, the framework activity plan of a commuter is listed in Table 7.1, where $U$ denotes uniform distribution and $N(\mu, \delta^2)$ denotes normal distribution with mean $\mu$ and variance $\delta^2$. As shown in

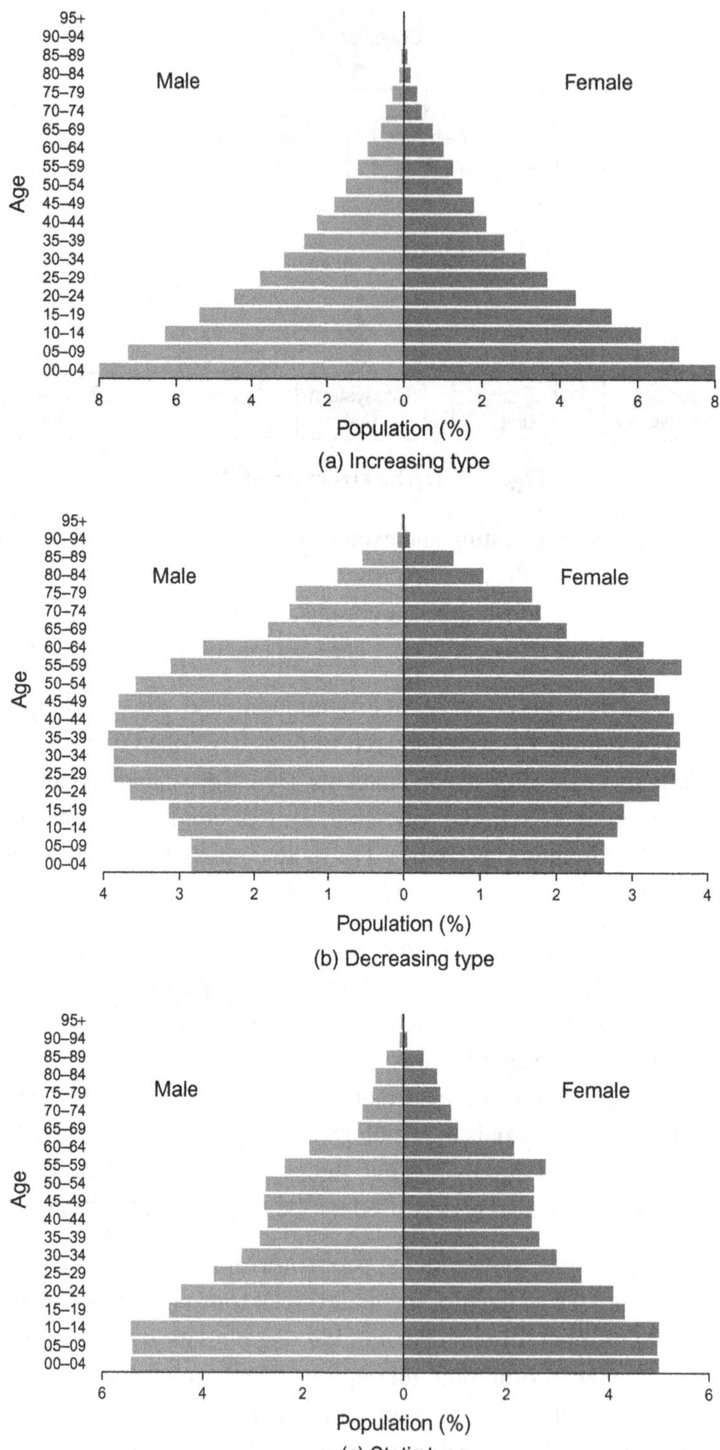


**Figure 7.2: Three types of population structures.**

**Table 7.1: The framework of an all-day activity plan for a commuter.**

| Sequence | Activity | Start Time | End Time | Duration (hour) | Location | Probability |
|---|---|---|---|---|---|---|
| 1 | Home | 00:00 | U(05:30–06:30) | $N(6,1^2)$ | Residential area | 1 |
| 2 | Dropping off children | U(06:30–07:00) | U(07:00–07:30) | $N(6,0.5^2)$ | Kindergarten | * |
| 3 | Work | U(07:30–09:30) | U(16:00–20:00) | $N(8,1^2)$ | Office | 1 |
| 4 | Business | U(10:00–15:00) | U(10:00–15:00) | $N(2,0.5^2)$ | Office | $\alpha_1$ |
| 5 | Picking up children | U(16:30–17:00) | U(17:00–17:30) | $N(0.5,0.5^2)$ | Kindergarten | * |
| 6 | Shopping | U(8:30–10:30) | U(17:30–22:00) | $N(1,0.25^2)$ | Shop | $\beta_1$ |
| | Sports | U(8:30–10:00) | U(19:30–23:00) | $N(2,0.5^2)$ | Gymnasium | $\gamma_1$ |
| | Eating | U(11:00–14:00) | U(19:30–23:00) | $N(1,0.25^2)$ | Restaurant | $\delta_1$ |
| | Entertainment | U(14:00–17:00) | U(19:30–23:00) | $N(2,0.5^2)$ | Leisure center | $\phi_1$ |
| | Party | U(18:00–19:00) | U(20:00–22:00) | $N(2,0.5^2)$ | Residential area | $\eta_1$ |
| 7 | Home | U(18:30–23:00) | 24:00 | $N(2,0.5^2)$ | Residential area | 1 |

*Value is 1, if one of his/her children is younger than 6 years and he/she drops off the child, else the value is 0.

**Table 7.2: Generating individuals' all-day activity plan using Monte Carlo simulation.**

| |
|---|
| Step 1: Utilizing the distribution information in the framework, calculating start time, end time, and duration of all subsistence and maintenance activities. |
| Step 2: Let the end time of the subsistence or maintenance activity before current activity be $t_b$ and the start time of the subsistence or maintenance activity after current activity be $t_e$, then the range of current activity is $[t_b, t_e]$. |
| Step 3: Denote the activities shopping, sports, eating, leisure, and party as $d_1, d_2, d_3, d_4$, and $d_5$, respectively. Let the probabilities of these activities be $\alpha_1, \alpha_2, \alpha_3, \alpha_4$, and $\alpha_5$ and define $\alpha_0 = 0$. $\alpha_1$, $\alpha_2$, $\alpha_3$, $\alpha_4$, and $\alpha_5$ satisfy $\sum \alpha_i \leq 1$. The properties of discretionary activities can be determined using the following steps: |
| Step 4: Let $S$ denote the result set of determined activities, $t_c$ denotes the current time. Initialize $S = \emptyset$, $t_c = t_b$. |
| Step 5: Generate random $r$ that are uniformly distributed in [0,1]. If $r > \sum_{i=1}^{5} \alpha_i$, the process is finished, return; otherwise, assume $r$ satisfying $\sum_{i=0}^{k-1} \alpha_k < r \leq \sum_{i=1}^{k} \alpha_k$, then $d_k$ is the optional activity to be executed in the next step. If $d_k \in S$, that is, $d_k$ is already included in the current activity plan, then reject $d_k$ and redo Step 5, otherwise, $S + d_k \rightarrow S$, that is, choose $d_k$ as the next activity. |
| Step 6: According to the distribution function of $d_k$, generate its duration $t_{kl}$. If $t_c + t_{kl} \geq t_e$, that is, the duration can be satisfied by the time budget, then set the time range of $d_k$ to be $[t_b, t_e]$, that is, start time to be $t_c$ and duration to be $t_e - t_c$, and assign $t_e \rightarrow t_c$; otherwise, set the time range of $d_k$ to be $[t_c, t_c + t_{kl}]$, and assign $t_c + t_{kl} \rightarrow t_c$. |
| Step 7: If $t_c + t_{min} \leq t_e$,* return to Step 5 and continue to select the next activity, otherwise the process is finished, return. |

*$t_{min}$ is the minimum duration that every activity must satisfy.

Table 7.1, there are at most 10 types of activities that can be performed by a commuter. Although the properties of subsistence and maintenance activities (including work, school, home, and dropping off/picking up children) can be easily determined using the information in the framework, the properties of discretionary activities (including shopping, sports, dining out, leisure, and party) cannot be obtained directly. We use Monte Carlo simulation to determine whether these activities occur, and if so, calculate their sequences, start time, end time, and so on. The algorithm of Monte Carlo simulation is shown in Table 7.2.

According to an individual's activity plan, when one person plans to execute one new activity and the activity's location is not in the current place, his/her travel demand from the current place to the destination is generated. Travel demand of an individual $k$ can be represented by one 6-tuple:

$$(\mathbf{A}_k, \mathbf{D}_k, \mathbf{P}_k, \mathbf{M}_k, \mathbf{ST}_k, \mathbf{ET}_k) = (A_{k1}, A_{k2}, ..., A_{kn}; D_{k1}, D_{k2}, ..., D_{kn}; P_{k1}, P_{k2}, ..., P_{kn}; \\ M_{k1}, M_{k2}, ..., M_{kn}; ST_{k1}, ST_{k2}, ..., ST_{kn}; ET_{k1}, ET_{k2}, ..., ET_{kn})$$

where $\mathbf{A}_k, \mathbf{D}_k, \mathbf{P}_k, \mathbf{M}_k, \mathbf{ST}_k$, and $\mathbf{ET}_k$ are vectors of the individual $k$'s activities, travel destinations, travel paths, travel modes, travel start time, and travel end time, respectively. Among these vectors, activities and their time can be directly obtained from the all-day activity plan; other activities need to be determined by the individual.

There are two main types of factors that influence the selection process. One is the internal properties of the options, such as the capacity of place and the length of travel path, and the other relates to the individual's psychology and behavior, such as his/her familiarity of travel path and the feeling of convenience of the travel mode. Designing and establishing appropriate models for the latter is one of the focuses of our work. These factors are closely connected with the social and behavioral characteristics of an individual, and their effects are usually expressed using natural language. Modeling the autonomous ability of one agent using discrete choice models (DCM) is a simple and compelling method that has been verified in various areas, especially in social and economic research, and has also been adopted in our work. Some of the examples are demonstrated below showing their usages in the decision process of one agent.

- Probability of performing an activity: The probability that an agent $i$ performs the $k$th activity in its complete all-day plan is calculated by a logistic model, as shown below:

$$P_{ik} = \frac{\exp(\alpha_k \cdot \text{gender}_i + \beta_k \cdot \text{age}_i + \gamma_k)}{1 + \exp(\alpha_k \cdot \text{gender}_i + \beta_k \cdot \text{age}_i + \gamma_k)},$$

  where $\text{gender}_i$ and $\text{age}_i$ are gender and age of agent $i$, $\alpha_k$ and $\beta_k$ are coefficients, and $\gamma_k$ is a constant. Typical values of $\alpha_k$, $\beta_k$, and $\gamma_k$ are listed in Section 7.4.
- Selecting the location of one activity: The locations of home, work, and school are basic properties of an individual. The locations of other activities will be selected in real time. Suppose that current place is $i$, the probability that individual $k$ will select place $j$ to perform the next activity is

$$P^k_{j|i} = \frac{\exp(\alpha_k D_{ij} + \beta_k \log(C_j) + \gamma_k)}{\sum_j \exp(\alpha_k D_{ij} + \beta_k \log(C_j) + \gamma_k)},$$

  where $D_{ij}$ is the distance between places $i$ and $j$, $C_j$ is the capacity of place $j$, and $\alpha_k, \beta_k$, and $\gamma_k$ are coefficients.

- Selecting travel mode: The probability of agent $k$ selecting travel mode $j$ is

$$P_m^k = \frac{e^{e_k/M_{km} + f_k/T_{km} + g_k R_m}}{\sum_m e^{e_k/M_{km} + f_k/T_{km} + g_k R_m}}$$

  where $M_{km}$ is the ratio of travel cost to individual $k$'s income, $T_{km}$ is travel time using mode $m$, $R_m$ is the degree of convenience (from 1 to 10) of mode $m$, and $e_k$, $f_k$, and $g_k$ are coefficients.
- Selecting travel path: There are three sources of travel path: habitual, shortest, and minimum cost. Habitual path is one of the properties of an individual. The shortest path is a global variable and will be kept constant until the road network is modified. Minimum cost path is generated by the system using real-time traffic information, which will be updated in fixed intervals. The probability of agent $k$ selecting path $l$ is

$$P_l^k = \frac{e^{c_k/L_l + d_k F_{kl}}}{\sum_l e^{c_k/L_l + d_k F_{kl}}},$$

  where $L_l$ is the length of link $l$ and $F_{kl}$ is the degree of the individual $k$'s familiarity of path $l$. $F_{kl}$ is a fuzzy variable and ranges between 0 and 10. $c_k$ and $d_k$ are coefficients.

Besides providing feasible ways for modeling the decision process of an agent, it is worth pointing out that there are many other advantages of modeling transportation systems from the bottom up. For example, both Cyber-Physical Systems (CPS) and Cloud Computing are natural and are embedded in this approach. As a matter of fact, CPS, as well as Cyber-Physical-Social Systems (CPSS), is a special case of intelligent spaces and an extension of Intelligent Transportation Spaces (ITSp); both were developed in our previous studies. As for cloud computing, it has already been used since the late 1990s in our work on agent-based control and management for networked traffic systems and other applications under the design principle of "Local Simple, Remote Complex" for high intelligence but low-cost smart systems.

## 7.3 The Evaluation Method

Essentially, the evaluation based on ATS is accomplished by observing and analyzing the traffic status that emerges while the TSCS is interacting with the environment. In this approach, ATS serve mainly as platforms for conducting computational experiments and for systematic, continuous application of computer simulation programs to analyze and predict behaviors of actual transportation systems in different situations, as shown in Fig. 7.3.

To obtain reliable results, the evaluation needs to be carried out in various experimental scenarios. The scenarios will not be limited in a transportation system, as traffic is heavily impacted by the environment. From microcosmic individuals' psychology and driving behavior to macrotravel gross and travel distribution, all are impacted by environmental

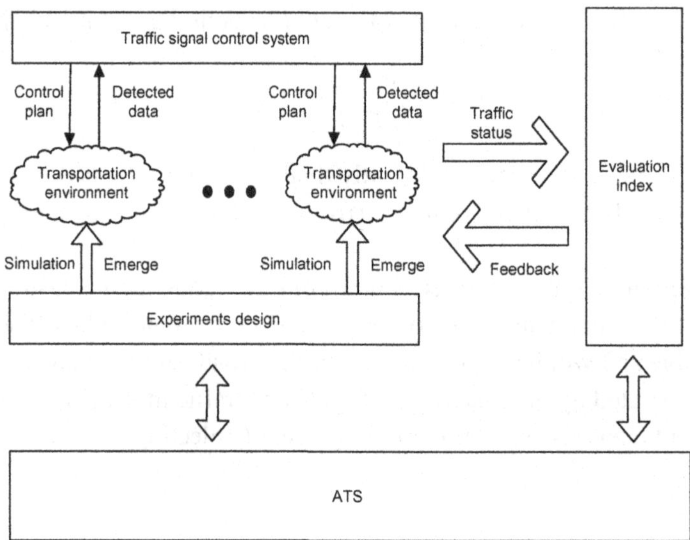


**Figure 7.3: Evaluating TSCS through experiments based on ATS.**

factors, such as economics and weather. The mechanisms by which the environment influences the traffic are very complex and there are still many disputes about how to model them as a whole (Hranac, Sterzin, Krehmer, Rakha, & Farzaneh, 2006; Koetse & Rietveld, 2009). However, for simple artificial objects, most of the current conclusions about the influences that they received from the environment are in agreement. So if simple objects and local behavior are modeled using these widely approved conclusions, the complex integrative phenomena that emerge are also expected to be understandable and agreeable. The idea for addressing this problem can be abstracted as simple-is-consistent, which was proposed in ATS research and its effectiveness has been proved by numerous experiments (Wang, 2008; Wang & Tang, 2004b).

To guarantee the generality of the evaluation method, we also designed an open interface for the communication between TSCS and ATS. There are many TSCSs that are quite different in their implementation, especially of the optimization models and the communication protocols. A well-designed interface hides the implementation details of TSCS and forms the basis for establishing a standard evaluation platform in the future. In the following, we will analyze the working procedure of a general TSCS and demonstrate the main ideas of the interface.

The working procedure of TSCS is represented in Fig. 7.4. TSCS computes the control plan using optimization model and sends it to traffic signal controllers that regulate traffic by controlling traffic lamps. At the same time, traffic signal controllers collect traffic information using various sensors in the field, and send it to TSCS through the communication network. When TSCS receives the uploaded real-time information, it continues to optimize the control

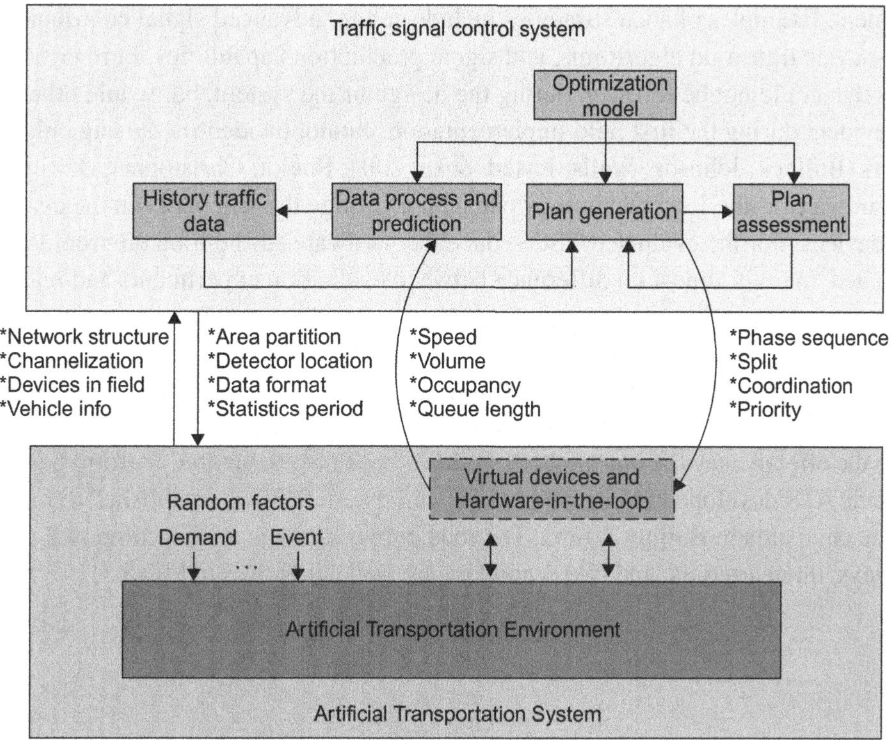


**Figure 7.4: The working procedure of TSCS.**

plan using historical data (stored in database), current data (collected in real time), and future data (predicted by itself). All these steps continue indefinitely. From this procedure, we can see that there are mainly two types of factors that can influence traffic environment. One type includes the uncontrollable random factors described by experimental conditions, such as accidents and traffic demands, and the other type includes control plans of TSCS. In the optimization process, TSCS tries to keep the fluctuations of the traffic low in a controllable range by generating control plans and interacting with signal controllers in the field continuously.

To increase the authenticity of the evaluation results, real devices instead of virtual controllers can be integrated into ATS. The traffic signal controller is the main operating unit in TSCS. It acquires traffic flow information through various detectors, such as loops, cameras, and so on, and controls the traffic lamps in the intersection. In traditional simulations, the traffic signal controller is usually implemented as a shared memory Dynamic Link Library (DLL) that may be called from any application that needs to access detector and phase data. Although many models that include emulation for actuated signals and basic signal coordination have been implemented, some types of advanced signal control may be difficult or impossible to implement. Examples of such strategies include certain advanced signal coordination strategies, cycle transition algorithms, and signal preemption capabilities. Furthermore, problems that could not be foreseen during the design of the system, but would otherwise become evident during the first field implementation, cannot be identified using only virtual controllers (Bullock, Johnson, Wells, Kyted, & Li, 2004; Poelof, Christopher, & Kevin, 1999). Hardware-in-the-loop technology can help us bridge the gap between the evaluated and real devices. For the evaluated TSCS, once the hardware-in-the-loop environment is established, there is almost no difference between evaluation experiments and real applications.

## 7.4 Building the Evaluation Platform

To verify the effectiveness of our method, we have built the evaluation platform based on a specific ATS developed in our laboratory. This specific ATS is established to model Zhongguanchun area in Beijing, China. The road network in this area is composed of four expressways, three arterials, and two secondary arterials, as shown in Fig. 7.5.

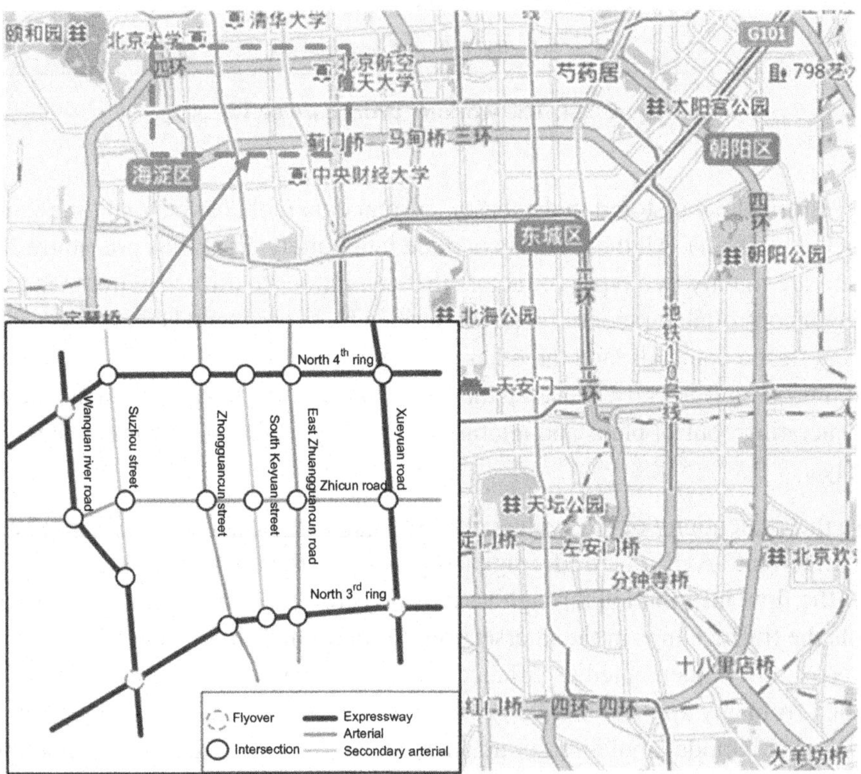


**Figure 7.5: Zhongguanchun area and its location on a map of Beijing.**

While the ATS is running, it provides multiscale friendly GUIs to show the traffic status in real time. Macro-GUIs show the traffic status from a global perspective and, in this interface, the road is drawn using different colors to represent various degrees of congestion, such as serious congestion (red), congestion (yellow), and expedite (green). On the contrary, micro-GUIs can only show local traffic scenes, but can depict the movement of vehicles in detail.

The sites of activity in the platform include 40 residential areas, 74 office buildings, 47 restaurants, 12 schools, 7 hospitals, 27 shopping centers, 11 leisure centers, and 8 sports centers. Travel demands are generated from an individual's activity plan, which serves as the foundation of ATS. Before carrying out computational experiments, the rationality of the individual's activity plan must be verified.

In ATS, we classify a person's activities into seven types: work, school, hospital, shopping, sports, eating (out), and entertainment. Start time, end time, and duration are the three basic attributes for a specific activity. We assume that they all obey a normal distribution, although their means and standard deviations are different. One shortcoming of a normal distribution is that its value range is infinite, which may generate meaningless values, for instance, negative values for start time. So, we use a bounded normal distribution (BND) instead of a common normal distribution, as shown below:

$$\begin{cases} x \sim N(u, \sigma), & \text{and} \\ \text{if } x < u - 4\sigma, & \text{then } x = u - 4\sigma, \text{ and} \\ \text{if } x > u + 4\sigma, & \text{then } x = u + 4\sigma \end{cases}$$

The global attributes of these activities on a working day and at the weekend for Jinan ATS are calculated according to BND and listed in Tables 7.3 and 7.4, respectively. Note that the time range is used to represent the start time and the end time. Tables 7.3 and 7.4 also list the parameters for calculating the probabilities of activities, which are explained in Section 7.2.

**Table 7.3: Attributes of activities (working day).**

| | Time Range (HH:MM) | Duration (minute) | Probability | | |
|---|---|---|---|---|---|
| | | | $\alpha_k$ | $\beta_k$ | $\gamma_k$ |
| School | (6:00–17:30) | $N(450, 20)$ | 0.1 | 0.01 | 12 |
| Work | (6:30–20:00) | $N(480, 40)$ | 0.1 | 0.01 | 10 |
| Hospital | (6:30–17:00) | $N(60, 10)$ | −0.25 | 0.02 | −1.65 |
| Shopping | (10:00–20:30) | $N(90, 20)$ | −0.91 | 0.01 | 0.56 |
| Sports | (9:00–20:00) | $N(90, 10)$ | 0.13 | 0.02 | −1.19 |
| Eating | (16:00–19:00) | $N(60, 10)$ | 0.25 | 0.01 | −1.68 |
| Entertainment | (15:00–20:00) | $N(90, 10)$ | 0.57 | 0.03 | −2.19 |

**Table 7.4: Attributes of activities (weekend).**

| | Time Range (HH:MM) | Duration (minute) | Probability | | |
|---|---|---|---|---|---|
| | | | $\alpha_k$ | $\beta_k$ | $\gamma_k$ |
| School | (6:00–17:30) | $N(450, 20)$ | 0.1 | 0.01 | −2.4 |
| Work | (6:30–20:00) | $N(480, 40)$ | 0.1 | 0.01 | −2.2 |
| Hospital | (6:30–17:00) | $N(320, 80)$ | −0.18 | 0.02 | −1.72 |
| Shopping | (10:00–20:30) | $N(240, 60)$ | −0.73 | −0.01 | 0.64 |
| Sports | (9:00–20:00) | $N(120, 40)$ | 0.13 | 0.02 | −1.25 |
| Eating | (9:00–19:00) | $N(90, 30)$ | 0.36 | 0.01 | −1.81 |
| Entertainment | (9:00–20:00) | $N(320, 80)$ | 0.63 | 0.03 | −2.56 |

Based on the preconditions listed in Tables 7.3 and 7.4, each individual will generate his/her specific travel demand using the discrete choice models described in Section 7.2. Then, the macro results will emerge naturally while numerous individuals are performing their activities. For example, Fig. 7.6 presents the distributions of persons performing different activities from 5:00 P.M. to 11:00 P.M. in ATS. Figure 7.6(a) shows the distributions on one working day, where the distributions of persons performing "work" and "school" are more regular than those of persons performing other activities. In addition, most people are performing "work" or "school" only in daytime, and very few people are participating in other activities until 6:00 P.M. Figure 7.6(b) shows the distributions at the weekend, and it has markedly different features when compared with Fig. 7.6(a). In Fig. 7.6(b), because more people are participating in discretionary activities (including shopping, sports, eating out, and entertainment), not only are the frequencies of these activities increased sharply but their time spans are also extended.

Clearly, the results in Fig. 7.6 match the real situation very well. Intuitively, school and work are regular activities and their times are usually limited between 8:00 A.M. and 6:00 P.M., whereas other activities are more flexible and individuals have more freedom to schedule them. It is worth mentioning that Fig. 7.6 shows the macrophenomena arising while individuals are pursuing their activity plans independently. As the environment is modeled using basic rules and each individual can adjust their activities as they wish, reasonable travel demands under various situations can be easily generated by changing experimental conditions.

## 7.5 GreenPass System

The evaluated TSCS is the GreenPass system (Fig. 7.7) that has been developed in our laboratory. The GreenPass system is equipped with many intelligent control algorithms, among which the most powerful is the networked adaptive control based on mobile agents.

Since the Agent-Based Control (ABC) technology was first proposed by Wang in the 2000s (Wang, 2000, 2005; Wang & Wang, 2002; Wang & Liu, 2008), it has been followed by researchers and engineers in many areas, and numerous successes have been

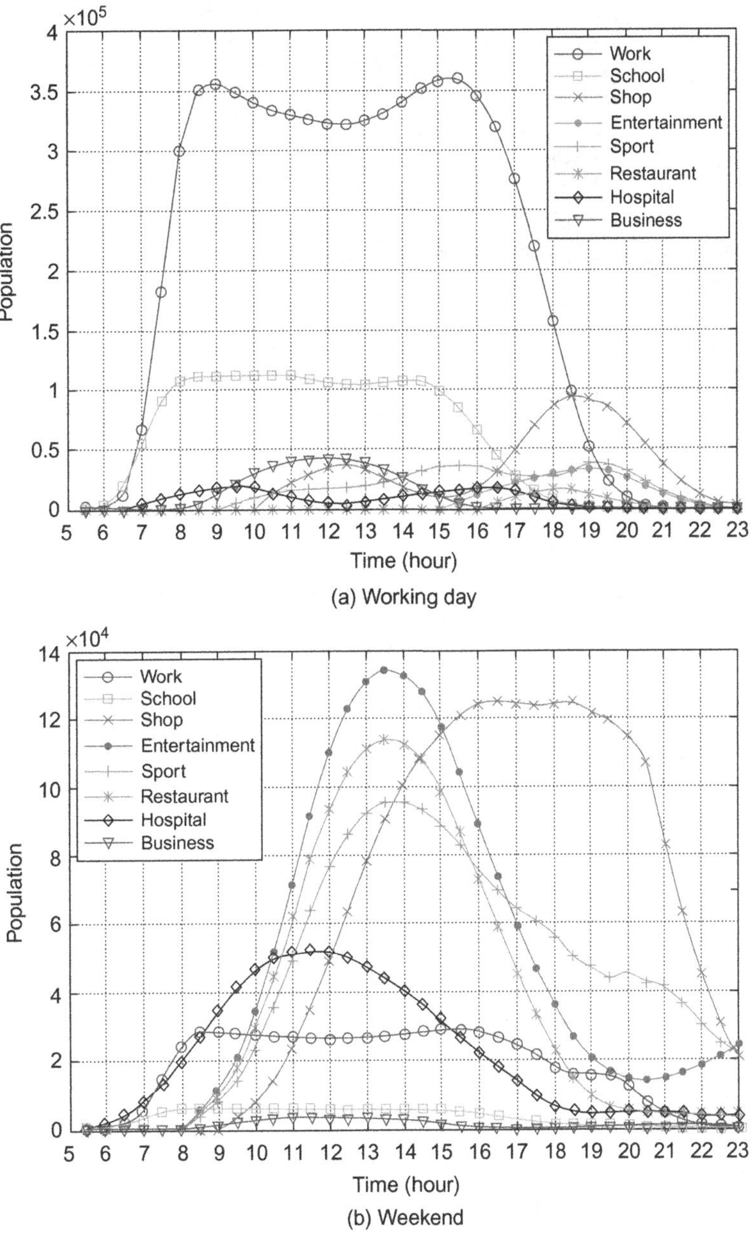


**Figure 7.6: Population distributions performing different activities in 1 day: (a) for a working day; (b) at the weekend.**

achieved. ABC decomposes an integrated control algorithm into many simple task-oriented agents, which are distributed over a network, generally a wide area network, and they can migrate from one node to another with their codes and states. Network-enabled devices only need to support the operation of the control agent for the current situation,

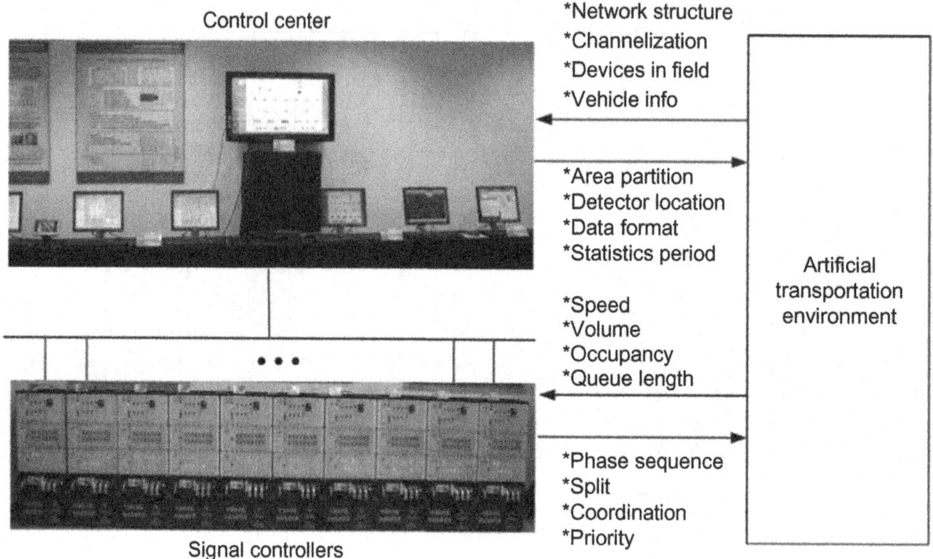


**Figure 7.7: The evaluated GreenPass system.**

and control agents can be easily deployed and replaced over the network as operating conditions vary. Hence, computing demand and storage consumption of the local devices decrease dramatically when compared with traditional devices controlled by algorithms.

A traffic control algorithm is modeled as an agent, and its structure model is shown in Fig. 7.8. It includes environment interaction module, task-solving module, knowledge base, internal state set, constraints, and routing strategy. Through the environment interaction module, a traffic control agent can communicate with the outside world including facilities, detectors, and other agents. The task-solving module contains execution module and associated reasoning methods and rules, and it generates control action that can act on the traffic lamps directly, based on traffic information obtained from detectors and other information. The knowledge base is the world model perceived by the traffic control agent, which acquires knowledge during the migration of the traffic control agent and implementation process. The internal state set, constraints, and routing strategy are the three basic modules interacting with the task-solving process. The internal state is the current state of agent. The constraints are set by constructors of the traffic control agent to ensure its basic behavior and performance, such as back time, stop time, completion degree of task, and so on. The routing strategy decides the path of the traffic control agent, which can be a static list of services and facilities or a rule-based dynamic routing to facilitate the solution of complex and nondeterministic tasks. Design of the traffic control agent embodies customizable and

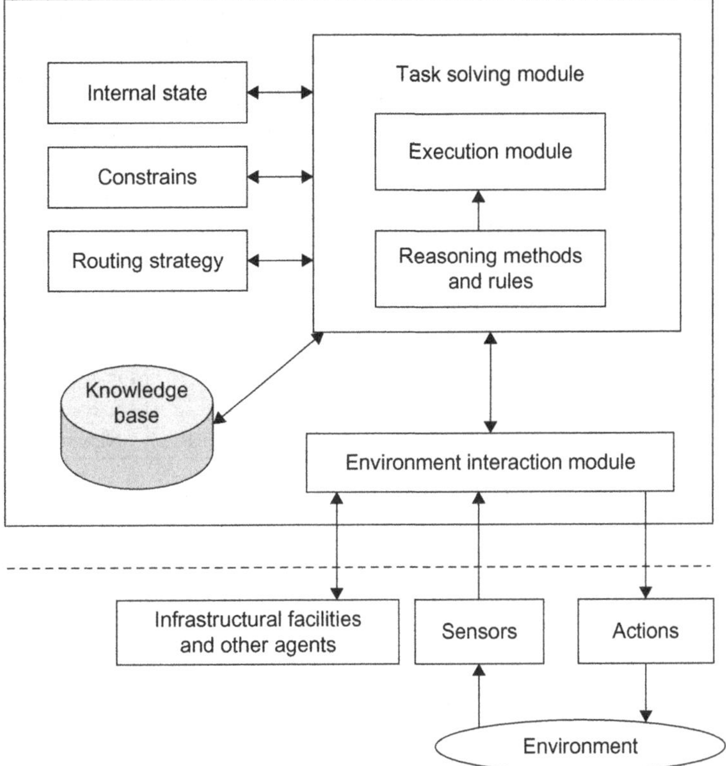


**Figure 7.8: Structure model of traffic control agent.**

reusable ideas. All the traffic control agents use the same structure, whereas some specific modules in the model can be customized to implement different control algorithms.

Agent-based control algorithms have been implemented in the GreenPass 2070 Traffic Signal controller, which conforms to the ATC 2070 hardware specification. GreenPass controllers communicate with the control center using the NTCIP center-to-field communication protocol. The controller is developed on a Samsung S3C6410 ARM11 processor and Linux OS, which can satisfy computation and communication demands for advanced networked control algorithms. We evaluate the performance of GreenPass by comparing it with the default condition, where all intersections are controlled by isolated traffic signal controllers.

In the implementation of the communication interface, web services are preferred over other middleware technologies for reasons of interoperability and portability. Some functions that are included in one TSCS are listed in Table 7.5. All these functions can be directly transferred to web services using Simple Object Access Protocol (SOAP), which is a standard for web services transportation.

Table 7.5: Function of TSCS.

| Function | Input | Output |
|---|---|---|
| Cycle adjusting to decrease the maximum saturation level | • Speed<br>• Occupancy<br>• Volume | • Cycle time<br>• Splits<br>• Offset |
| Phase sequence adjusting for quicker servicing of emergency vehicles and reducing stop and delay | • Volume<br>• Queue length<br>• Preempt signal | • Phase sequence<br>• Transition mode |
| Phase time adjusting to minimization of risk of oversaturation and queue spillback | • Queue length<br>• Link capacity | • Extend current phase or not<br>• The time of next phase |
| Provide priority to public transport vehicles (PTV) | • PTV priority<br>• PTV schedule<br>• Current signal status | • Green extension<br>• Red omit<br>• New phase |

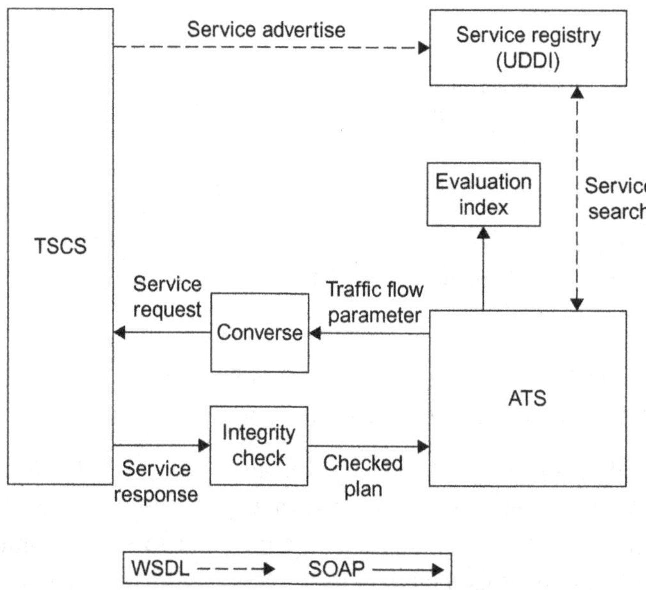


Figure 7.9: Communications between TSCS and ATS.

On the basis of the communication interface, we have constructed a distributed evaluation platform, on which a TSCS developed on any platform and deployed anywhere on the internet can communicate with ATS. The evaluation platform is composed of three parts: ATS (the service requestor), TSCS (the service provider), and the service registry. TSCS advertises its services in a service registry. ATS find suitable services from the service registry and subsequently interact with the associated TSCS. This architecture of the evaluation platform is shown in Fig. 7.9.

## 7.6 Experiments

Plenty of experiments have been carried out on ATS to evaluate the performance and reliability of the GreenPass system. Due to limitations of space, we show only two of them, one in adverse weather and the other after the building of a shopping mall. Although the GreenPass system has been evaluated by both FoTs and simulation software, before the following experiments are introduced, there are no evaluation results in these special scenarios.

### 7.6.1 Scenario 1: Adverse Weather

The first experiment is carried out to evaluate the GreenPass system in adverse weather. For each individual in the transportation system, the influences he/she received from adverse weather include two aspects. One aspect is the individual's experience and behavior, including road conditions, driver's psychology, driving behavior, the possibility of an accident, and the extent of damage. The other aspect concerns the arrangement of the activity plan. In adverse weather, people will adjust their activities on their own initiative and avoid unnecessary activities. In particular, discretionary activities will clearly be reduced. Using medium rainfall (the precipitation is between 10 mm and 20 mm) as an example, we have modeled its influences using five rules, as shown in Table 7.6. Clearly, all these rules are expressed from the individual's perspective and can be easily integrated into ATS.

The simulated period in our experiment is from 4:00 P.M. to 9:00 P.M., during which the period of rainfall is from 6:00 P.M. to 8:00 P.M. Figure 7.10 shows the influence of this adverse weather (medium rainfall) by comparing the traffic flow in normal conditions with that in adverse weather, both in the default condition, that is, all intersections are controlled separately. Because it rains during peak hours, according to our experience, it usually causes serious congestion in road networks. As expected, this phenomenon is seen in our experiment on ATS, that is, the number of vehicles that are blocked in the network increases rapidly and the average speed drops to zero in a short time.

**Table 7.6: The rules modeling the influences of medium rainfall.**

| | |
|---|---|
| Rule W1: Activity arrangement | Commuting activities are delayed 10–30 minutes and the probability of discretionary activity is reduced 50% |
| Rule W2: Accident | The probability of accident increases 30% and the severity of accident increases 20% |
| Rule W3: Driver's psychology | Irritability level increases 30% |
| Rule W4: Driving behavior | Free speed decreases 30%, time headway decreases 20%, and minimum safe distance decreases 20% |
| Rule W5: Vehicle performance | Acceleration decreases 20% and deceleration decreases 20% |

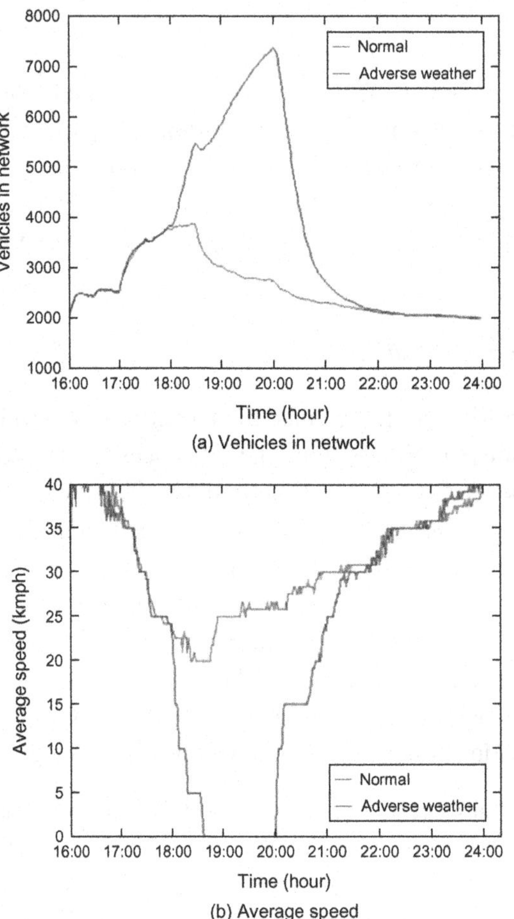


**Figure 7.10: Comparing traffic flow in normal weather with that in adverse weather, both in the default condition.**

After modeling the adverse weather successfully on ATS, we continue to evaluate the GreenPass system by putting it into the artificial scenario and observing and analyzing the resulting phenomena. The experiments with and without GreenPass are carried out six times, and the evaluation results are shown in Fig. 7.11 by comparing the average of six experimental results under the control of GreenPass system with that in the default condition. In Fig. 7.11, under GreenPass's optimization, the congestion caused by adverse weather is clearly alleviated. Besides comparative curves, plenty of quantitative data can be calculated, as ATS modeled every individual's movement in detail. For example, the average of the number of vehicles in the network from 4:00 P.M. to 12:00 P.M. decreases from 3587 to 3205, that is, the decrease is about 10%, and the temporal average of average speed increases from 24.4 kmph to 28.8 kmph, that is, an increase of about 15%.

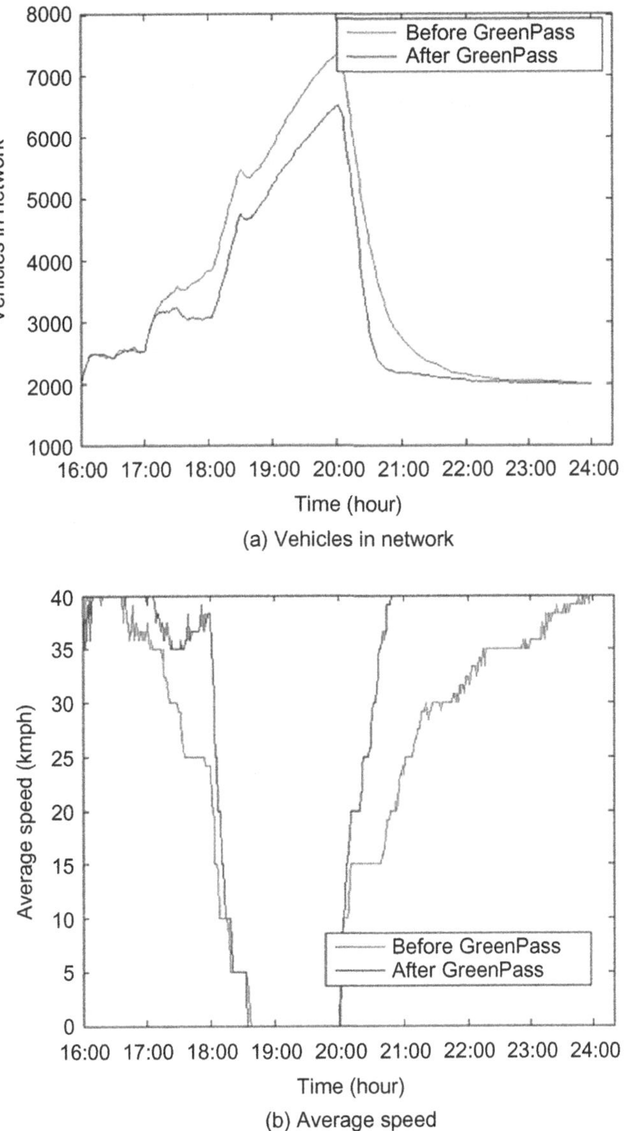


(a) Vehicles in network

(b) Average speed

**Figure 7.11: Experimental results in adverse weather.**

### *7.6.2 Scenario 2: After the Building of a Shopping Mall*

Suppose a shopping mall is planned to be built at the lower right corner of the network (Fig. 7.12); it is reasonable to assume that the surrounding traffic flow will be changed significantly. If an artificial scenario can be established beforehand, it can be of great benefit to both the traffic management and the land use administration.

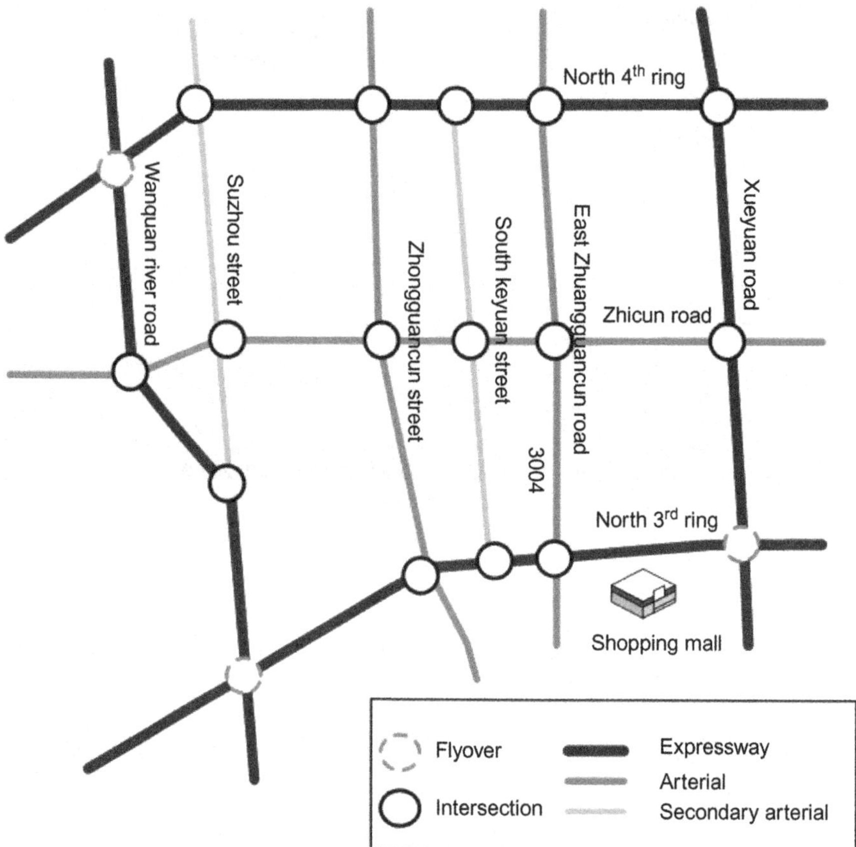


**Figure 7.12: The location of the shopping mall.**

In ATS, the processes where individuals select activity locations and travel paths have been modeled using basic rules, so this scenario can be constructed in our experiment by simply adding one activity location. It turns out that the influences are only evident on several links around the newly built shopping mall, so our evaluation is carried out just in a limited area. The traffic flow on link 3004, which is a section of a directed road in reality, is expected to increase notably in the afternoon peak hour, because this link is on the main path for people working in the left part of the network to go to the shopping mall. We will use this link as an example to illustrate our results.

We have verified the influences brought by the new shopping mall by comparing the traffic flow before and after the construction, both in the default condition. Figure 7.13 shows the influences brought by the new shopping mall by comparing the traffic flow before and after the construction, both in the default condition. As expected, serious congestion is caused on link 3004 from 5:30 P.M. to 7:00 P.M. Based on this artificial scenario, we

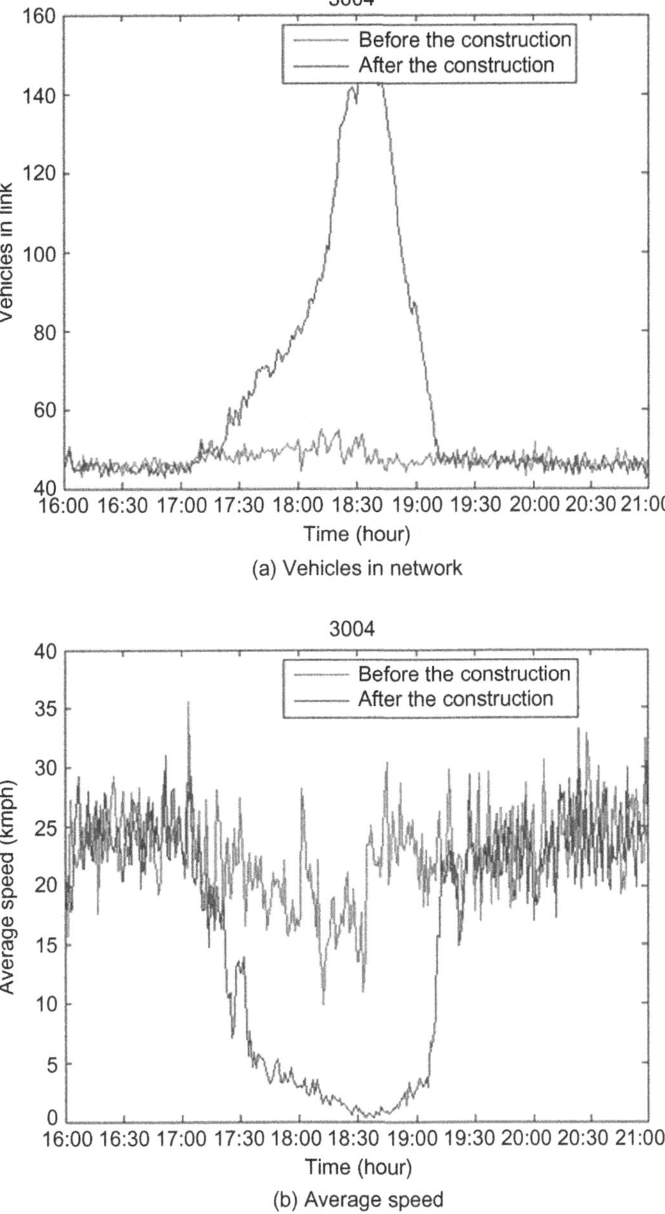


**Figure 7.13: Comparing traffic flow before the construction with that after the construction, both in default condition.**

also conduct evaluation experiments by integrating GreenPass with this scenario and running the experiments with and without GreenPass, respectively. Figure 7.14 compares the result under the control of the GreenPass system with that in the default condition. It

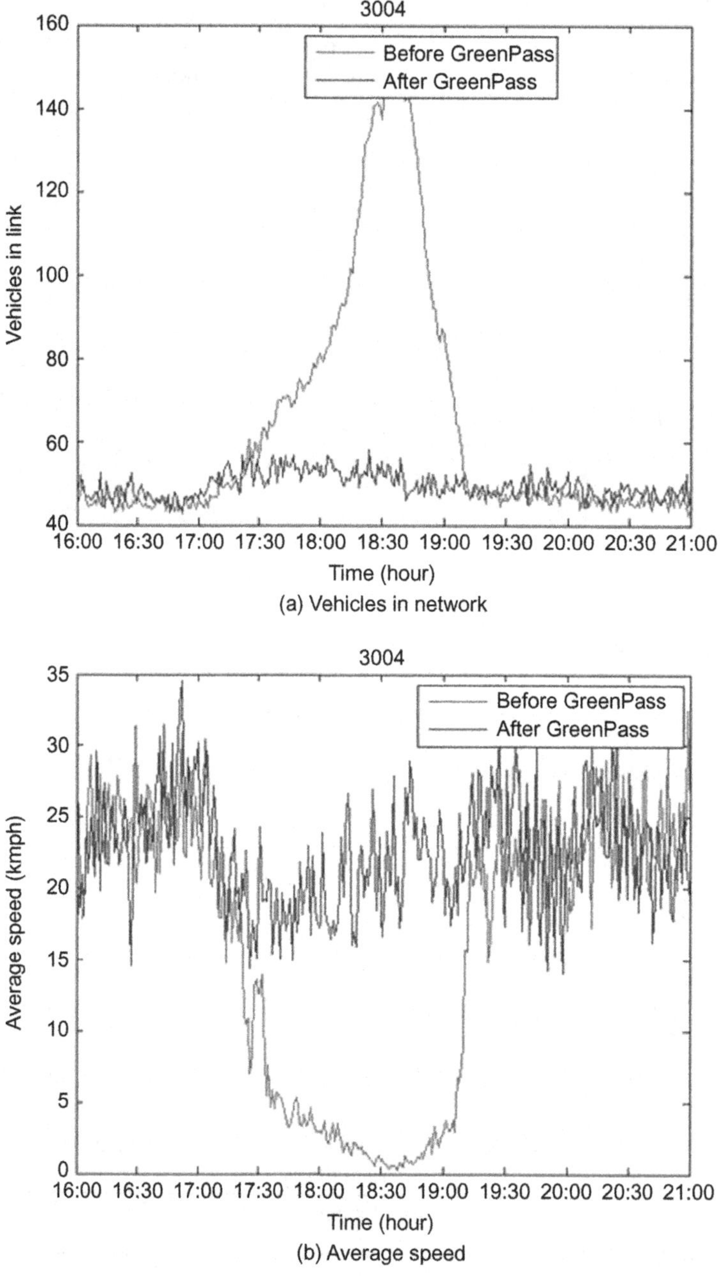


**Figure 7.14: The results of evaluation after the construction of a shopping mall.**

is unexpected that the effect of GreenPass is such that the congestion almost disappears completely. After an in-depth study of the experiment, we think the result is reasonable. In the experiment, the influence of the shopping mall is very limited and the congestion only happens in a few surrounding links. The traffic demand of most links in the network is very low, so there is enough space for GreenPass to exert its optimization function. As GreenPass decreases the number of vehicles flowing into link 3004 by adjusting the time plan of the controller located in the upstream intersection, the traffic status of this link is almost restored to its original level.

## 7.7 Conclusions

The main difficulty of evaluating TSCS lies in the ability to reproduce an authentic transportation environment within the laboratory, as real-world traffic scenarios are both too huge and too complex to be described by traditional simulation methods. ATS aim at exploring feasible approaches to reproducing traffic scenarios in a laboratory; thus they provide a new method to evaluate TSCS. Using this method, the evaluation can be carried out before the installation on site, so the cost is decreased enormously. Furthermore, as various traffic demands can be generated from individuals' activities, the reliability of evaluation results can be guaranteed. The effectiveness of this method is also verified by two examples evaluation in adverse weather and evaluation after a new shopping mall is built. Both experiments are very difficult, if not impossible, to carry out using traditional methods.

The work presented here is the first step in our plan to set up an evaluation theory and method based on ATS. One of our future projects is to design and carry out abundant experiments in various environments, such as other adverse weather conditions, economic activity conditions, and so on, and set up an evaluation index system for TSCS by combining and trading off comprehensive metrics (Miao et al., 2011; Zhu, Li, Li, Chen, & Wen, 2011).

## References


Bhat, C. R. & Koppelman, F. S. (1999). Activity-based modeling of travel demand. In R. W. Hall (Ed.), *The handbook of transportation science* (pp. 35–61). Norwell, MA: Kluwer Academic.

Bullock, D., Johnson, B., Wells, R. B., Kyted, M., & Li, Z. (2004). Hardware-in-the-loop simulation. *Transportation Research Part C: Emerging Technologies, 12*, 73–89.

Davidson, W., Donnelly, R., Vovsha, P., Freedman, J., Ruegg, S., Hicks, J., … Picado, R. (2007). Synthesis of first practices and operational research approaches in activity-based travel demand modeling. *Transportation Research Part A: Policy and Practice, 41*, 464–488.

Hranac, R., Sterzin, E., Krechmer, D., Rakha, H., & Farzaneh, M. (2006). Empirical Studies on Traffic Flow in Inclement Weather. F. H. Administration, Washington, DC: Publication No. FHWA-HOP-07-073.

Kitamura, R. & Fujii, S. (1998). Two computational process models of activity-travel behavior. In T. Garling, T. Laitila, & K. Westin (Eds.), *Theoretical foundations of travel choice modeling* (pp. 251–279). Oxford: Elsevier Science.

Koetse, M. J. & Rietveld, P. (2009). The impact of climate change and weather on transport: An overview of empirical findings. *Transportation Research Part D: Transport and Environment, 14*, 205–221.

Kosmatopoulos, E., Papageorgiou, M., Bielefeldt, C., Dinopoulou, V., Morris, R., Mueck, J., … Weichenmeier, F. (2006). International comparative field evaluation of a traffic-responsive signal control strategy in three cities. *Transportation Research Part A: Policy and Practice, 40*, 399–413.

Metropolitan Travel Forecasting Current Practice and Future Direction (Special Report 288) (2007). Washington, D.C.: Committee for Determination of the State of the Practice in Metropolitan Area Travel Forecasting, The National Academy.

Miao, Q., Zhu, F., Lv, Y., Cheng, C., Chen, C., & Qiu, X. (2011). A game-engine-based platform for modeling and computing of artificial transportation systems. *IEEE Transactions on Intelligent Transportation Systems, 12*(2), 343–353.

Mirchandani, P. & Head, L. (2001). A real-time traffic signal control system: architecture, algorithms, and analysis. *Transportation Research Part C: Emerging Technologies, 9*, 415–432.

Moore, J. E., II, Jayakrishnan, R., McNally, M. G., & MacCarley, C. A. (1999). Evaluation of the Anaheim Advanced Traffic Control System Field Operational Test: Introduction and Task A: Evaluation of SCOOT Performance. Final report prepared under project RTA-65V313 for the US Department of Transportation, Federal Highway Administration, Washington, DC.

Poelof, E. R., Christopher, P. M., & Kevin, B. N. (1999). Development of a distributed hardware-in-the-loop simulation system, for transportation networks. Transportation Research Board Annual Meeting, National Research Council, Washington, DC. Preprint #990599.

Tate, J. F. & Bell, M. C. (2000). Evaluation of a traffic demand management strategy to improve air quality in urban areas. *Tenth International Conference on Road Transport Information and Control,* London, UK (pp. 158–162).

Wang, F.-Y. (2000). *ABCS: Agent-based Control Systems* (SIE Working Paper). University of Arizona, Tucson, AZ.

Wang, F.-Y. (2005). Agent-based control for networked traffic management systems. *IEEE Intelligent Systems, 20*(5), 92–96.

Wang, F.-Y. (2008). Toward a revolution in transportation operations: AI for complex systems. *IEEE Intelligent Systems, 23*, 8–13.

Wang, F.-Y. (2010). Parallel control and management for intelligent transportation systems: Concepts, architectures, and applications. *IEEE Transactions on Intelligent Transportation Systems, 11*(3), 630–638.

Wang, F.-Y. & Lansing, S. J. (2004). From artificial life to artificial societies: New methods in studying social complex system. *Journal of Complex Systems and Complexity Science, 1*, 33–41.

Wang, F.-Y. & Liu, D. (2008). *Networked control systems: Theory and applications.* London, UK: Springer.

Wang, F.-Y. & Tang, S. (2004a). Artificial societies for integrated and sustainable development of metropolitan systems. *IEEE Intelligent Systems, 19*, 82–87.

Wang, F.-Y. & Tang, S. (2004b). Concept and framework of artificial transportation system. *Journal of Complex Systems and Complexity Science, 1*, 52–57.

Wang, F.-Y. & Wang, C.-H. (2002). On some basic issues in network-based direct control systems. *Acta Automatica Sinica 28* (Suppl. 1), 171–176. (in Chinese)

Wilson C., Millar G., & Tudge R. (2006). Microsimulation evaluation of the benefits of SCATS coordinated traffic control signals. *Proceedings of Transportation Research Board Meeting*, Washington, DC.

Zhang, N., Wang, F.-Y., Zhu, F., Zhao, D., & Tang, S. (2008). DynaCAS: Computational experiments and decision support for ITS. *IEEE Intelligent System, 23,* 19–23.

Zhou, H. (2008). Evaluation of route diversion strategies using computer simulation. *Journal of Transportation Systems Engineering and Information Technology, 8*, 61–67.

Zhu, F., Li, G., Li, Z., Chen, C., & Wen, D. (2011). A case study of evaluating traffic signal control systems using computational experiments. *IEEE Transactions on Intelligent Transportation Systems, 12*(4) 1220–1226.

CHAPTER 8

# An Approach to Optimize Police Patrol Activities Based on the Spatial Pattern of Crime Hotspots[1]

**Li Li, Zhongbo Jiang, Ning Duan, Weishan Dong, Ke Hu, and Wei Sun**

*IBM China Research Laboratory, Zhongguancun Software Park, Shangdi, Beijing, China*

## 8.1 Introduction

Criminal activities are a major risk factor for the well-being of society in many countries. Police patrol service is a critical instrument to combat the criminal activities with violent aspects (Reis, Melo, Coelho, & Furtado, 2006; Ruan, Meirina, Yu, Pattipati, & Popp, 2005). Because of the central role of the patrol activities in police functions, police patrol strategies are a topic of vigorous debate and have undergone many changes. Initially, patrol activities received few instructions and they were mostly instantaneous responses of the patrol officers toward criminal activities. Later, the policing concepts evolved and police officials generally regard the patrol function as having two major law enforcement objectives. First, it provides rapid responses to emergency calls through sector concentration. Second, it involves street surveillance to deter crimes (Bahn, 1974). An emerging policing philosophy is intelligence-led policing, which seeks to reduce crime through the combined use of crime analysis and criminal intelligence to determine crime reduction tactics that concentrate on the enforcement and prevention of criminal offender activities. Several tools are developed under this philosophy. These tools can be classified into two categories: random patrol and focused patrol. Random patrol aims to provide the public with a sense of security and deter crimes through the random appearances of police forces. The random patrol is easy to implement and is unpredictable. Focused patrol tends to focus patrolling resources to the locations where the police presence

[1] This chapter is adapted from Li, L., Jiang, Z., Duan, N., Dong, W., & Sun, W., Police patrol service optimization based on the spatial pattern of hotspots, in 2011 IEEE International Conference on Service Operations and Logistics, and Informatics.

Service Science, Management, and Engineering.


DOI: 10.1016/B978-0-12-397037-4.00008-9

is most needed, which often are locations with a cluster of crimes. The underlying reason for this patrolling strategy is that crime is not spread evenly across city landscape. Because of these observations, focusing limited resources on places with clusters of crime seems to be a rational choice.

Although the two different focuses of patrol approaches are meaningful, most studies have centered around the focused patrol. Algorithms have been developed to implement focused patrol activities (Chawathe, 2007; Chevaleyre, Sempe, & Ramalho, 2004). Previous work mainly focuses on determining the important scores of locations based on hotspot analysis or identifying important routes based on the topology of road networks (Chawathe, 2007). However, there are several problems in patrol route planning based solely on the locations of hotspots or the topology of road networks. First, hotspots are usually not randomly distributed and the patterns in the distribution of hotspots may affect the patrol effectiveness (Sherman, Gartin, & Buerger, 1989). Although the effect of such patterns on patrol route selections was highlighted, the patterns of such distributions have rarely been considered in patrol planning (Sherman et al., 1989). Second, some patrol optimization algorithms that maximize hotspot coverage result in a single (fixed) patrol recommendation (Chawathe, 2007). Fixed patrol routes can obviously lead to a predictable patroller, which may encourage potential criminals to predict the arrival patterns of patrol vehicles. The importance of randomized patrols has been recognized in law enforcement for some time, but the nature of the randomization is often ignored (Sherman, 2002). Nevertheless, since patrolling is a daily activity for police officers, it is important to identify an approach that can optimize the overall efforts of patrolling. If approaches are focusing on a single optimal patrol activity, patrol planning might not be able to produce an optimal and effective allocation plan. Finally, it is possible for some algorithms to comprehensively list all possible routes and calculate the weights of these routes based on the scores of locations. However, these solutions only work for small jurisdictions. For typical jurisdictions consisting of hundreds or thousands of street segments, more efficient algorithms are needed.

The police resource allocation for street patrolling is one of the most important tactical management activities. It is important to continuously improve patrolling strategies, and the goal of patrol route planning should be to recommend a set of suitable routes to optimize the overall performance of patrol activities. To achieve this goal, a thorough understanding of the nature of hotspots and its implications on patrol route optimization are required. For example, hotspots might not be spatially independent of each other and some hotspots are surrounded by other hotspots. The effect of this knowledge on patrol route optimization needs to be investigated, and algorithms that can efficiently identify suitable routes should be used.

To address the limitations of current patrolling strategies, this chapter aims to integrate a spatial pattern identification approach with an efficient route optimization algorithm to produce randomized optimal patrol routes. We note that our goal is not to provide a solution for the entire process of planning patrol routes, rather to propose an approach to determine sets of patrol routes, which can help to achieve the optimization of the collective efforts of multiple patrol activities. We will also address how to ensure randomness in optimal route identification with an efficient algorithm.

In Section 8.2, we provide background information on patrol strategies, hotspot identification methods, and patrol route optimization approaches. In Section 8.3, we propose an approach to incorporate the spatial pattern of hotspots and randomness in the patrol route optimization process. In Section 8.4, we provide a case study to illustrate our approach. Finally, the results and conclusions are presented in Section 8.5.

## 8.2 Background

Police patrolling has long been an important public service that helps to deter and prevent crimes. It also creates a sense of public security. Despite the importance of patrol service, police resources are limited and efficient patrol route planning is required. Spatial analysis approaches, such as hotspot analysis, have often been used to facilitate patrol planning. In this section, we first introduce common patrol strategies and typical spatial analysis approaches used in crime analysis.

### 8.2.1 Police Patrolling

The history of police can be dated back to Babylonian times in 2000 B.C. (Lee, 1901). Strictly speaking, the first police force, as a social control organization, was established in London in 1829. Rooted in military traditions, the newly established police force was cheaper than a military force and created less resentment (Monkkonen, 1992). Since the first establishment, the role of police forces has been through a series of reforms. Despite these reforms, patrolling is a major function of police forces. The first Metropolitan Police patrols went on to the streets on September 29, 1829, 3 months after the establishment of the police force (Metropolitan Police, 2010). Initially, the police patrol provided a wide range of services to citizens and gradually it became focused on enforcing law (Burrows & Lewis, 1988). Because of its central role in the police function, the police patrol was a subject of innovations and widespread experimentation in new practices. In the 1960s, random patrols became a core tactic in the "traditional model" of policing and continued to be a widely adopted policing approach. The following subsection introduces the random patrol.

#### 8.2.1.1 Random Patrolling

Random patrol is defined as "self-initiated patrol activities" (Schnelle, Kirchner, Casey, Uselton, & McNees, 1977). The patrol activities may include a variety of actions, such as traffic enforcement, business checks (convenience stores etc.), unoccupied vehicle checks, and surveillance of problem areas for criminal activity. The advantage of random patrol is the irregularity of patrol schedules, which increase the awareness of the general population that patrol is taking place. In addition, randomness of the patrol schedule discourages potential criminals, who find that they cannot predict the arrival patterns of patrol vehicles (Rosenshine, 1970). Because of its advantages, random patrolling was detrimental to the link between the police and the public (Kelling, Pate, Dieckman, & Brown, 1974).

The model of random patrolling has relied on the uniform provision of police resources and the law enforcement powers of the police to prevent crime and disorder across a wide array of crimes and across all parts of the jurisdictions that police serve (Weisburd & Eck, 2004). This, however, contradicts the common observations that crimes are not distributed evenly across jurisdictions. The Kansas City Preventative Patrol Experiment revealed that random patrolling had no effect on the crime rate (Kelling et al., 1974). Statistically speaking, a patrolling officer on a busy city street might hope to come across a street robbery in process once every 14 years by patrolling randomly (Reiss, 1967). These studies highlighted a problem of random patrolling: random patrol equals random results. As a result, numerous managerial and technological movements in police reform and crime prevention have surfaced since the late 1970s. Although random patrol across police jurisdictions is one of the most enduring of standard police practices, alternatives to random patrol have emerged (Weisburd & Eck, 2004). Most of these newly emerged techniques have emphasized the focused patrol.

#### *8.2.1.2 Focused Patrolling*

Focused patrol aims to focus police forces on locations where police presence is needed most. It often involves team patrolling and careful police resource allocation plans. The police resource allocation to perform preventive policing is one of the most important tactical management activities of police departments. The resource allocation plan often assumes that by knowing where the crime is happening and the reasons associated with this crime, it is possible to make an optimized allocation and consequently to decrease the crime rate. The police departments often collect extensive data in the belief that the locations of reported occurrences are good predictors of where incidents will occur in the future (Lobo, 2005).

Hazard formulas are widely used for allocating manpower by time and geography. This method takes all factors thought relevant to determine the need for patrol services, weighs each factor by its importance, and adds them up to arrive at a single hazard number (Kakalik, Wildhorn, & Housing USDo, Development U, Corporation R, 1971). The additive weighted combinations of the many hazard factors can neither reflect the highly complex interactions among the factors nor focus individual attention on any single factor. Most importantly, the weights of factors are often estimated in a subjective way. To deal with the limitation of hazard formulas, methods that provide unbiased estimation of crime occurrence are preferred and are often used in conjunction with the hazard formulas. Hotspot analysis is one such method.

## 8.2.2 Crime Hotspots

Hotspot analysis has been used to determine the spatial heterogeneity in the need for patrol services. Hotspot patrolling, which focuses patrolling resources at locations with clusters of crimes, has become a widely accepted patrolling strategy. It is well known that crime is not spread evenly across city landscapes and there are usually clusters of crime, or "hot-spots," that generate half of all criminal events (Braga, 2005). A reasonable choice would be to focus limited resources on places with clusters of crime. Some police departments have identified a list of "hotspots" and used that as a guidance for patrol activities. It was relatively easy for the experts to spot those points and then to allocate the teams to patrol mostly around them. The techniques used to detect hotspots belong to the family of point pattern analysis.

### 8.2.2.1 Point Pattern Analysis

The data collected by police departments often include crime events, such as event time and location. Such datasets make it possible to represent crime events as points located in geographical space, which are referred to as point pattern datasets and analyzed using point pattern analysis approaches. Point pattern analysis includes a family of statistics that allows for the identification and description of point patterns such as outliers, clusters, hotspots, cold spots, trends, and boundaries (Besag & Newell, 1991).

Point pattern analysis approaches can be global or local. Global statistics detect the existence of a spatial structure in the data. Local statistics complement the global statistics by identifying where the spatial structures are (Ord & Getis, 1995). Since the global statistics only identify, but do not necessarily quantify, the spatial heterogeneity in the police service local statistics seem to be more suitable for police service-related studies, especially the statistics that can identify crime hotspots.

Many versions of definitions exist for the term "hotspots" and there is no consensus on the best definition for crime hotspots. The major disagreement in the definition of "hotspot" is on whether to aggregate data or on how to aggregate them using different spatial units. In some studies, crime hotspots refer to the clusters of criminal events where the inter-distances between these events are shorter than would be expected in a random pattern and data aggregation is not necessary. Crime hotspot studies often have different contexts and data aggregation is often necessary for taking the contextual information into account. For strategic planning of law enforcement, which uses zoning as a tool, hotspots are often referred to as areas with higher level of criminal events than other areas. Kulldorff's spatial scan statistic is one of the most popular local statistics, which finds spatial regions where some quantity of point pattern is significantly higher than expected. It has been widely applied to identify crime hotspots (Anselin, Cohen, Cook, Gorr, & Tita, 2000; Ceccato, 2005; Kulldorff, 1997; Kulldorff, Huang, Pickle, & Duczmal, 2006).

#### 8.2.2.2 Kulldorff's Spatial Scan Statistic

Kulldorff's spatial scan statistic searches over a given set of spatial regions, finding those regions that maximize a likelihood ratio statistic and thus are more most likely to be generated under the alternative hypothesis of clustering rather than the null hypothesis of no clustering (Kulldorff, 1997). For each specified location, a series of circles of varying radii are constructed. All the events that fall into each circle are counted. Each circle is set to increase continuously from zero until a threshold is reached. The spatial scan statistics are usually implemented using two models: a Bernoulli model and a Poisson model.

In the Bernoulli model, cases and controls in each circle are compared to Boolean variables. In the Poisson model, the number of cases is compared to the background population data and the expected number of cases in each unit is proportional to the size of the population at risk. A commonly used likelihood function based on the Bernoulli model is the following (Kulldorff, 1997):

$$L(z,p,q) = p^{cz}(1-p)^{nz-cz}q^{C-cz}(1-q)^{(N-nz)-C-cz} \tag{8.1}$$

where $z$ represents the cell, $p$ is the probability of a disease case being inside the cell, $q$ is the probability of a disease case being outside the cell, $cz$ denotes the number of disease cases in the cell, $nz$ denotes the number of individuals in the cell, $C$ denotes the total number of cases in the region, and $N$ denotes the total number of individuals in the region.

Despite its popularity, the spatial scan statistics have two major limitations (Duczmal, Cançado, Takahashi, & Bessegato, 2008). First, they use a circular scanning window, which decreased the likelihood of the statistic detecting noncircular clusters. Second, they are computationally extensive as they usually search across the entire study area. These two limitations also affect the application of spatial scan statistics in patrol-related studies.

#### 8.2.2.3 A Modified Kulldorff Spatial Scan Statistic

Within the context of police patrol, hotspot identification using the street segment as basic unit seems to be the most relevant as more than 80% of patrol activities are motor patrol, which is constrained by the street networks (Menton, 2008). Street segments have been used in cluster analysis as basic units. Shekhar et al. proposed a "mean street" approach to identify streets or routes that have the highest level of crimes based on spatially aggregated crime data (Shekhar, Vatsavai, & Celik, 2008). A benefit of spatially aggregating data is that it may reduce the calculation difficulties. However, the success of this approach requires sophisticated approaches to define the basic unit, which has a large influence on the outcomes of aggregation. For point level data, Yamada and Thill proposed an approach to identify network constrained clusters. To apply this approach, we have to assume that crimes are located on the streets, since this approach uses network distances to calculate the interactions between points.

Li modified Kulldorff's spatial scan approach to identify clusters for points influenced by linear features (Li, 2010). Li's approach incorporates the characteristics of linear features in the spatial scan analysis using ellipses. In this approach, a vector map showing the distribution of linear features is first obtained. If the linear features have a complicated structure, it may be necessary to simplify their structures first to improve the efficiency of the algorithm. For example, connected segments with similar orientations could be merged. Each observed event location is assumed to be a potential center of a cluster. Therefore, it is not necessary to scan the entire area to locate possible clusters. Around each possible center of clusters, an ellipse with varying orientations, sizes, and shapes is drawn. The orientation of this ellipse could be inherited from a nearby linear feature. For example, the ellipse could be either parallel or perpendicular to the nearest linear feature. The size and shape of ellipse can be specified based on prior knowledge of the study area. For example, the biggest ellipse should cover at least one-fourth of the study area.

Note that this approach is different from the original elliptical cluster analysis that does not consider the influence of other features on points. This approach, though not previously applied to crime data, reflects the true nature of crimes, of which the distribution is influenced, but not necessarily constrained, by the distribution of streets.

### 8.2.3 Estimators of Spatial Autocorrelation

Using hotspot detection approaches, it is fairly easy to identify areas with elevated levels of crimes. Some police departments provide a list of the locations of hotspots to their patrol officers and instruct them to visit these locations. The identification of hotspots often brings analysis of crime to a close and it is often assumed that hotspots represent locations that demand police officers' attention. Any interactions that may exist between hotspots have often been ignored. We know that there are informal and formal interactions between crimes. Clusters of crimes may indicate a concentration of victims or criminals, which may interact with each other and complicate the dynamic nature of criminal activities. Therefore, it is important to examine the relationships between hotspots and one important method of doing this is spatial autocorrelation.

#### 8.2.3.1 Definition of Spatial Autocorrelation

Spatial autocorrelation is a fundamental concept in spatial analysis (Getis, 2008). It is the correlation among values of a single variable strictly attributable to their relatively close locational positions on a two-dimensional surface, introducing a deviation from the independent observation assumption of classical statistics (Diniz, Bini, & Hawkins, 2003). Positive spatial autocorrelation means that geographically nearby values of a variable tend to be similar on a map: high values tend to be located near high values, medium values near medium values, and low values near low values.

Spatial autocorrelation in crime data has often been observed (Ratcliffe, 2002). Demographic characteristics, for example population density, often exhibit positive spatial autocorrelation. Some demographic characteristics are often related to the level of crimes in different regions, since they explain the opportunities for criminals.

Spatial autocorrelation is often measured to avoid violating some fundamental statistical assumptions of certain statistical methods (Lichstein, Simons, Shriner, & Franzreb, 2002). It has also been quantified to support spatial prediction. Spatial autocorrelation analysis consists of a set of statistics describing how a variable (e.g., crime likelihood) is autocorrelated through space (Hardy & Vekemans, 1999). This variable can also be an attribute of street segments. The autocorrelation in attributes of a street network implies that the attributes of each edge in a network can be predicted in part from knowledge of the attributes of related edges. For patrol route planning, inclusion of an edge in a route may lead to impacts that are disproportional to the crime likelihood of this edge, since the neighbors of this edge that are unavoidably added in the route may also have high/low values (Dow, 2007). There are a variety of methods that have been used to quantify the spatial autocorrelation in a variable (Cliff & Ord, 1970). Getis-Ord Gi* is one of the most popular approaches to quantify the spatial autocorrelation.

#### *8.2.3.2 Getis-Ord Gi**

Getis-Ord Gi*, known as hotspot analysis, is commonly used to detect spatial clusters of the magnitude of an attribute (Anselin, 1995). It is a multiplicative measure of overall spatial association of values that fall within a critical distance of each other. The following equation is for general Getis-Ord Gi* test (Anselin, 1995):

$$G(d) = \frac{\sum_{i=1}^{K} \sum_{j=1}^{K} w_{i,j}(d) x_i x_j}{\sum_{i=1}^{K} \sum_{j=1}^{K} x_i x_j} \tag{8.2}$$

where $x_i$ is the value of the $i$th point and $w_{i,j}(d)$ is the weight for points $i$ and $j$ for distance $d$.

The general Getis-Ord Gi* only measures the characteristics of spatial clusters and does not test the statistical significance of the obtained measurements. Gi* values indicate whether certain values are likely to occur in one location or are equally likely to occur at any location. Positive values indicate that neighboring features are more like each other than distant features. Negative values indicate that neighboring features are unlike each other.The z-score of Getis-Ord Gi* provides an estimate of the statistical significance of the Gi* score. Depending on the assumption about the data, different functions can be used to calculate the z-score. A typical formulation of the z-score can be found in Anselin (1995).

## 8.2.4 Route Optimization

The Getis-Ord Gi* can be used to quantify the interrelationship between crime clusters. However, how to utilize this information to direct the patrol is still unknown and many questions regarding the patrol effectiveness still remain unanswered. For example, some police departments have identified a list of "hotspots" and used that as guidance for patrol activities. It was relatively easy for the experts to spot those points and then to allocate the teams to patrol mostly around them. However, it is unclear how patrol officers got to these points and whether these points were indeed visited more often than other locations. It is quite possible for a police officer to travel across several streets with a low level of crimes to reach a location with high crimes.

In addition to directing patrol officers to visit certain locations, an alternative way to plan patrol activities is to plan patrol routes. The direct benefit of planning patrol routes is that it provides an opportunity to optimize the collective efforts of patrol activities. When on duty, police officers often remain mobile to increase their chance to be seen and they are mostly moving along streets. The patrol directions given as a set of routes can be followed in a natural manner. This may also reduce the chances that the patrol performance is affected by personal biases of police officers, as the routes are decided in a subjective manner.

The routing problem is a well-studied problem in the operations research literature. Many algorithms have been developed to solve this problem. These algorithms can be classified into three categories: (1) direct tree search methods; (2) dynamic programming; and (3) integer linear programming (Laporte, 1992). The goal of the routing problem is often to optimize delivery capability. One of the frequently used measurements for delivery capability is the travel time. The routing problems focusing on minimizing travel time are often dealt with using shortest path approaches.

### 8.2.4.1 Shortest Path Problem

Shortest path problems aim to find a path between two locations in a graph (i.e., vertex) such that the sum of the weights of the segments in this path are minimized. The weights of the segments (i.e., edge) could be the travel distance, travel time, travel cost, or other attributes. Many algorithms have been developed to deal with the shortest path problem (Dreyfus, 1969). Among these algorithms, Dijkstra's algorithm is widely adopted and it is considered as the theoretically most efficient algorithm (Dijkstra, 1976). The following describes Dijkstra's algorithm.

Step 1: Label the starting vertex (could be arbitrarily selected) as 0 and label the rest of the vertices as infinite.

Step 2: Change labels of each vertex that is connected directly to the starting vertex to the distance between this vertex and the starting vertex.

Step 3: Find the vertex with the smallest temporary label, and select it as the next vertex of the path.
Step 4: From this vertex, find the total distance to each directly connected vertex and write as temporary labels. If a longer distance is already at a vertex, cross it out and replace it with the shorter distance. Then find the smallest temporary label and box it to make it permanent.
Step 5: Repeat Step 4 until all vertices are assigned to the path.

Dijkstra's algorithm is a successive approximation scheme. Therefore, it is a greedy search approach and it works as long as there is no negative weight for path segments. Dijkstra's algorithm is also a greedy algorithm. The greedy algorithms try to make good local choices and hope that a series of good local choices can lead to an optimal global solution. It is proven to be the best solution for the "single-source shortest path" problem, that is, a shortest path from a single vertex to all other possible vertices (Davis & Impagliazzo, 2004). However, it is not suitable for constraint path problems, for example the travel salesman problem, where certain vertices must be included in the path (Korkmaz, Krunz, & Tragoudas, 2002). These characteristics limit its application in patrol-related studies, since there may be negative weights associated with edges and constraints for path selection depending on the purposes of the patrol activities and constraints. For constraint path problems, Markov decision processes have been used (Briggs, Detweiler, Scharstein, & Vandenberg-Rodes, 2004).

In the Markov decision process, a route is defined as a series of connected edges and whether to travel from one edge to another is defined by transition probability. When the transition probability is unknown, the Markov decision process models are often solved using reinforcement learning, through which an agent learns the behavior of the system through trial and error with an unknown dynamic environment (Sutton & Barto, 1998). One problem of reinforcement learning is that the majority of algorithms developed based on it are computationally inefficient. For police patrol planning, fast computational means are needed. Recently, some fast learning algorithms were developed based on the Cross-Entropy (CE) method, which has become a standard tool in Monte Carlo estimation (Rubinstein, 1997). Section 8.2.4.2 introduces the CE method.

#### *8.2.4.2 Cross-Entropy Method*

The CE method provides a simple, efficient, and general method for solving complicated optimization problems. It has been used to solve the traveling salesman problem, combinatorial optimization problem, as well as patrol route optimization (Chen & Yum, 2010; Chepuri & Homem-De-Mello, 2005; Kroese, Porotsky, & Rubinstein, 2006; Rubinstein, 2001; Rubinstein & Kroese, 2004). The CE method translates the "deterministic" optimization problem into a related "stochastic" one. It has been applied to optimization problems concerning a weighted graph and introduces randomness in either the nodes or the edges of the graph (Kroese et al., 2006). In the case of patrol optimization, street networks are often simplified as a weighted graph.

In general, the CE method consists of two phases: (1) generate a random data sample according to a specific mechanism and (2) update the parameters of the random mechanism to produce a "better" sample in the next iteration. For patrol route estimation, the first phase of the CE method involves randomly generating patrol routes by selecting edges based on an initial transition matrix. This matrix defines the benefits of traveling between any two nodes. The initial matrix often assumes that traveling between different edges renders equal benefits. In the second phase, the transition matrix will be updated based on the calculated true benefits of traveling between the selected edges (Chen & Yum, 2010). This approach can consider the dynamic nature of criminal events. However, it uses aggregated crime data and does not consider the distribution of hotspots and their spatial pattern. The transition matrix was updated only based on the simulation process and it does not consider the spatial relationship between hotspots.

## 8.3 Strategy for Near-Optimal Patrol Route Planning

In this section, we introduce the proposed approach for incorporating the spatial autocorrelation scores of hotspots in a patrol route optimization procedure. We first provide an overview of our approach. After that, details on the patrol route model are provided. Finally, the route optimization procedure is explained.

### 8.3.1 Overview

In our approach, the CE method is used for the route optimization and randomization. The procedure of our approach is illustrated in Fig. 8.1. In our approach, we first calculated a crime likelihood value for each basic unit in a patrol route model (the details of this model will be given in the next subsection). Crime likelihood values are typically used to calculate patrol rewards for a patrol route. We are adding a critical step to calculate the spatial autocorrelation values for the likelihood values. As illustrated in Fig. 8.1, we used Gi* values as patrol rewards to take into account the spatial patterns of crime hotspots. Using assigned patrol rewards, the CE approach helps to update the transition matrix, which helps the identification of optimal routes. To ensure the randomness of the generated patrol routes, we randomize the starting points of routes and the convergence rule of the CE approach is relaxed. The updating procedure for the transition matrix is stopped before it reaches its optimal state. Finally, a set of near-optimal routes is obtained. For comparison purposes, the original crime likelihood values are also used for the same procedure and another set of near-optimal routes is obtained. To examine if these two sets of near-optimal routes help to achieve a collective optimal solution for patrol service, the $R_2$ score (proportion of variability in a data set that is accounted for by a statistical model) is used to test whether the frequency of each basic unit in these routes is proportional to the crime likelihood of this unit.

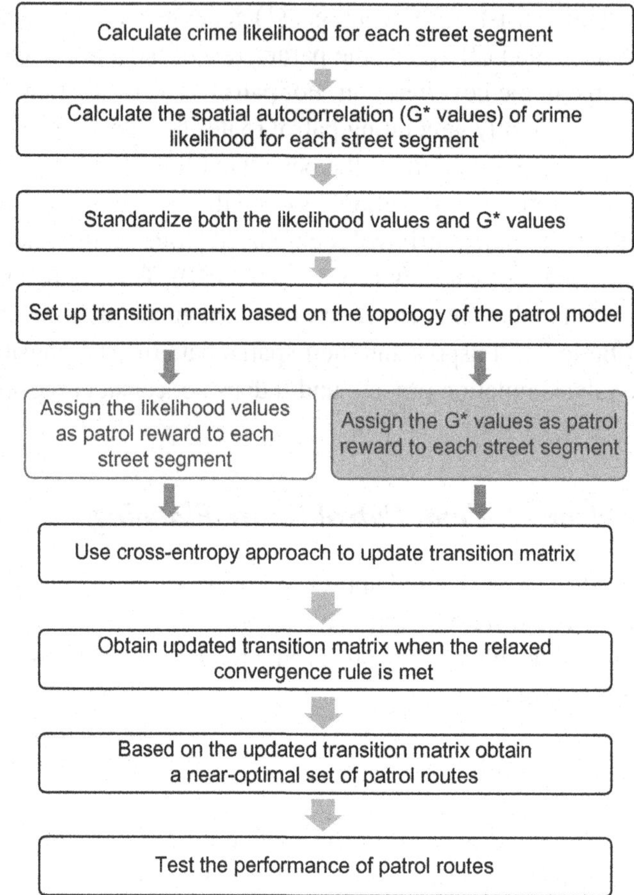


**Figure 8.1: Flowchart explaining the procedures in the proposed approach.**

## 8.3.2 Street Network-Based Patrol Route Model

### 8.3.2.1 Route Model Establishment

Currently, patrol planning activities, including foot and motor patrols, are mainly planned using street network models, such as street central line maps. To ensure the usability of the proposed algorithms, we first introduce a street network-based patrol route model that enables patrol route planning as follows:

- $\{1, 2, \ldots, N\}$ denotes edges in a street network model for an area to be patrolled. In street data, an edge $e_{(n)}$ usually represents the smallest street segment connecting two neighboring intersections.
- $T_n$ denotes the traveling time for the patrol unit on edge $n$.
- $C_n$ denotes the crime likelihood of edge $n$.
- $G_n$ denotes the spatial autocorrelation value of the crime likelihood for edge $n$.

- $P_n$ denotes transition probability of traveling on edge $n$. The initial value of weight for an edge is zero and it is updated using the CE method, which will be introduced in Section 8.3.2.2.

#### 8.3.2.2 Patrol Planning Problem Formulation

For this route model, the patrol reward for a route $r$ is:

$$r = \sum_{n=1}^{M} CnI_{n \in r} \qquad I_{Cn \in r} \in \{0,1\} \tag{8.3}$$

where $M$ is the total number of edges in a route.

Given the patrol reward equation, the patrol planning problem is formulated as:

$$\max \sum_{n=1}^{N} CnI_{n \in r} \qquad r \in \Omega \tag{8.4}$$

where $\Omega$ is a complete set of all possible suitable routes.

### 8.3.3 Route Optimization Procedure

Based on this route model, a random route ($R_n$) that can be completed within time $T$ can be generated using the following algorithm:

- Select a random edge $e_0$
- Stop until $(t_0 + t_1 + ... + t_n) > T$
  - Randomly select the neighboring edge of $e_k$
  - Check if $(t_0 + t_1 + ... + t_n) > T$
- Calculate the patrol rewards.

When selecting the neighboring edge, if the one with highest transition probability is selected, the results would give a local optimal route determined by a greedy algorithm. To obtain an optimized route, an informative transition matrix is needed.

#### 8.3.3.1 Cross-Entropy Algorithm

To obtain an optimal transition matrix, the CE algorithm can be used. The CE algorithm is as follows:

- Generate $M$ number of random routes by the algorithm described at the beginning of Section 8.3.3
- Sort these routes by patrol rewards
- Select top $g\%$ of routes with highest importance values
- Update transition matrix
- Repeat the above steps until convergence.

The transition matrix can be updated using Eq. (8.5) (Rubinstein, 1997). Using the street network model defined above, the transition probability of edge $i$ is:

$$P_i = \frac{\sum_{n=1}^{NR} N_{\{E_i \in R_n\}}}{\sum_{m=1}^{NR} N_{\{E_i \in R_m\}}} \tag{8.5}$$

where $NR$ is the total number of randomly generated routes, $ER$ is the total number of selected routes with highest importance values, $E_i$ is the edge $i$, $N_{\{E_i \in Rn\}}$ denotes the number of times $E_i$ appeared in randomly generated routes, and $N_{\{E_i \in Rm\}}$ denotes the number of times $E_i$ appeared in selected routes.

### 8.3.3.2 Convergence Rule Relaxation

The convergence rule is usually defined based on the patrol reward of the top performing route. In each iteration, the patrol reward of the generated route is recorded. Using cross-entropy minimization, this procedure stops when the patrol reward of the newly generated routes continuously equals the highest record of patrol rewards. This procedure would favor the best solution obtained over all previous iterations, up to and including the current iteration. In this way, the CE algorithm ensures a global optimal solution and it stops when the results converge to the optimal one.

In the context of patrol planning, it is important that a near-optimal set of solutions (here, a solution is referred to as a route) is found. One global single solution for patrolling would lead to blind points, which should be avoided in patrol plans. One way to achieve this goal is to relax the convergence rule. For example, the procedure can stop when an iteration reaches a near-optimal solution and its several subsequent iterations produce similar optimal solutions. The similarity of patrol reward can be defined using a threshold value. Using this convergence rule, the algorithm can recommend different patrol routes and these routes could be local optimal solutions. It is important to select an appropriate threshold value to ensure both the identification of near-optimal solutions and the near-comprehensive coverage of patrol plans.

The selection of the threshold can be based on the distribution of patrol reward values. For example, it can be assumed that the patrol reward values have a normal distribution. A simulation program can be used to estimate a theoretical distribution range for patrol reward values. The threshold can be arbitrarily selected (e.g., a threshold can be set to be larger than 80% of possible patrol reward values). Since the purpose of having this threshold is to allow some randomness in the resulting solutions, it can also be selected to ensure that a certain percentage of routes are different from each other.

## 8.4 A Case Study

Our approach has been applied to a study site in the United States. This section provides details of this application. We first describe the study site and then explain our choice of parameter values. In this study, we emphasize the importance of real-time application. We also compare our approach with an existing one.

### 8.4.1 Study Site

The method defined above was demonstrated using crime data along with patrol boundaries from the Police Department of a city in Washington State. This Police Department deploys 16 patrol teams working 11-hour days. Among these teams, eight teams each consisting of eight patrol officers are on duty each day. The city is divided into eight precincts and each patrol team is assigned to one of the precincts. Each precinct is further divided into two patrol neighborhoods. A patrol group usually consists of two police officers. Therefore, at any time, at least four police officers would be assigned to a precinct and a patrol group would be assigned to a patrol neighborhood. To demonstrate the proposed method, the downtown area, the $0.9 \times 1.1$ miles patrol neighborhood with the highest crime rate, was selected to be the study site. The boundary of the downtown area is shown in Fig. 8.2.

**Figure 8.2: Downtown area (dashed line), crime incidents in 2008 (black dots), and street network (gray line).**

According to Cordner and Kenney's study, patrol activities are often interrupted and patrol officers spend 60% of their time on duties other than patrolling (Cordner & Kenney, 1999). Patrol planning should select small time units to accommodate this problem. In this study, the length of recommended routes is controlled and a recommended patrol route should be finished within 10 minutes. Assuming patrol officers spend 40% of their shift time on patrolling, which is 5 hours, a patrol group can complete at least 30 patrol routes in each shift.

### 8.4.2 Data and Analysis

In the context of police patrol planning, selecting an appropriate time range to aggregate crime data ensures the reliability of crime likelihood estimation. The frequency and type of crimes varies at different times of day or year (Ratcliffe, 2002). However, the temporal aspect is not within the scope of this study. It is common practice to use yearly data to estimate crime likelihood. To test the efficiency of the proposed approach, crime data from the entire year 2008 (black dots in Fig. 8.2), which include 2175 crime cases, are used. The types of crimes include arson, assault, burglary, drugs, malicious mischief, robbery, theft, and murder. Li's approach is applied to the crime data to estimate crime likelihood for each basic patrol unit (Li, 2010). In Li's approach, centers of clusters were selected from locations with observed point events. To fit into the context of patrol route optimization, the centers of clusters in this study are centroid of basic patrol units. The Gi* value of the crime likelihood values is calculated and an inverse distance approach is used to determine the spatial proximity of the basic patrol units. To evaluate the temporal generality of crime data in 2008, the crime likelihood and Gi* values are calculated for crime data obtained in 2009 and the relationship between these two sets of data is examined using the $R_2$ score.

Also, as shown in Fig. 8.3, street network data are obtained from the GIS unit of the city and they are further broken into basic patrol units using ArcGIS (http://www.esri.com). The street network data consist of 320 basic patrol units. The average length of the basic unit is 250 ft and the total length is 15 miles.

### 8.4.3 Results

In this section, results of our case study are presented. One of the important contributions of this approach is that we introduced the Gi* score for the patrol route calculation. At the beginning of this section, we summarized the crime likelihood values and Gi* scores. We also explained the parameter settings in the convergence threshold. Finally, the performance of our approach has been accessed and compared by the cumulative patrol rewards of recommenced routes and running time of our program.

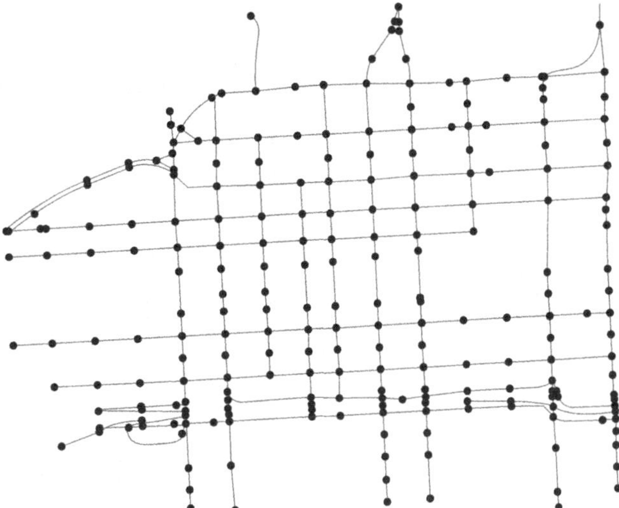

**Figure 8.3: The street network (the line represents edges/patrol units and each dot represents a vertex).**

### *8.4.3.1 Crime Likelihood Values and Gi* Scores*

Using the crime likelihood method explained in Section 8.3, the crime likelihood ratio for each basic patrol unit is calculated using crime data in 2008 and displayed as the size of pie charts in Fig. 8.4. The crime likelihood ratio values range from 0 to 1.51 with an average of 0.03. Based on the calculated crime likelihood ratio, Gi* score is calculated for each edge using ArcGIS with spatial analysis extension. Figure 8.3 displays Gi* z-scores as the size of patrol units. The range of Gi* z-score is from –0.860 to 6.91 with an average of –0.01. The relationships between the crime likelihood values/Gi* z-score in 2008 and 2009 are evaluated using $R_2$ score. The $R_2$ score for the crime likelihood is 0.785 and is 0.797 for the Gi* z-scores.

The z-score rescaling is applied to both the crime likelihood ratio and the Gi* z-score values. After rescaling, the crime likelihood ratio values range from 0 to 1 with an average of 0.3 and the Gi* z-score values range from 0 to 1 with an average of 0.6.

For comparison purposes, the CE method is applied to both the rescaled crime likelihood and the Gi* z-score using the process described in Section 8.3. The random routes are generated using a python program and the igraph module of python is used to construct a patrol route model. The CE algorithms are implemented in python code. Each process includes 500 iterations. This number of iterations is considered sufficient, since the total length of the simulated patrol routes in these iterations is 20 times the total length of the entire street network. The average convergence time increases and the possibility for each edge appearing in a set of patrol routes decreases when the convergence threshold is reduced.

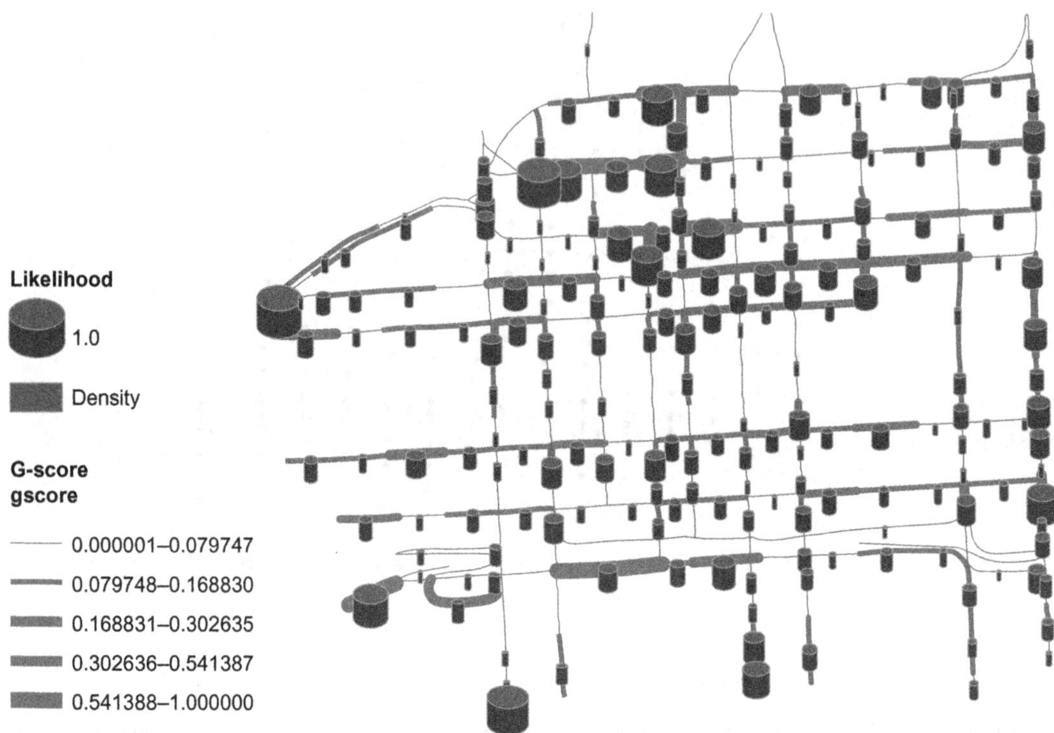


**Figure 8.4: Crime likelihood and Gi* scores (the size of the pie chart represents the likelihood and the size of the patrol unit represents G* score).**

### *8.4.3.2 Convergence Threshold*

To select a threshold that ensures a reasonable convergence time and an ideal patrol coverage, five thresholds were tested. These thresholds are 0.05, 0.1, 0.15, 0.2, and 0.25 multiply the average value of the rescaled crime likelihood and Gi* z-score. Using these test thresholds, we tested the calculation time and the percentage of overlapping routes. Our criterion to select the threshold is the balance between a reasonable calculation time and a small percentage of overlapping routes. Finally, convergence thresholds are set to 0.05 multiply the average values of the rescaled crime likelihood and Gi* z-score.

### *8.4.3.3 Performance of the Proposed Approach*

Using the selected thresholds, the CE methods are run 50 times on a computer with Dual-Core 2.53 GHz. The average running times for both the likelihood values and the Gi* values are recorded in Table 8.1. As shown in Table 8.1, the time that is required for the program to find a set of solutions is about 13 seconds. There is no apparent difference in the calculation times for the algorithm using the original crime likelihood value and the Gi* value.

**Table 8.1: Running time of cross-entropy algorithm on the crime likelihood values and the Gi* values of the crime likelihood values.**

| Approach | Average Running Time (second) | *R*-Score | Value Range |
|---|---|---|---|
| Crime likelihood | 13.02 | 0.1 | 1.25–2.47 |
| Gi* value | 12.1 | 0.3 | 2.17–3.48 |

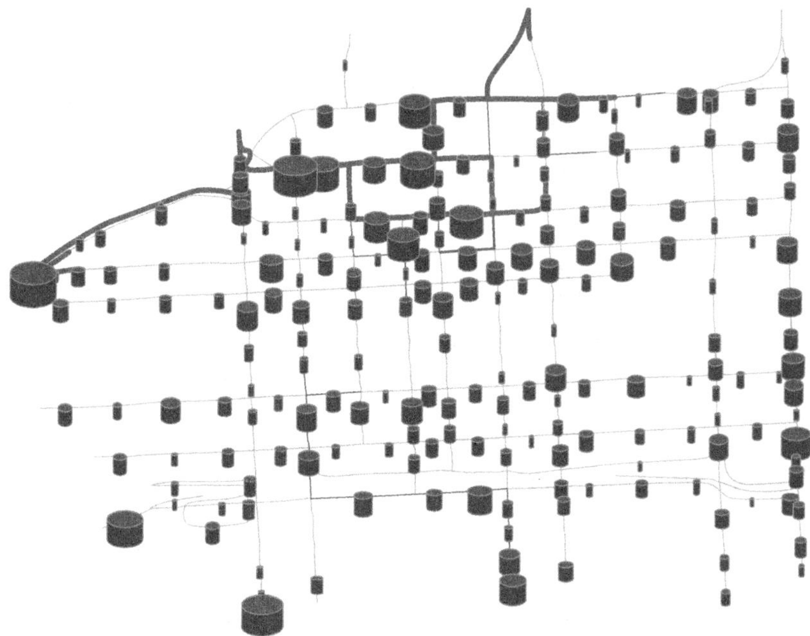

**Figure 8.5: Fifteen near-optimal routes created by the proposed approach using the original likelihood values (the size of the patrol routes represents the patrol rewards).**

Figures 8.5 and 8.6 show examples of near-optimal routes created using the proposed approach based on the original likelihood values and Gi* scores, respectively. In Fig. 8.5, the patrol route at the upper left corner was recommended because it has high reward values. It is apparent that the high reward value of this route can be attributed to the high reward value of a small patrol unit at its left side (the one with large pie chart). This may be caused by some random effects that may not happen again. The patrol rewards for other patrol units on this route are low. In Fig. 8.6, the recommended routes are reasonably connecting patrol units with a high level of patrol rewards.

**Figure 8.6: Fifteen near-optimal routes created by the proposed approach using the G* scores (the size of the patrol routes represents the patrol rewards).**

## 8.5 Conclusions and Prospects

Our study of the patrol route planning approach has been motivated by the widespread use of random patrolling, where concentrations of crimes have often been encountered. In this context, spatial autocorrelation in crime cases is often observed and spatial measurements such as Gi* scores have become the most relevant.

### 8.5.1 Conclusions

In this chapter, we described an approach to plan optimal randomized patrol routes. This approach is based on the Getis-Ord Gi* and the CE algorithms. In this approach, the basic street unit was used as the basic analysis unit instead of hotspots, which are often used in patrol planning. We also examined the spatial autocorrelation of the crime likelihood defined using a hotspot identification approach.

Suggesting routes instead of point locations of crime hotspots may be technically difficult. As most patrol cars are now equipped with sophisticated navigation equipment such as GPS, planning patrol activities at route level is feasible. In terms of running time, our approach shows similar performance with patrol route planning strategies proposed by other studies. For example, in Chen and Yum's study (2010), the running times for several major patrol route planning approaches were around 20 seconds, which is close to the average running time of our approach (13 seconds).

In addition, the results show that the patrol routes suggested by our approach can help to achieve a comprehensive and near-optimal coverage of the patrol area (Figs 8.5 and 8.6). The advantage of using Getis-Ord Gi* values for patrol rewards is that they consider the spatial similarity of crime likelihood of nearby street units and have a good temporal generality. Most importantly, this method helps to identify routes that lead to the improvement of collective patrol rewards of multiple routes. If this approach is used continuously, the collective performance of patrol activities can be improved. Results also show that our approach is effective and can be used for real-time solutions.

### 8.5.2 Limitations and Future Directions

We have confirmed in our approach that the use of Gi* scores in the patrol route selection procedure leads to improvements in patrol planning. In terms of real-life applications, our approach can be improved in several ways. First, when calculating the crime likelihood, different crime types can be assigned with different weights based on their emergency levels, which is usually defined by local authorities. For example, it is common practice to give violent crimes higher weight than nonviolent crimes, such as theft. It is also important to separate response priority from the patrol needs. Fast response for some crimes, such as burglary, is critical for the deterrence of criminals. For other crimes, such as robbery, patrol presence may prevent their occurrence. Our approach can also be improved by better defining an appropriate time range to calculate crime likelihood. For example, burglary tends to cluster in the temporal domain and a shorter time range might provide more insights into crime likelihood identification. Flexible scheduling for police officers is another area inviting further research, as it has been found that response rates become lower during the change of shifts than at other time periods. Finally, the police patrol shift and disturbances in their patrol services often need to be considered in the implementation of the recommended patrol plans. The proposed approach can recommend multiple routes for a patrol team to cover an entire shift. To take the work interruptions into account, it is possible to generate route recommendations based on the current location of the patrol team and the patrol history.

## References


Anselin, L. (1995). Local indicators of spatial association-LISA. *Geographical Analysis, 27*(2), 93–115.

Anselin, L., Cohen, J., Cook, D., Gorr, W., & Tita, G. (2000). Spatial analyses of crime. *Criminal Justice, 4*, 213–262.

Bahn, C. (1974). The reassurance factor in police patrol. *Criminology*, 12, 338–352.

Besag, J. & Newell, J. (1991). The detection of clusters in rare diseases. *Journal of the Royal Statistical Society Series A (Statistics in Society), 154*, 143–155.

Braga, A. A. (2005). Hot spots policing and crime prevention: A systematic review of randomized controlled trials. *Journal of Experimental Criminology*, *1*(3), 317–342.

Briggs, A. J., Detweiler, C., Scharstein, D., & Vandenberg-Rodes, A. (2004). Expected shortest paths for landmark-based robot navigation. *The International Journal of Robotics Research*, *23*(7–8), 717.

Burrows, J. & Lewis, H. (1988). *Directing patrol work* (Vol. 99). Home Office Research Study. London: HMSO.

Ceccato, V. (2005). Homicide in Sao Paulo, Brazil: Assessing spatial-temporal and weather variations. *Journal of Environmental Psychology*, *25*(3), 307–321.

Chawathe, S. S. (2007). Organizing hot-spot police patrol routes. In *2007 IEEE international conference on intelligence and security informatics (ISI)*, New Brunswick, NJ, May 23–24 (pp. 79–86).

Chen, X. & Yum, T. S. (2010). Cross entropy approach for patrol route planning in dynamic environments. In *2010 IEEE international conference on intelligence and security informatics (ISI)*, Vancouver, British Columbia, Canada, May 23–26 (pp. 114–119).

Chepuri, K. & Homem-De-Mello, T. (2005). Solving the vehicle routing problem with stochastic demands using the cross-entropy method. *Annals of Operations Research*, *134*(1), 153–181.

Chevaleyre, Y., Sempe, F., & Ramalho, G. (2004). A theoretical analysis of multi-agent patrolling strategies. In *Proceedings of the third international joint conference on autonomous agents and multiagent systems* (Vol. 3, pp. 1524–1525). Silverspring: IEEE Computer Society Press.

Cliff, A. D. & Ord, K. (1970). Spatial autocorrelation: A review of existing and new measures with applications. *Economic Geography*, *46*, 269–292.

Cordner, G. W. & Kenney, D. J. (1999). Tactical patrol evaluation. In *Police and policing: Contemporary issues* (pp. 127–155). Westport, CT: Greenwood.

Davis, S. & Impagliazzo, R. (2004). Models of greedy algorithms for graph problems. In J. I. Munro (Ed.), *Proceedings of the fifteenth annual ACM-SIAM symposium on discrete algorithms*, New Orleans, LA (pp. 381–390). New York: ACM.

Dijkstra, E. W. (1976). The problem of the shortest subspanning tree. *Operational Research Quarterly, 17*, 269–279.

Diniz, J. A. F., Bini, L. M., & Hawkins, B. A. (2003). Spatial autocorrelation and red herrings in geographical ecology. *Global Ecology and Biogeography*, *12*(1), 53–64.

Dow, M. M. (2007). Galton's problem as multiple network autocorrelation effects. *Cross-Cultural Research*, *41*(4), 336–348.

Dreyfus, S. E. (1969). An appraisal of some shortest-path algorithms. *Operations Research, 17*, 395–312.

Duczmal, L., Cançado, A. L., Takahashi, R. H., & Bessegato, L. F. (2008). *A comparison of simulated annealing, elliptic and genetic algorithms for finding irregularly shaped spatial clusters*. Vienna: Simulated Annealing I-Tech Education and Publishing.

Getis, A. (2008). A history of the concept of spatial autocorrelation: A geographer's perspective. *Geographical Analysis*, *40*(3), 297–309.

Hardy, O. J. & Vekemans, X. (1999). Isolation by distance in a continuous population: Reconciliation between spatial autocorrelation analysis and population genetics models. *Heredity*, *83*(2), 145–154.

Kakalik, J. S., Wildhorn, S., Housing USDo, Development U, Corporation R. (1971). *Aids to decision making in police patrol*. Santa Monica, CA: Rand.

Kelling, G. L., Pate, T., Dieckman, D., & Brown, C. E. (1974). *The Kansas City preventive patrol experiment: A summary report*. Washington, DC: Police Foundation.

Korkmaz, T., Krunz, M., & Tragoudas, S. (2002). An efficient algorithm for finding a path subject to two additive constraints. *Computer Communications*, *25*(3), 225–238.

Kroese, D. P., Porotsky, S., & Rubinstein, R. Y. (2006). The cross-entropy method for continuous multi-extremal optimization. *Methodology and Computing in Applied Probability*, *8*(3), 383–407.

Kulldorff, M. (1997). A spatial scan statistic. *Communications in Statistics: Theory and Methods*, *26*(6), 1481–1496.

Kulldorff, M., Huang, L., Pickle, L., & Duczmal, L. (2006). An elliptic spatial scan statistic. *Statistics in Medicine*, *25*(22), 3929–3943.

Laporte, G. (1992). The vehicle routing problem: An overview of exact and approximate algorithms. *European Journal of Operational Research*, *59*(3), 345–358.

Lee, W. L. (1901). *A history of police in England*. London, UK: Methuen & Co.

Li, L. (2010). Detecting local clusters in the data on disease vectors influenced by linear features. In *2010 18th international conference on geoinformatics*, Beijing, June 18–20 (pp. 1–6).

Lichstein, J. W., Simons, T. R., Shriner, S. A., & Franzreb, K. E. (2002). Spatial autocorrelation and autoregressive models in ecology. *Ecological Monographs*, *72*(3), 445–463.

Lobo, V. (2005). One-dimensional self-organizing maps to optimize marine patrol activities. In *Oceans'05 IEEE conference and exhibition*, Brest, France, 20–23 June (Vol. 1, pp. 569–572).

Menton, C. (2008). Bicycle patrols: An underutilized resource. *Policing: An International Journal of Police Strategies and Management, 31*(1), 93–108.

Metropolitan Police (2010). *History of the Metropolitan Police*. http://www.met.police.uk/history/archives.htm

Monkkonen, E. H. (1992). History of urban police. In M. Tonry & N. Morris (Eds.), *Crime and justice* (pp. 547–580). Chicago: University of Chicago Press.

Ord, J. K. & Getis, A. (1995). Local spatial autocorrelation statistics: Distributional issues and an application. *Geographical Analysis, 27*(4), 286–306.

Ratcliffe, J. H. (2002). Aoristic signatures and the spatio-temporal analysis of high volume crime patterns. *Journal of Quantitative Criminology*, *18*(1), 23–43.

Reis D., Melo, A., Coelho, A., & Furtado, V. (2006). GAPatrol: An evolutionary multiagent approach for the automatic definition of hotspots and patrol routes. In Sichman, Coelho, & Oliveira (Eds.), *Proceedings of IBERAMIA/SBIA* (pp. 118–127). Berlin: Springer-Verlag.

Reiss, A. J. (1967). The challenge of crime in a free society. *A report of a research study submitted to the President's commission on law enforcement and administration of justice*. London, UK: Methuen & Co.

Rosenshine, M. (1970). Contributions to a theory of patrol scheduling. *Operational Research Quarterly, 21*, 99–106.

Ruan, S., Meirina, C., Yu, F., Pattipati, K., & Popp, R. L. (2005). Patrolling in a stochastic environment. *Online Information for the Defense Community*, *10*, 28–34.

Rubinstein, R. Y. (1997). Optimization of computer simulation models with rare events. *European Journal of Operational Research*, *99*(1), 89–112.

Rubinstein, R. Y. (2001). Combinatorial optimization, cross-entropy, ants and rare events. *Stochastic Optimization: Algorithms and Applications*, *54*, 303–363.

Rubinstein, R. Y. & Kroese, D. P. (2004). *The cross-entropy method: A unified approach to combinatorial optimization, Monte-Carlo simulation, and machine learning*. New York: Springer-Verlag.

Schnelle, J. F., Kirchner, R. E., Jr., Casey, J. D., Uselton, P. H., Jr., & McNees, M. P. (1977). Patrol evaluation research: A multiple-baseline analysis of saturation police patrolling during day and night hours. *Journal of Applied Behavior Analysis*, *10*(1), 33.

Shekhar, S., Vatsavai, R. R., & Celik, M. (2008). Spatial and spatiotemporal data mining: Recent advances. In H. Kargupta, J. Han, P. S. Yu, R. Motwani, & V. Kumar (Eds.), *Next generation of data mining*. CRC Press.

Sherman, L. W. (2002). *Evidence-based crime prevention*. London, UK: Routledge.

Sherman, L. W., Gartin, P. R., & Buerger, M. E. (1989). Hot spots of predatory crime: Routine activities and the criminology of place. *Criminology*, *27*, 27–35.

Sutton, R. S. & Barto, A. G. (1998). *Reinforcement learning: An introduction*. Cambridge, MA: The MIT Press.

Weisburd, D. & Eck, J. E. (2004). What can police do to reduce crime, disorder, and fear? *The Annals of the American Academy of Political and Social Science*, *593*(1), 42.

CHAPTER 9

# Chemical Emergency Management Research Based on ACP Approach

**Wei Wei[*], and Changjian Cheng[†]**

[*]State Key Laboratory of Management and Control for Complex Systems, Institute of Automation, Chinese Academy of Sciences(CASIA), Beijing 100190, China, and Academy of Equipment Command and Technology, Beijing 102249, China

[†]State Key Laboratory of Management and Control for Complex Systems, Institute of Automation, Chinese Academy of Sciences(CASIA), Beijing 100190, China

## 9.1 Introduction

Throughout the 20th century and beyond, all kinds of emergencies have been occurring frequently. These incidents are so paroxysmal that they become significant threats to people's lives, their property, and society. Frequent emergencies have caused a number of significant tragedies. To prevent the occurrence of emergencies effectively and to control the losses caused by emergencies, research on emergency management and the establishment of a reliable emergency response system are very necessary.

We analyze emergency management by taking chemical accidents as the entry point. Large chemical installations are defined as complicated, nonlinear, sociotechnological systems (Rasmussen & Gronberg, 1997). Chemical accidents occurring during operations may result in substantial consequences both inside and outside the premises of an installation. The initiating event of an accident in such a system is a failure, which may, and usually does, lead to more failures and finally to uncontrollable hazardous events. Much attention has been given, so far, to developing and implementing Safety Management Systems to reduce the probability of a major chemical accident. On the other hand, chemical emergency management has been given considerably less attention, although many related software packages are currently being used for this purpose. In this presentation, a new methodology is proposed, based on an ACP (artificial systems, computational experiments, parallel execution) approach (F.-Y. Wang, 2004), which is a new theory on modeling, analyzing, managing, and controlling complex systems, aiming at improving emergency management.

Service Science, Management, and Engineering.


DOI: 10.1016/B978-0-12-397037-4.00009-0

Emergency management is a process with the characteristic of time series, including four stages: emergency early warning, emergency response plans evaluation, emergency implementation, and emergency ex-post treatment.

In the early stage of emergencies, the affected department and on-site officers shall actively adapt necessary measures so as to prevent it becoming more serious. Building an effective early-warning system will predict and prevent accidents as early as possible. X.-Z. Wang (2005) introduced the early-warning management system on poisonous and harmful gas in the chlorine industry, which can provide basic data intuitively and play an important part in environmental protection, security monitoring, and early-warning analysis. W. Zhang (2008) used the emergency early-warning instant instruction technology to analyze the time when anomalies were first noticed based on the Automatic Emergency early-warning Information Platform. Chang (2003) developed an emergency early-warning software system based on the geographic information system by combining the accident model, the best path algorithm, and the GIS system, and by providing decision support for accident prevention and emergency rescue. Meanwhile, the Mine Fire Command Expert System, developed by British Coal Company Technology Department and the University of Nottingham, can simulate the rescue considerations of experts in emergencies and put forward disposal measures by analyzing a variety of related data. Similar research in China has produced the Industrial Accident Analysis Expert System developed by the China University of Geosciences and the Coal Mining Accident Decision Support System developed by the Xi'an Mining College (Zhai, 1998).

The emergency response plan is a plan formulated in advance to ensure fast, orderly, and effective emergency rescue so as to reduce the damage caused by the accidents. To ensure the expected effects of the emergency response plan, ex-ante evaluation is required. The evaluation before plans are implemented is usually developed by building an evaluation index system (Huang, 2006; Yu, 2007; Y. Zhang, 2004). Chemical plants often carry out dynamic evaluation through a simulation drill (Boppana, 2005), because the drill can analyze the rationality of the operational process and the problems that may arise during the emergency implementation (Burns, Robinsonk, & Lowe, 2007; Kaji & Lewis, in press). However, many factors restrict the application of this method, such as long period, high costs, and unsuitability for some plans due to legal or moral issues. In recent years, computer simulation evaluation (Lanrian & Berke, 2004; Mark & Westhues, 1989) has been widely used because of its efficiency and safety.

The two key problems, emergency early warning and emergency response plan evaluation, are studied in the presentation to assist in taking rational decisions in chemical emergency management.

## 9.2 Problems and Challenges

An emergency response system is a complex system with respect to social and behavioral dimensions. In the same way, the emergency management is also a complex system. Chemical plants are typical complex systems with both engineering and social components. Once the accident happens, the treatment of operators, the decisions of managers, and the change of external environment will have an effect on the emergency response. The research of complex systems is a well-developed field since it began when humans started to explore nature and acquire knowledge. However, it is also a new field because people have begun to realize that many problems will not be solved by existing concepts and methods from the last century. Thus, we must seek new ideas and introduce new concepts and methods to establish a new theoretical system aiming at complex systems and complexity science. Recently, Chinese AI researchers have been working on ways to transform systems and methods along this direction. Some scholars focus on the ACP approach. The approach involves modeling with artificial systems, analysis with computational experiments, and operation through parallel execution for control and management of complex systems. Moreover, the ACP approach has extensive adaptability for all kinds of complex system. In the social research field, it has been applied in emergency management and e-commerce, forming the cutting-edge theory of social computing. In the engineering technology research field, it has also been applied in traffic, electric power, population, ethylene production, and so on (F.-Y. Wang, 2007b, 2008).

Emergency response is essentially the interaction between operators, equipment, and environment. A person is the main operator in the emergency rescue and any function of plans must be realized by the operators' subjective behaviors. Although the general theory deems that people are completely rational, practical behaviors provide indisputable evidence for Simon's so-called "limited rationality" (Simon, 1957). This limited rationality leads people to often make choices that can satisfy themselves according to their recognition of the environment and their own limited thinking, not pursuing "the most utility". At present, related research on emergency management mostly takes into account the objective factors in emergency scenes, while consideration of subjective factors is lacking. In fact, people often have limitations when dealing with emergency situations, so the error probability of operation will be far higher than usual (B.-J. Liu, 1994). The emergency treatment will have a variety of possible outcomes due to the subjective complexity of personnel behaviors. So how to evaluate the operational effectiveness of the selected emergency response plan in the rescue process is a pending problem to be solved, which will be very necessary and meaningful.

In a people-oriented system, if we consider the behaviors of people, the complexity of the model will obviously increase. Many approaches have been proposed to try to handle this problem. Andersen and Rasmussen (1987) discuss the management science approach that is based on decision making by using the outcomes of utility theory and focus on decision making from a prescriptive point of view. It is argued by the authors that the utility theory has a difficulty in incorporating human reliability and in representing the decision-making process followed by humans, which is, in fact, more complex than previously thought and cannot be easily represented by economic and management models. Social science has also been used to comprehend the human part of emergency management systems. However, social science by itself is not enough to describe the whole system, because it does not take into account the technological aspects of chemical emergencies. Thus, effective emergency management requires the formulation of a new structure, a new sociotechnological system capable of coping with the situation at hand. The ACP approach can not only consider all kinds of factors comprehensively but also describe the multiple possibilities of personnel behaviors fundamentally in emergency rescue, reflecting the influence of human behaviors on emergency implementation.

## 9.3 A Research Framework: Parallel Emergency Management System

The ACP approach is applied in emergency management by Professor F.-Y. Wang (2007a), forming the framework of PeMS (Parallel Emergency Management System). This framework involves modeling emergencies using an artificial social method, analyzing the emergency disposal strategies with computational experiments, and implementing the emergency response plans using parallel execution. The core idea of PeMS is to integrate the emergencies analysis and the cultivation of emergency psychology and ability, making the simulation of emergency process and emergency implementation unified, real time, intelligent, and finally making the emergency management scientific and practical.

Figure 9.1 presents the ACP-based framework of PeMS for the control and management of complex emergency systems. This framework prescribes that an artificial emergency system can be constructed by modeling personnel behaviors and equipment technological processes, respectively, in an actual emergency system, in order to realize emergency early warning and emergency response plans evaluation. To construct the framework, we utilize solid, effective mathematical methods and analytical tools for modeling, analysis, and synthesis developed by researchers in control theory. For complex systems and the corresponding decision-making problems, we must take into account social and behavioral dimensions beyond traditional mathematical analysis. AI techniques can play a critical role in this process, especially in the construction of agent-based artificial systems and in issues related to social computing, behavioral modeling and prediction, and intelligent decision-support systems.

Using artificial systems, we can treat computers as social laboratories, developing computational experiments, using emerging methods and various statistical means to analyze the effectiveness of emergency strategies or plans, and putting forward corresponding measures of modification. Emerging concepts will play a key role in the computational experiments because various complicated phenomena can be developed easily based on artificial systems.

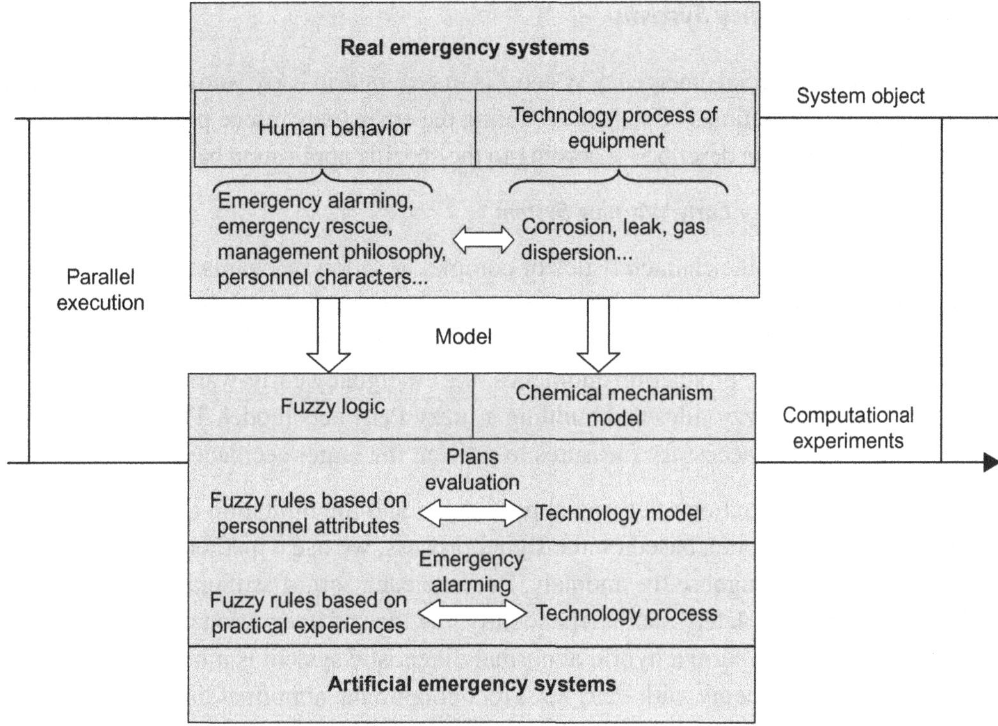


**Figure 9.1: The framework of the Parallel Emergency Management System based on ACP.**

In the framework, the actual system and its artificial counterparts can be connected in various modes for different purposes, such as learning and training, experimentation and evaluation, and control and management. We can design and conduct controllable experiments that are easy to manipulate and repeatable. By comparing and analyzing real and simulated behaviors, we can learn and predict systems' future actions and accordingly plan and modify control and management strategies for their operations.

In many aspects, this approach is driven by new developments in agent technology, computing architectures, and networked operational environments. It is also motivated by the desire to develop, organize, and apply AI methods and techniques for complex systems according to the decision-making structures and procedures in cybernetics and modern control theory.

We study two key problems in chemical emergency management: emergency early warning and emergency response plans evaluation. In view of the influence of social factors being seldom considered in current emergency management research, the presentation adds the factor of human behaviors into chemical emergency early warning and emergency response plans evaluation. In emergency early warning, fuzzy diagnostic rules including human behaviors are used to construct an artificial emergency early-warning system, which will provide forecasts for the anomaly's future status and looking for possible causes of the anomaly. In plan evaluation, the impacts of human behaviors on emergency implementation are fully evaluated.

### 9.3.1 Artificial Emergency Systems

The main function of artificial emergency systems is to generate an early warning of an accident and the evaluation of operational effectiveness during the emergency rescue process. The construction process will be described according to the specific application background.

#### 9.3.1.1 Artificial Emergency Early-Warning System

Chemical production has the characteristics of complex reaction mechanisms and higher operation risk. If an anomaly is seen, it may trigger a chain reaction and disrupt the normal operation of the whole chemical process, perhaps causing a serious accident (Z.-J. Liu, 2008). Based on finding chemical production anomalies, the emergency early-warning mechanism is established by creating fuzzy rules and building a fuzzy Petri nets model. Through the early warning, we can take the necessary measures to prevent the larger accidents.

Considering the actual situation of chemical production and the limitation of the methods based on the analytical model, based on the signal process, we use a method based on artificial intelligence to diagnose the anomaly. Because each sort of artificial intelligence cannot be diagnosed completely and independently, the idea of integration of several intelligent technologies to form a hybrid abnormal diagnostic system is introduced. This chapter integrates fuzzy theory with Petri nets, recognizing the abnormal diagnosis in early chemical accidents. The chemical process often produces a longer time lag: the next anomalies caused by the preceding anomaly usually have a time delay; therefore, the time factor is considered in the early-warning diagnosis to build the time fuzzy Petri nets model.

The fuzzy diagnostic rules are established as follows. First, the abnormality is confirmed, such as draught fan or forced draught fan anomaly. Second, factors relating to the anomaly are collected from the people, object, and management, such as the unsafe behavior of people, the insecure state of the object, and the defects of management. Third, the relationship between the various factors and their own importance is analyzed, determining the confidence level, delay time, and actual probability of rules according to expert opinion. Finally, fuzzy rules are modeled by the fuzzy Petri nets, in combination with a Matlab Stateflow module to construct an artificial emergency early-warning system, capable of making abnormal diagnoses.

Abnormal diagnoses based on the artificial emergency early-warning system include:

a. Abnormal backward reasoning, namely, find the root cause of the anomaly, provide reasonable measures to remove the anomaly, and return the system back to normal as soon as possible.
b. Abnormal forward deduction, namely, for the original anomaly, give the possible propagation paths and forecast the subsequent anomalies that the initial anomaly may cause.

### 9.3.1.2 Artificial Emergency Plans Evaluation System

An Artificial Emergency Plans Evaluation System is constructed by hierarchical Petri nets and Matlab Stateflow module, involving a technological physical model of equipment and a behavioral model of operators. The hierarchical Petri nets model is used to describe the implementation process of emergency response plans, which is the dynamic interactive process between the equipment and the operators. According to the status and function of each layer, different expanded Petri nets are used. The emergency response process is divided into four layers, namely system layer, operational layer, inferential layer, and basal layer, as shown in Fig. 9.2. The system layer is described with stochastic Petri nets, indicating the successful shift process of the chemical plant according to the emergency response plans depicted (Ai, 2008). The operational layer is described with basic Petri nets, which is the expansion of the abstract transition in the system layer. The abstract transition in the system is the operational link in the emergency response process. Different operators have different

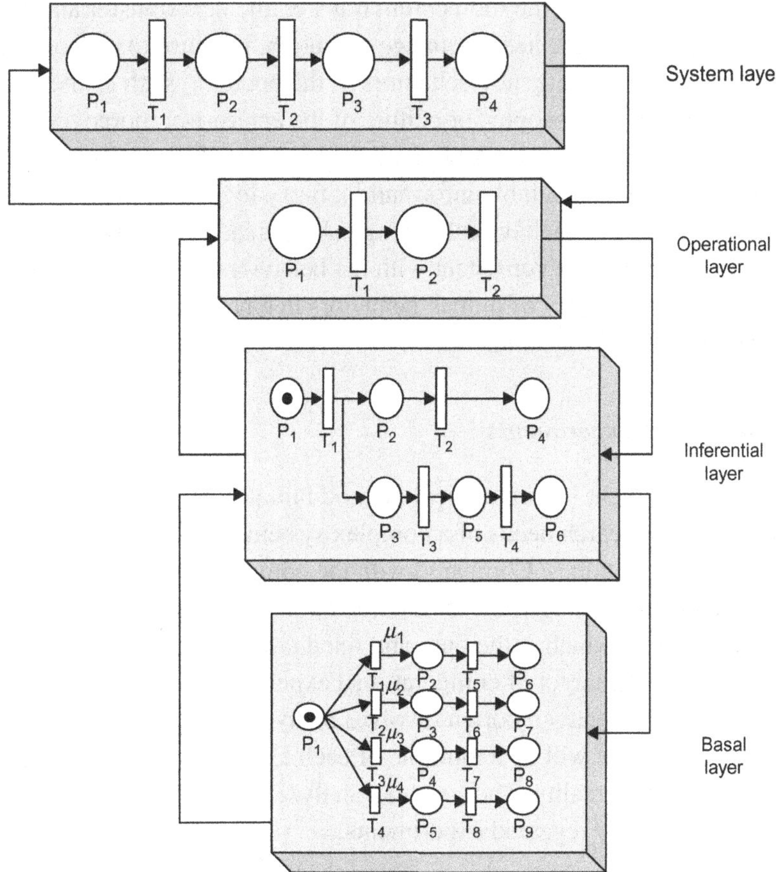


**Figure 9.2: Hierarchical model of emergency procedure.**

behaviors because of their own attributes. After the firing of abstract transition, there will be a variety of possible behaviors determined by fuzzy logic and a behavioral distinguishing matrix. These behaviors will be displayed in the form of multibranches in the operational layer. The inferential layer is described with basic Petri nets, which models the diagnostic process of system abnormality due to an operational human error. The process of behaviors inference and validation will involve more rules and is more difficult in contrast to the operational layer. Finally, various possible diagnostic behaviors are displayed in the form of multibranches in the inferential layer. The basal layer is constructed by fuzzy Petri nets, which models the fuzzy reasoning process in the inference layer and diagnoses based on fuzzy rules. The four layers are connected through a Combining Terminal.

The technological model is established based on chemical mechanisms, which is usually the simulation of natural processes using the differential and difference equations. In addition, the technological model will be selected based on the operational requirements of the emergency response plans, such as the corrosion model and the gas dispersion model, that model need to be established in leakage plans, the momentum transfer, the heat transfer and the mass transfer model need to be selected in the heat exchange operation. Because many social factors will also have an impact on the emergency behaviors of the operator, such as the attributes of operator, the management philosophy, or culture of the enterprise; moreover, these influencing factors mostly involve non-numeric and lingual descriptions. To transform the static text information into dynamic computable information, fuzzy logic is used to model human emergency behaviors, and the behavioral distinguishing matrix is defined to judge whether the actual emergency behaviors are consistent with the behaviors the emergency response plans require. We will collect a variety of human behaviors in emergency situations by the Artificial Emergency Plans Evaluation System.

### 9.3.2 Computational Experiments

A computational experiment is a theoretical method initiated by the emergence of an artificial social method and the research needs of a complex system, which is a natural extension of computer simulation techniques. Compared with the computer simulation, computational experiments simulate the realities. The simulation target of computer simulation is the objective natural process, which is the only and fixed target, almost having no room to play subjectively. However, the target of computational experiments often lacks the uniqueness and the occurrence of its behaviors often involves many subjective factors. The realities that an artificial system presents will be a multiple of each experiment. The realities are not the only reality, but a possible reality. Thus, we can analyze and evaluate the various aspects of a system comprehensively by repeated experiments.

In complex production systems, especially those involving human behaviors and social organizations, the simulation results can be considered as an alternative to reality or a possible reality. Therefore, the method of computer simulation in computational experiments is put forward based on this concept. Using artificial systems, we can treat computers as social laboratories. We can design and conduct controllable experiments that are easy to manipulate and repeat and then evaluate and quantitatively analyze various factors in the emergency response process. The basic problems of the calibration, design, analysis, and verification of computational experiments are discussed in the literature (F.-Y. Wang, 2004).

Human behaviors and the formulation of emergency response plans will bring more uncertainty due to their own social subjectivity. Therefore, human behaviors and the factors constraining them, such as corporate culture, management philosophy, and emergency response plans, will be the main content of computational experiments. Through computational experiments, we are able to analyze the production anomalies that result from the emergency behaviors of the operator, forecast and evaluate the behavioral effect of the operator in the emergency rescue process, and then judge whether the development of plans is reasonable and the postconfiguration of personnel in emergencies is rational. The basic considerations are shown in Fig. 9.3.

The computational experiments are conducted through artificial systems. Artificial emergency early-warning systems recognize an abnormal diagnosis based on the established fuzzy diagnostic rules, which cover a wide range of rules parameters, such as confidence level, delay time, and initiating probability. However, these parameters are set according to the judgment of experts, with different judgments leading to different diagnostic results. To compare these experimental results, we develop the computational experiments and explore the laws between these results. Concerning the same experimental environments, validated computational experiments are utilized, which is not the statistical technology, but the analytical method to investigate the repeatability of the experimental results under the same experimental conditions. Through validated experiments, we consider the results deduced by the most experienced experts as the final diagnostic results.

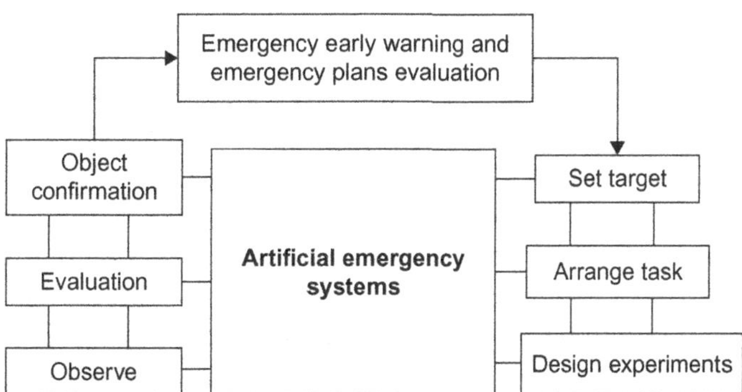


**Figure 9.3: Computational *experiments* scheme based on *artificial systems*.**

An Artificial Emergency Plans Evaluation System evaluates the operational effectiveness in the emergency rescue process. People with different attributes will produce different emergency behaviors, which will affect the practice effects of emergency response plans. To analyze the behavioral difference resulting from the personnel attributes, the computational experiments are used to present every possibility. Among the personnel attributes, the main attribute is different in different operations. For a specific emergency operation, we need to use optimized computational experiments to determine the main influence attribute. The computational experiments are divided into two types: single-factor and multiple-factor experiments. Single-factor experiments analyze the impact of a single attribute on the emergency operational reliability, whereas multiple-factor experiments confirm the main attribute affecting operation most. In these experiments, each attribute is defined as $N$ level, and through the multiple-factor experiment the optimal attribute combination will be explored, providing the direction for the personnel postconfiguration in emergency rescue.

### 9.3.3 Parallel Execution

By parallel execution, the artificial system running in parallel with the actual system, the comparison and analysis between the behaviors of the artificial system and the actual system can provide a reference and estimate of the future situation and then adjust their own management and control method.

The artificial system tries to simulate the actual system as much as possible to estimate various possible behaviors and responses of the actual system. Through parallel running, the difference of the two systems' evaluation states will be observed and error feedback signal will be generated to realize the correction of the evaluation model and parameters of the artificial system. Because of the multiple realities of the artificial system, realizing the early-warning analysis and the evaluation of the emergency operator's behaviors in the reality of "practice" require parallel execution between the actual system and the artificial system.

The fuzzy diagnostic rules are established in the emergency early warning, where various parameters need to be assigned by parallel execution. The personnel attributes in the emergency rescue are also assigned in the same manner, providing the input variables for the plans evaluation. Figure 9.4 describes the parallel interactive process between the two systems. Processes ①, ②, ③, and ④ represent the constraints of emergency response plans to human behaviors, the mapping from personnel attributes to actual behaviors, the emergency operating behaviors acting on chemical equipment directly, and the self-correction of the system according to the error feedback signal, respectively. Through parallel running between the two systems, the actual system provides the equipment status parameters and on-post personnel attributes for the artificial system. A variety of possible emergency scenarios are deduced during the continuous interaction between operators and equipment.

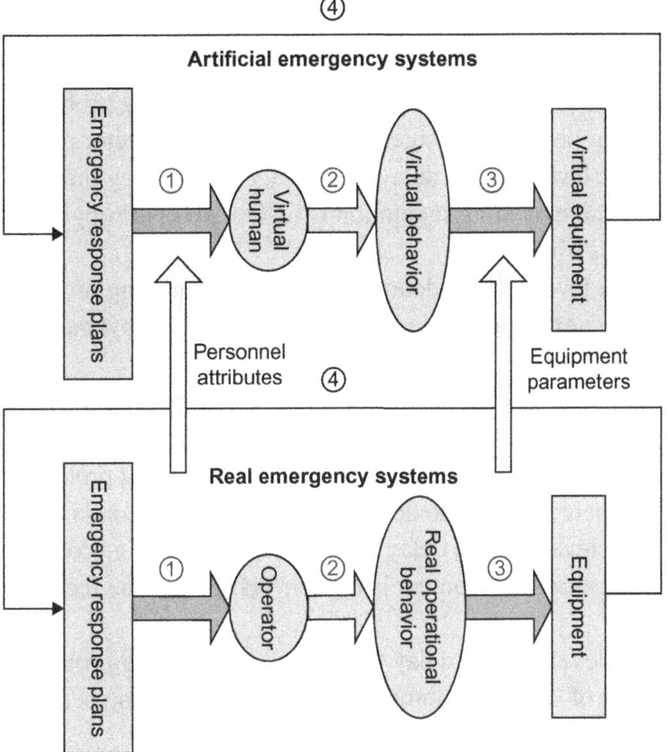


**Figure 9.4: Parallel interface between artificial system and actual system.**

## 9.4 Case Study

To illustrate the human behaviors in our framework, we present two case studies that come from the use cases of a chemical emergency response system. The cases describe the specific process of emergency early warning and emergency plans evaluation clearly.

### 9.4.1 Case Study 1: Chemical Early-Warning Research

Taking the polypropylene reactor process from the literature (Z.-J. Liu, 2008) as an example, we describe the abnormal reasoning and propagating process containing the behaviors of operators in emergency early warning. It is a suspension polymerization reaction by the action of the catalyst and activator, using liquid hexane as solvent.

The fuzzy diagnostic rules are established based on the analysis of process engineering, which are related to the reaction temperature and pressure. Among the rules, the rules marked with "*" contain personnel behaviors, describing the causal relations between the operator's improper operational behaviors and the production anomaly. The confidence level, delay time, and initiating probability of the rules are set according to expert advice by parallel execution. Part of the rules is as follows:

Rule 1 (confidence level is 0.85, delay time is 4, and initiating probability is 0.5)
IF Propylene feeding percentage content is higher THEN Reactor pressure is higher
Rule 2* (confidence level is 0.93, delay time is 3, and initiating probability is 0.6)
IF Propylene feed valve opening regulation is higher THEN Propylene feeding percentage content is higher
Rule 3* (confidence level is 0.94, delay time is 3, and initiating probability is 0.5)
IF Hexane feed valve opening regulation is lower THEN Propylene feeding percentage content is higher
Rule 4 (confidence level is 0.8, delay time is 4, and initiating probability is 0.65)
IF Reaction kettle level is higher THEN Reactor pressure is higher
Rule 5 (confidence level is 0.92, delay time is 1, and initiating probability is 0.55)
IF Hexane feeding percentage content is higher THEN Reaction kettle level is higher
Rule 6*(confidence level is 0.95, delay time is 3, and initiating probability is 0.7)
IF Hexane feeding percentage content is higher, THEN Hexane feeding percentage content is higher
Rule 7 (confidence level is 0.97, delay time is 2, and initiating probability is 0.4)
IF Hexane flow control valve is abnormal, THEN Hexane feeding percentage content is higher
Rule 8 (confidence level is 0.9, delay time is 1, and initiating probability is 0.65)
IF Propylene feed is higher, THEN Reaction kettle level is higher
Rule 9*(confidence level is 0.95, delay time is 3, and initiating probability is 0.7)
IF Propylene feed valve opening regulation is higher, THEN Propylene feed is higher
Rule 10 (confidence level is 0.97, delay time is 2, and initiating probability is 0.4)
IF Propylene feed valve is abnormal, THEN Propylene feed is higher
Rule 11 (confidence level is 0.99, delay time is 2, and initiating probability is 0.75)
IF Reactor temperature is higher, THEN Reactor pressure is higher.

The time fuzzy Petri nets model is established based on the above rules, as shown in Fig. 9.5. Then, we can develop the abnormal backward reasoning and forward deduction based on the model.

Figure 9.6 describes the abnormal backward reasoning process. The confidence level of the proposition for the places P1, P4, P12, P18, and P21 in Fig. 9.5 is set as 0.7, 0.65, 0.6, 0.8, and 0.9, respectively, by parallel execution. The confidence level of the proposition "the reaction kettle pressure is higher" initiated by the five causes is calculated. The results show that the confidence level calculated by the initial proposition "catalyst A feed is higher" is the highest, 0.796. So, the proposition "catalyst A feed is higher" is determined as the most likely initiated reason and the final reasoning path is determined as $P_{11}$–$P_{24}$–$P_{23}$–$P_{21}$, identified by the dotted line, as shown in Fig. 9.6.

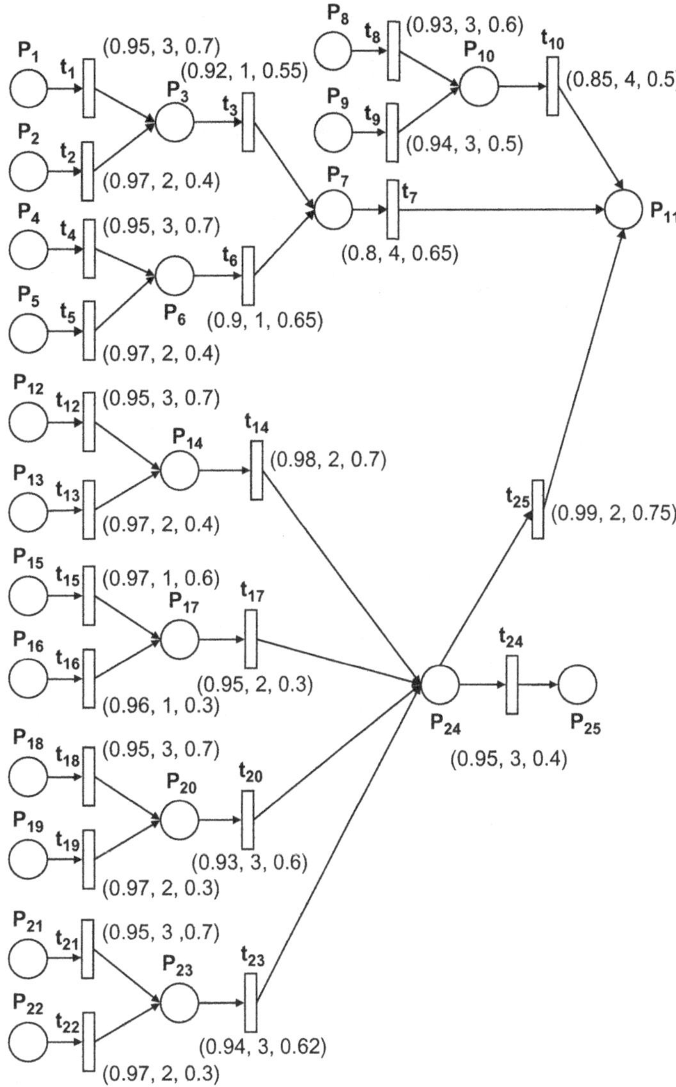


**Figure 9.5: Time fuzzy Petri nets model of "the reaction kettle process of polypropylene."**

Figure 9.7 describes the abnormal forward deduction process. The deduced results show that the proposition "cooling valve opening regulation is smaller" will trigger the proposition "jacket cooling water flow is smaller" after three unit time, the proposition "reaction kettle temperature is higher" after five unit time, and the proposition "reaction kettle pressure is higher" after seven unit time. And the credibility of the proposition "reaction kettle pressure is higher" is 0.74 when the confidence level of the initial proposition "cooling valve opening regulation is smaller" is 0.8.

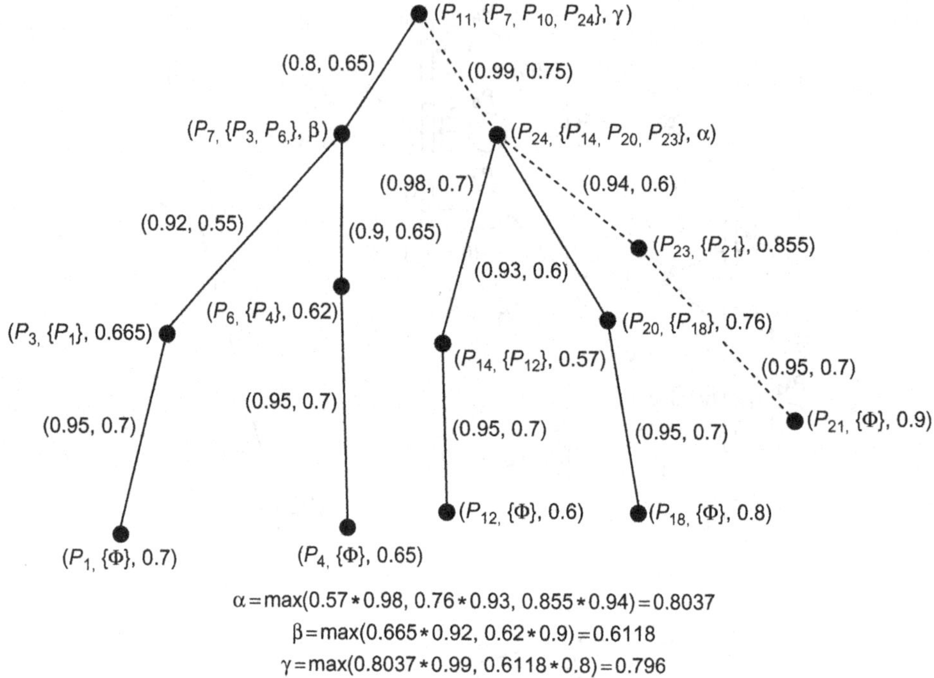


**Figure 9.6: The calculation of confidence level in abnormal reasoning.**

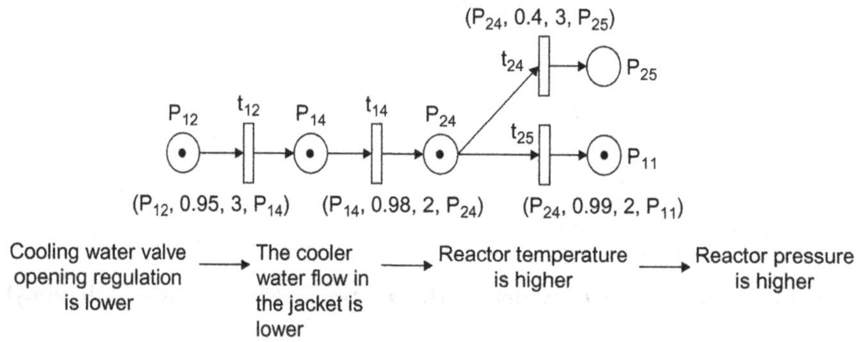


**Figure 9.7: The Petri model of the "cooling valve opening adjustment is smaller."**

The anomalous propagation process is simulated using the Matlab Stateflow module, as shown in Fig. 9.8. The green part on the simulation interface indicates that the initial anomaly is spreading, without the conflict and the deadlock, having the same propagating laws as the theoretical analysis.

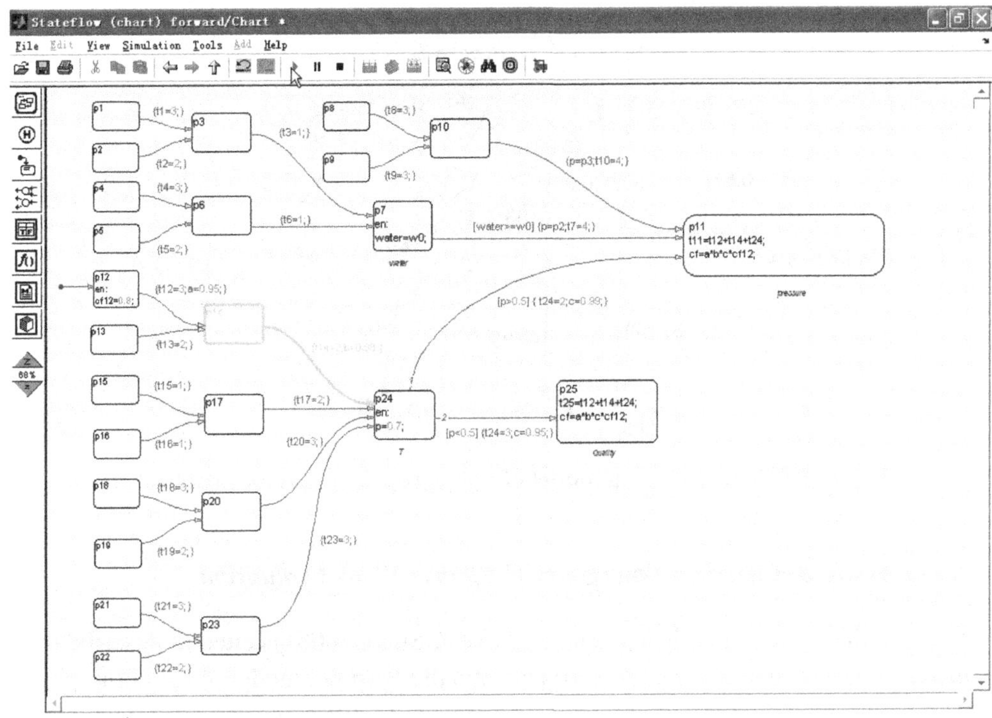


**Figure 9.8: The propagation process of an anomaly.**

**Table 9.1: The results of validation computational experiments for four sets of expert rules.**

| Expert | The Result of Abnormal Reasoning | Confidence Level | Reasoning Path |
|---|---|---|---|
| A | Catalyst A feed quantity is higher | 0.796 | $P_{11}$-$P_{24}$-$P_{23}$-$P_{21}$ |
| B | Propane feed valve open regulation is higher | 0.672 | $P_{11}$-$P_{7}$-$P_{6}$-$P_{4}$ |
| C | Catalyst A feed quantity is higher | 0.738 | $P_{11}$-$P_{24}$-$P_{23}$-$P_{21}$ |
| D | Catalyst A feed quantity is higher | 0.642 | $P_{11}$-$P_{24}$-$P_{23}$-$P_{21}$ |

We still use the example "polypropylene reaction kettle process" for discussion, aiming at the proposition "reaction kettle pressure is higher", setting four sets of expert rules, and developing the validation computational experiments based on artificial emergency early-warning systems. The experimental results are shown in Table 9.1, indicating that the reasoning results of the three sets of expert rules are the same, namely, the proposition "Catalyst A feed quantity is higher" is confirmed as the most likely cause to trigger the proposition "reaction kettle pressure is higher", although their reasoning credibility is a little different. The corresponding reasoning model is displayed in Fig. 9.9, the dotted line in which is the final reasoning path.

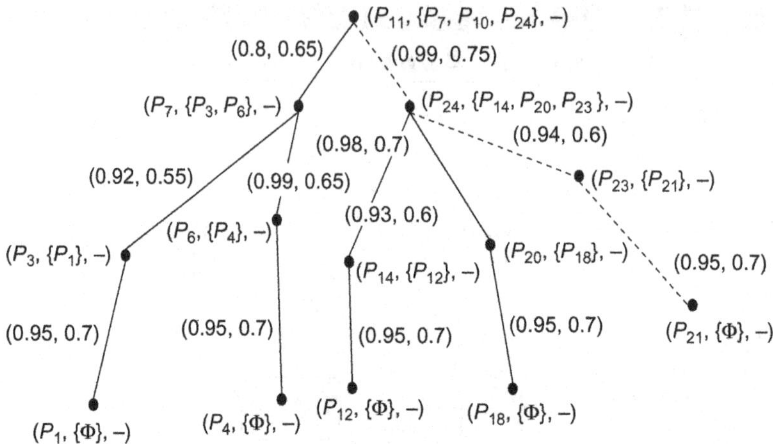


**Figure 9.9: The reasoning model of "polypropylene reactor process."**

### *9.4.2 Case Study 2: Chemical Emergency Response Plans Evaluation*

In this chapter, the switching of heat exchanger (Ye & Shao, 2005) is cited to describe the modeling and analysis process. Figure 9.10 presents the Petri net model of the entire process of switching operation, in which eight types of people with different attributes are given by parallel interface and eight branches display respectively the path of emergency scenario evolution for each type.

#### *9.4.2.1 Modeling of Human Behaviors*

People with different attributes often have different behaviors even in the same emergency scenario. Some personnel attribute–behavior mapping rules in the switching of heat exchanger established on the basis of fuzzy logic are listed below.

- If personality traits = excellent, professional knowledge = solid, operational capability = strong and physical fitness = good, then check = correct and drive = normal and operation of heat exchange = good
- If personality traits = excellent, professional knowledge = solid, operational capability = average and physical fitness = good, then check = correct, drive = normal and operation of heat exchange = average
- If personality traits = excellent, professional knowledge = solid, operational capability = poor and physical fitness = good, then check = correct, drive = normal and operation of heat exchange = bad
- If personality traits = excellent, professional knowledge = solid, operational capability = strong and physical fitness = bad, then check = wrong and drive = abnormal
- Operational capability = strong, physical fitness = bad and drive = abnormal then troubleshooting = rapid and operation of heat exchange = good

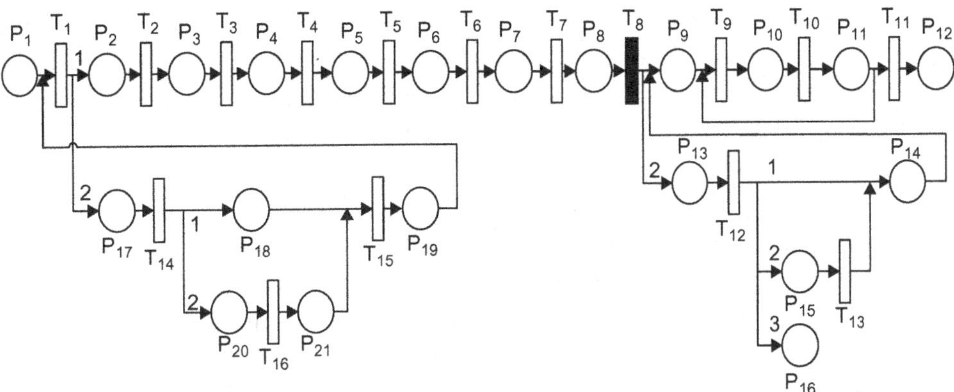


Figure 9.10: Petri nets model of emergency operation of switching heat exchanger.

Table 9.2: Twelve behavioral modes of DISC.

| a | Accuracy | d | Enthusiasm | g | Patience | j | Self-motivation |
|---|---|---|---|---|---|---|---|
| b | Cooperativeness | e | Friendliness | h | Persistence | k | Sensitivity |
| c | Efficiency | f | Independence | i | Self-confidence | l | Thoughtfulness |

Table 9.3: The relationship strength level between operator attribute and operating procedures.

| Language | Weaker | Weak | General | Strong | Stronger |
|---|---|---|---|---|---|
| Range of values | (0, 0.2) | (0.2, 0.4) | (0.4, 0.6) | (0.6, 0.8) | (0.8, 1) |

The mapping rules from attributes to natural behaviors only describe the behaviors that the specific attributes may map, but whether the mapped behavior is the one prescribed by the emergency response plan requires further judgment. The behavioral matrix is defined as $\varphi(W,A,B) * \lambda > T$, where A represents the personnel attributes matrix because the same post is held by more than one person. By referring to the operation evaluation factors for the petrochemical posts proposed by M.-Y. Wang (2006) and considering the required personnel attributes for the switching operation of the heat exchanger, only four factors, namely personality traits, professional knowledge, operational capability, and physical fitness, are adapted. Personality traits are quantified by the character strength of DISC. Table 9.2 displays 12 types of character modes of DISC. Character strength is assumed as general, average, and high, which is denoted by numbers from 0.5 to 1. The other three factors are also denoted by numbers from 0 to 1, representing strong or weak ability, rich or poor knowledge, good or bad physique, respectively.

B is the relationship strength matrix between personnel attributes and emergency response plans. It is divided into five levels in accordance with experience: weaker, weak, general, strong, and stronger. The responding numeric scope is as shown in Table 9.3.

W is the weight matrix of attributes, denoting the weights of attributes in a specific operation. Then behavioral matrix $\varphi$ is calculated. If conditions meet $\varphi(W,A,B) * \lambda > T$, behaviors as prescribed by the plan will be assumed to occur, in which $\lambda$ is a random number between (0,1) and $T$ is a behavioral constant assigned by experts.

### 9.4.2.2 Constructing Artificial System

The emergency operation of switching the heat exchanger consists of two steps: start-up and heat exchange. The process and equipment model to be established is a one-dimensional dynamic equation describing the heat exchange process, as shown in Eq. (9.1). The heat exchange operation is mainly aimed to cool the potassium phosphate solution from 65° to 32°, and warm the cooling water at a flow rate of 18,441 kg/h from 20° to 34.728°. The behavior in the operation of heat exchange is usually produced under the constraints of operational plans, personnel attributes, and dynamic parameters calculated by Eq. (9.1). The personnel model only illustrates the impact of emergency response plans and attributes on behaviors.

$$F_1 \mathrm{Cp}_1 (T_{1\mathrm{out}} - T_{1\mathrm{in}}) + \mathrm{Cp}_1 M_1 \frac{dT_{1\mathrm{out}}}{dt} = KA\Delta T_{\mathrm{m}}$$

$$F_2 \mathrm{Cp}_2 (T_{2\mathrm{out}} - T_{2\mathrm{in}}) + \mathrm{Cp}_2 M_2 \frac{dT_{2\mathrm{out}}}{dt} = -KA\Delta T_{\mathrm{m}}$$

$$\Delta T_{\mathrm{m}} = \frac{(T_{2\mathrm{in}} - T_{1\mathrm{out}}) - (T_{2\mathrm{out}} - T_{1\mathrm{in}})}{\ln \frac{(T_{2\mathrm{in}} - T_{1\mathrm{out}})}{(T_{2\mathrm{out}} - T_{1\mathrm{in}})}} \tag{9.1}$$

In this formula, overall heat transfer coefficient $K$ is 924.8 kJ/(m$^2$ h/°), thermal transfer area $A$ is 13.88 m$^2$, and specific heat capacities $\mathrm{Cp}_1$ and $\mathrm{Cp}_2$ are 4.183 and 3.9 kJ kg$^{-1}$ K$^{-1}$, respectively.

Thus, we must establish a hierarchical Petri nets model of the emergency response plan to provide a carrier for the interaction between human and equipment.

1. System layer: This models the standard operating procedures of switching heat exchanger, as shown in Fig. 9.11. Table 9.3 describes the meanings of places and transitions in Fig. 9.11. $T_1$ and $T_8$ stand for abstract transitions in Fig. 9.11, which are identified with black vertical boxes that represent the checking of equipment before start-up and high-point exhaust, respectively. Based on the possible branches, the operation layer model is constructed. Rhombic places $Q_{11}$ and $Q_{12}$ realize the information exchange between the system layer and the operation layer.
2. Operation layer: Figures 9.12 and 9.13 show the operation layer model of checking and exhaust operations. Input place $P_1$ conducts elementary behavior inference and validation. After the firing of transition $T_1$, the branches produced are the possible actions. For

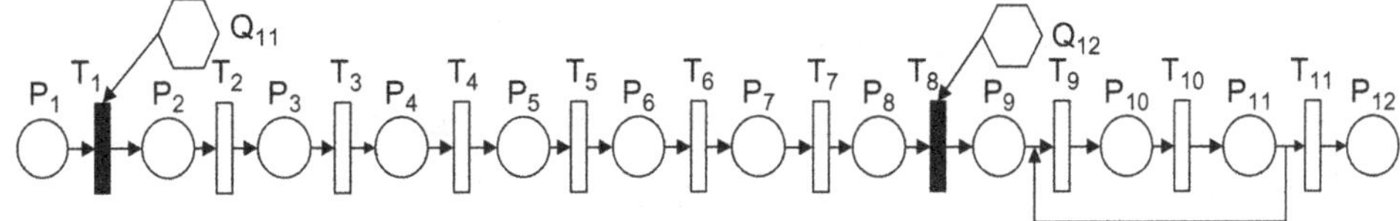


**Figure 9.11: Petri nets model of system layer.**

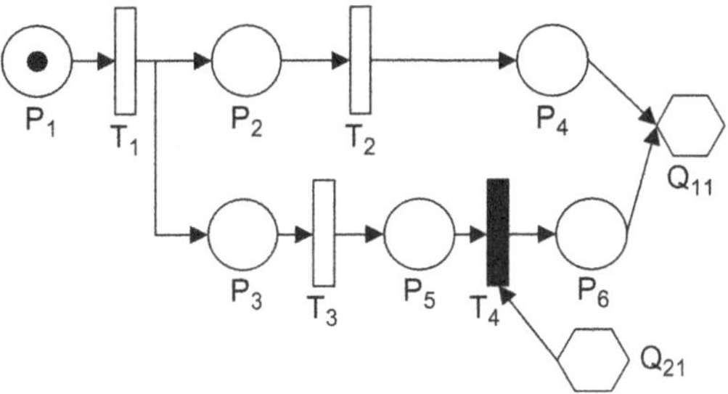


**Figure 9.12: Operation layer model for transition $T_1$ in Fig. 9.11.**

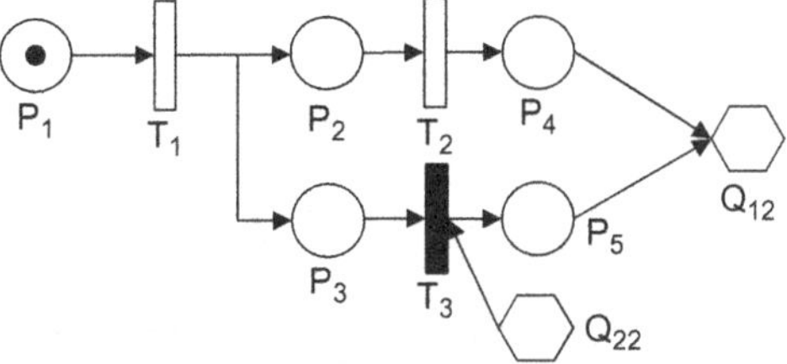


**Figure 9.13: Operation layer model for transition $T_8$ in Fig. 9.11.**

example, branches $P_1$–$P_2$–$P_4$ and $P_1$–$P_3$–$P_5$–$P_6$ illustrate whether the result of checking is correct or wrong, respectively. Abstract transitions $T_4$ and $T_3$ stand for equipment status diagnosis, which will be unfolded in the inference layer. Rhombic places $Q_{21}$ and $Q_{22}$ realize the information exchange between operation layer and inferential layer.

3. Inference layer: This models the process of equipment status diagnosis. The abstract transitions in Figs 9.12 and 9.13 unfold as shown in Figs 9.14 and 9.15, which illustrate the diagnostic process of system abnormalities resulting from the heat exchanger leak and omission of high-point exhaust, respectively. Input place $P_1$ of the inference layer completes behavior inference and validation. However, inference involves more rules and is more difficult. In the inference of input place $P_1$ in Fig. 9.14, for example, the equipment state is that the outlet temperature of the heat exchanger is higher within the normal flow. After the firing of transition $T_1$, the branches produced are the possible actions. Branches $P_1$–$P_2$, $P_1$–$P_3$–$P_2$, and $P_1$–$P_4$ stand for the behaviors where the fault is resolved by one

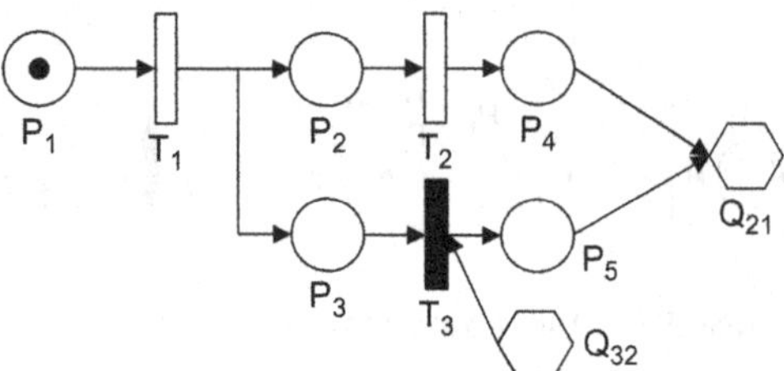


**Figure 9.14: Inference layer model for transition $T_4$ in Fig. 9.12.**

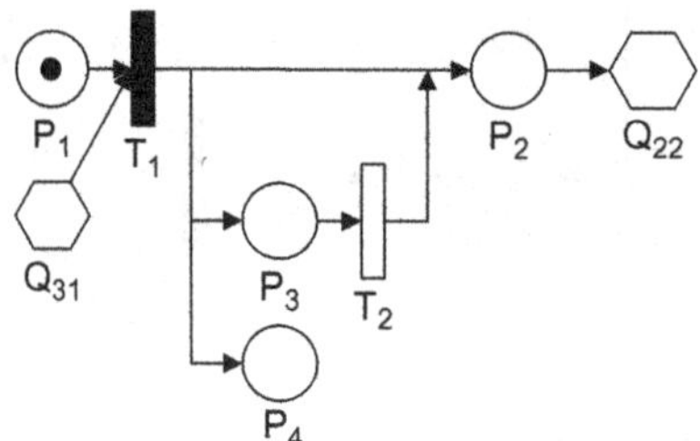


**Figure 9.15: Inference layer model for transition $T_4$ in Fig. 9.12 and transition $T_3$ in Fig. 9.13.**

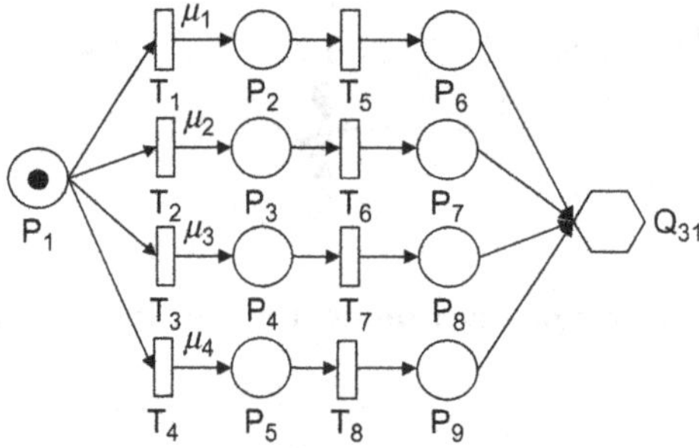


**Figure 9.16: Base layer model of transition $T_1$ in Fig. 15.**

diagnosis, two diagnoses, and remains unresolved, respectively. Abstract transitions $T_3$ and $T_1$ stand for the fuzzy reasoning process unfolded in the base layer. Rhombic places $Q_{31}$ and $Q_{32}$ realize the information exchange between the inference layer and the base layer.

4. Base layer: A fuzzy Petri nets model is constructed for abstract transition $T_1$ in Fig. 9.15 according to the rules as follows. Rule confidence is noted by $\mu_i$, as shown in Fig. 9.16.

If the exhaust valve is not open, then the outlet temperature is higher, $\mu_1 = 0.8$
If the heat exchange leaks, then the outlet temperature is higher, $\mu_2 = 0.7$
If the pump is faulty, then the outlet temperature is higher, $\mu_3 = 0.6$
If the temperature controlling device is faulty, then the outlet temperature is higher, $\mu_4 = 0.4$

Table 9.4: The attributes of eight types of operator in Table 9.3.

| Category | Attribute | | | |
|---|---|---|---|---|
| | Personality Traits | Professional Knowledge | Operational Capability | Physical Fitness |
| I | Excellent | Solid | Strong | Good |
| II | Excellent | Solid | Strong | Bad |
| III | Excellent | General | Common | Good |
| IV | Excellent | General | Common | Bad |
| V | Good | Solid | Common | Good |
| VI | Good | Solid | Common | Bad |
| VII | Good | General | Strong | Good |
| VIII | Good | General | Strong | Bad |

Every transition in the basic layer is an atom transition that is not unfolded. They stand for the basic judgments of the operator, such as alarm confirmation, obtaining parameter values, parameter variation confirmation, and equipment operation status confirmation.

In artificial systems, human behaviors and the development of emergency response plans are featured with obvious social subjectivity of complex systems, which cause greater uncertainty. Therefore, human behavior and its constraints including corporate culture, management philosophy, and emergency response plans will be the main content of computational experiments. Through computational experiments, we are able to analyze the typical emergency behaviors of the operator and determine whether the postconfiguration in emergencies is rational, thus providing a reference for developing reasonable plans.

The model of human behaviors only estimates whether the operating behavior is what the plans require, but whether the emergency response is successful mainly depends on whether the response time is within the limit time $T$. Thus, computational experiments on emergency response processes for each type of people are required to be done on the Artificial System Simulation Platform to compute the average response time for each type of people to complete the entire operation. After running $N$ times, if the response time of $j$ is smaller than limit time $T$, the emergency response is deemed successful and $m{+}=1$ ($m$ is the frequency of success). Otherwise, the emergency response fails. Finally, the emergency response reliability $R$ for each type of people is calculated using the formula $m/N$.

Then, SPSS is used to design the orthogonal experiment and analyze the results. Compared with conventional methods, orthogonal design can significantly reduce the number of experiments and complexity of analysis, and a combination of optimal factors can be obtained. The orthogonal experimental design Table L8($2^7$) not considering interaction is shown in Table 9.5, with each factor defining two levels.

Table 9.5: The results of orthogonal computational experiment.

| No. | Type | A | B | C | D | | | | Emergency Reliability |
|---|---|---|---|---|---|---|---|---|---|
| 1 | I | 1 | 1 | 1 | 1 | 1 | 1 | 1 | 98 |
| 2 | II | 1 | 1 | 1 | 2 | 2 | 2 | 2 | 95 |
| 3 | III | 1 | 2 | 2 | 1 | 1 | 2 | 2 | 70 |
| 4 | IV | 1 | 2 | 2 | 2 | 2 | 1 | 1 | 72 |
| 5 | V | 2 | 1 | 2 | 1 | 2 | 1 | 2 | 65 |
| 6 | VI | 2 | 1 | 2 | 2 | 1 | 2 | 1 | 64 |
| 7 | VII | 2 | 2 | 1 | 1 | 2 | 2 | 1 | 76 |
| 8 | VIII | 2 | 2 | 1 | 2 | 1 | 1 | 2 | 77 |

## 9.5 Conclusions

An important issue in emergency management is how to ensure the effectiveness of emergency rescue. This chapter proposes a novel emergency management framework of Parallel Emergency Management, based on an ACP approach. Taking chemical emergency management as an entry point, we apply the framework in chemical emergency early warning and emergency plans evaluation. We have made three important contributions to this work. First, we have identified a general set of key features that can be used to characterize the emergency behaviors in chemical emergency management. Second, we have presented a new causal link between human behaviors and equipment anomaly in emergency early-warning diagnosis. Third, we have proposed a dynamic technique to evaluate the operational effectiveness of plans in the emergency rescue process.

To validate the effectiveness of this method, we build up a corresponding artificial emergency system about typical chemical emergency operations, illustrating and analyzing the process of chemical early-warning and plans evaluation. In plans evaluation, we set a variety of personnel attributes to carry out the rescue actions, evaluating the operational effectiveness of the plan by the operational reliability and optimizing the personnel post configuration. All these results show the rationality and feasibility of the method.

## References

Ai, H.-W. (2008). *Study of railway emergency rescue flow based on stochastic Petri nets* (Master's thesis, Beijing Jiao tong University).

Andersen, V. & Rasmussen, J. (1987). Decision support systems for emergency management. In H. B. F. Gow & R. W. Kay (Eds.), *Emergency planning for industrial hazards* (pp. 127–148). Elsevier Applied Science, Ispra Establishment.

Boppana, V. R. (2005). Disaster management plan for chemical process industries case study: Investigation of release of chlorine to atmosphere. *Journal Loss Prevention in the Process Industries, 13*(1), 57–62.

Burns, K. J., Robinson, K., & Lowe, E. G. (2007). Evaluation of responses of an air medical helicopter program during a comprehensive emergency response drill. *Air Medical Journal, 26*(3), 139–143.

Chang, D.-Q. (2003). *The GIS research of major accident warning and emergency system in petrochemical enterprise* (Master's thesis, Northeast University).

Huang, D.-J. (2006). The evaluation method research of emergency plans in petrochemical enterprise. *Oil Chemical Safety Technology, 22*(5), 17–19.

Kaji, A. H. & Lewis, R. J. (2008). Assessment of the reliability of the Johns Hopkins/Agency for Healthcare Research Quality Hospital Disaster Drill Evaluation Tool. *Annals of Emergency Medicine, 52*(3), 204–210.

Lanrian, D. M. & Berke, P. (2004). Evaluating plan implementation. *Journal of Planning Association, 70*(4), 86–94.

Liu, B.-J. (1994). The error factors of human in emergency operation. *Nuclear Power Project, 15*(3), 199–203.

Liu, Z.-J. (2008). *The research of industrial process condition monitoring method based on Petri nets* (Ph.D. thesis, Beijing Chemical University).

Mark, S. & Westhues, A. (1989). A developmental stage approach to program planning and evaluation. *Journal Evaluation Review, 13*(1), 56–77.

Rasmussen, B. & Gronberg, C. D. (1997). Accidents and risk control. *Journal of Loss Prevention in the Process Industries, 10*, 325–332.

Simon, H. A. (1957). *Models of man*. New York, NY: John Wiley and Sons.

Wang, F.-Y. (2007a). The parallel system method and its application research for PEMS of Emergency management. *Chinese Emergency Management, 1*(12), 22–28.

Wang, F.-Y. (2004). Parallel systems methods for control and management of complex systems. *Journal of Control and Decision, 19*(5), 485–489.

Wang, F.-Y. (2007b). Toward a paradigm shift in social computing: The ACP approach. *IEEE Intelligent Systems, 22*(3), 65–67.

Wang, F.-Y. (2008). Toward a revolution in transportation operations: AI for complex systems. *IEEE Intelligent Systems, 23*(3), 1–8.

Wang, M.-Y. (2006). Study on the operator competence model of petrochemical enterprises. *Journal of Petroleum Chemical Technology, 22*(1), 58–62.

Wang, X.-Z. (2005). The application of security warning system in the chemical production. *Alkali Industry, 3*(3), 43–45.

Ye, B. & Shao, J.-F. (2005). Chemical emergency response plans of Maoming Ethylene. Maoming Ethylene plants internal data.

Yu, Y.-Y. (2007). The operable research of emergency plans based on network plan. *Journal of Public Management, 4*(2), 100–107.

Zhai, X.-M. (1998). The review and prospect of emergency research. *System Engineering Theory and Practice, 2*(7), 17–24.

Zhang, W. (2008). The research and application of Security Warning System in Coal Chemical Enterprise. *Shandong Coal Technology, 5*(3), 83–85.

Zhang, Y. (2004). The evaluation of emergency plans based on Comprehensive Fuzzy evaluation method. *China Management Science, 12*(2), 154–156.

CHAPTER 10

# Bus Arrival Prediction and Trip Planning for Better User Experience and Services

**Feng Li[*], HongBin Lin[*], Yu Yuan[*], ChangJie Guo[*], Wanli Min[†], and XinYu Zhao[‡]**

*[*]IBM Research – China, ZGC Software Park, Beijing, China, 100193*
*[†]IBM Singapore, Changi Business Park, Singapore, 486072*
*[‡]Puhua Guaranty Limited Company, Beijing, China, 100032*

## 10.1 Introduction

Traffic plays an important role in modern urban society. Based on the report from the United Nations Population Division, almost 64 billion people will live in urban areas in 2050 and the annual growth rate is around 1.5% (World Urbanization Prospects, 2007). Obviously, it will cause an increase in trip demand. Because of the limitations of traffic resources, this increment will lead to urban traffic congestion. Traffic congestion, which cost USD 87.2 billion in 2007, is only getting worse, according to a new report from the Texas Transportation Institute (Urban Mobility Report and Appendices, 2010). According to some other studies (Horizon Research Consultancy Group, 2009), traffic congestion costs USD 50.3 per month for each person in Beijing, USD 39.8 in Guangzhou, and USD 38.0 in Shanghai. In order to relieve the congestion, governments all around the world provide funding and support to develop public traffic systems and build traffic applications such as subway systems, signal control systems, traffic information management systems, and electronic toll collection systems.

Public transport is a shared passenger transportation service that is available for use for the public. Public transport modes include buses, trolleybuses, trams and trains, "rapid transit" (metro, subways, undergrounds, etc.), and ferries. In large cities, public transport has already become an extremely important part of people's daily life. For example, most office workers choose to take the bus in peak time, e.g., 8 AM and 6 PM. In 2010, the contribution rate of public transport in China is about 20% (= public transport trip number / total trip number). Because cars removed from the road through public transport options translate to less congestion and faster speeds for remaining motorists, the general public can potentially reduce total transport costs by improving public transport. Besides relieving traffic congestion, public transport use benefits the environment, saves energy, and increases economic development.

Service Science, Management, and Engineering.


DOI: 10.1016/B978-0-12-397037-4.00010-7

A report provided by the Brookings Institution and the American Enterprise Institute mentioned that public transport in the United States used approximately half the fuel required by cars, SUVs, and light trucks. A 10% increase in public transportation use in the top five cities of the United States would save more than 85 million gallons of gasoline per year. Using public transport can reduce an individual's carbon footprint. If an individual were to use public transport rather than making a 20-mile trip by car, this would result in a net reduction of 4,800 lbs/year $CO_2$ emission(American Public Transportation Association, 2010).

Considering the above benefit, governments, public transport authorities around the world, and research departments are developing public transport systems. A key problem in the development process is to provide an adequate public transport service or enhance the service level. This problem is also faced by public transport planners and operators. As we know, a good service can attract more people to take public transport for their trips. For example, an accurate prediction of vehicle arrival times and the timely dissemination of the information to transit users may reduce their waiting time, thus improving the service quality. With the advent of advanced public transportation systems (APTS), traveler information systems (TIS) should have the capability to provide timely updated information such as predicted vehicle arrival and departure times (e.g., TRAVELINK, PA/CIS, and AZTech). This information can be disseminated to travelers before trips or en route through a variety of media (e.g., cell phone, kiosks, and Internet), allowing them either to arrange their departure times or to coordinate transfers at a lower cost in terms of travel time (Abdelfattah & Khan, 1998).

Bus trip planning services are another way to improve public transport services. In fact, there already exist many bus trip planning services such as Google Maps, Bing Maps, Baidu Maps, and so on, but the problem with these services is that they are based on static data and have few objectives for trip planning. When we search from origin A to destination B in Google Maps, the planning objective is to minimize the traveling time or transfer times. Bus routes generated by this map are marked with gray thick lines in Fig. 10.1. If traffic congestion happens in the circled area, these routes will just push more people into congestion. In this traffic case, the red route may be the best one. Other maps also have the following problems: user objective is limited to sort by length or transfers and the given routes are not based on real-time traffic data. In recent years, automatic vehicle location (AVL) systems have been universally popular. With global position system (GPS), general packet radio service (GPRS) or radio frequency identification (RFID), the bus location can be found and published in real time. Figure 10.2 shows a typical scenario for bus arrival time and traffic data in real time. Based on these collected data, many researchers have studied traffic and bus arrival time prediction (Seema & Alex, 2010; Sun, Zhang, & Yu, 2006). With reasonable prediction accuracy, these data can be used as the basis for a bus trip planning service.

For the aforementioned reasons, we will present a statistical approach to predict public bus arrival times and a system of bus trip planning to improve the public bus service

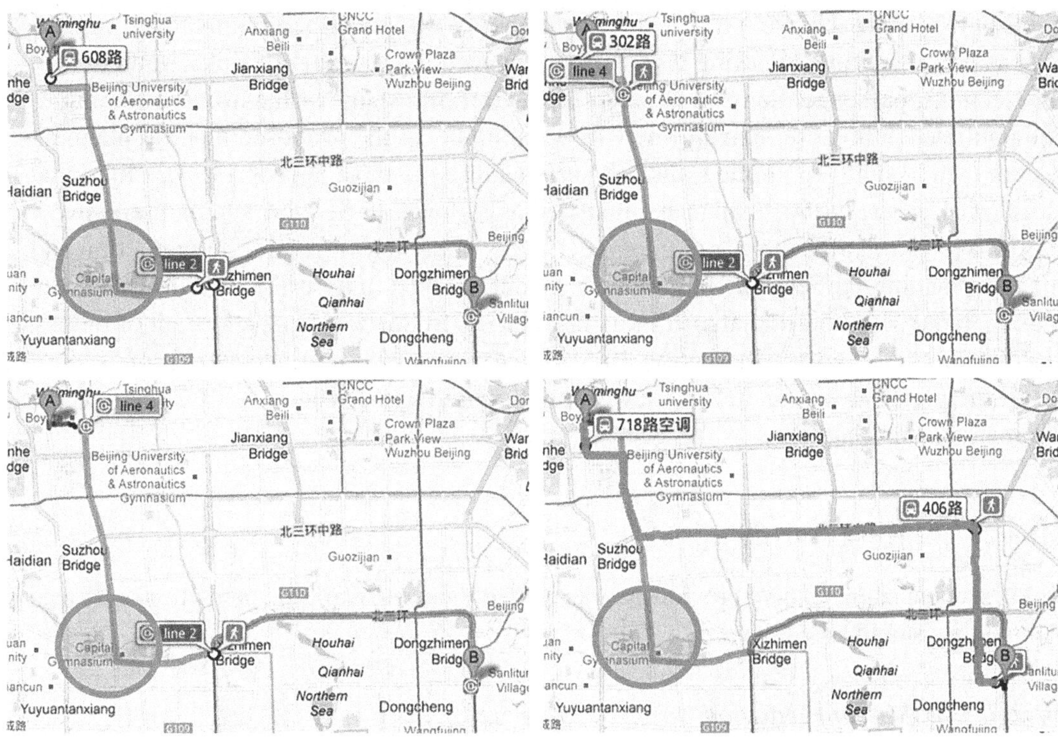


Figure 10.1: Results by Google Maps.

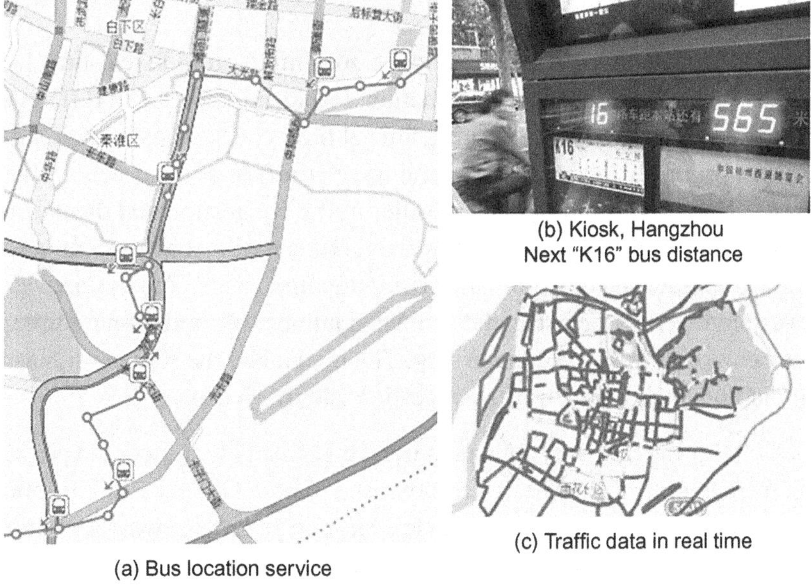

秦淮区
(b) Kiosk, Hangzhou
Next "K16" bus distance
(c) Traffic data in real time
(a) Bus location service


Figure 10.2: Bus arrival time and traffic data in real time.

level. In this approach, we will consider a number of factors affecting bus travel time, such as departure time, working day, current bus location, number of links, number of intersections, passenger demand at each stop, and traffic status of the urban network. The predicted public bus arrival time will be an input for the proposed bus trip planning system, which can help public transport users choose the most appropriate bus lines and transfers. The rest of this chapter is organized as follows. In Section 10.2, a literature review will summarize the current approaches for public bus arrival time prediction and bus trip planning. In Section 10.3, a linear model is given to describe the bus arrival time and a parameter calibration algorithm is introduced. In Section 10.4, a system of bus trip planning is proposed. In Section 10.5, a prototype and experiments are proposed to verify the practicability and efficiency of the proposed approaches. Finally, Section 10.6 summarizes this chapter.

## 10.2 Literature Review

In this section, some relative previous works related to bus arrival time prediction and bus trip planning are summarized.

### 10.2.1 Bus Arrival Prediction

The bus arrival time prediction models can be classified into the following three items: mathematical algorithms, Kalman Filter model with historical data, and artificial neural network (ANN) model.

In 1999, Lin and Zeng developed a mathematical algorithm to provide real-time bus arrival information. They considered schedule information, bus location data, the difference between scheduled and actual arrival time, and waiting time at time-check stops in their algorithm. The algorithm was primarily developed for rural traveler information systems where no congestion exists. Their algorithm did not consider traffic congestion and dwell time at stops. In the same year, Ojili (1999) developed a bus arrival time notification system in College Station. The model divides the bus route into 1-minute time zones. The bus arrival time at a given stop was predicted by counting the estimated number of the 1-minute time zones between the current location and the given stop. The model had the same issues as that of Lin and Zeng's. It did not consider the traffic congestion and dwell time.

Wall and Dailey (1999) were the first authors to use the Kalman Filter model to predict the bus arrival time. In their algorithm, they used a combination of both GPS data and historical data; they used a Kalman filter model to track a vehicle's location and a statistical estimation technique to predict travel time. It was found that they could predict bus arrival time with less than 12% error. However, they did not explicitly deal with dwell time as an independent variable. In 2003, Shalaby and Farhan proposed another bus travel time prediction model by using a Kalman filtering technique. In their model, they considered the passenger information at each bus stop. However, they predicted the dwell time only at time check points and not at every bus stop.

Due to its ability to solve complex nonlinear relationships, artificial neural network (ANN) models have been used to model the transportation problems. The ANN models had shown better results than the existing link travel time techniques, including a Kalman filtering model, historical average model, auto-regressive integrated moving average (ARIMA) model, and exponential smoothing model. In 2002, Chien et al. developed an artificial neural network model to predict dynamic bus arrival time in New Jersey. Considering the back-propagation algorithm unsuitable for on-line application, the authors developed an adjustment factor to modify their travel time prediction by using recent observed real-time data. However, the dwell time and the scheduled data were not considered in their model (Steven, Ding, & Wei, 2002). In 2004, Jeong and Rilett provided a historical data-based model, regression models, and ANN models to predict bus travel time by considering traffic congestion, schedule adherence and dwell times at stops. In 2006, Ramakrishna et al. proposed a multiple linear regression and an ANN model to predict bus travel times. In their model, they considered real-time GPS data of bus locations (Ramakrishna, Ramakrishna, Lakshmanan, & Sivanandan, 2006). In 2009, Suwardo et al. proposed a statistical neural network model to predict the bus travel time in mixed traffic, while considering bus travel time, distance, average speed, number of bus stops, and traffic conditions. In their paper, they assessed these factors and studied their mode of relationship between the factors and bus travel time (Suwardo, Madzlan, Ibrahim, & Oyas, 2009).

Among the aforementioned models, the artificial neural network model and statistical neural network model have shown greater promise than other models such as the Kalman Filter model, historical average model, auto-regressive integrated moving average (ARIMA) model, and exponential smoothing model (Li, Lin, Yuan, & Min, 2011). However, the parameters for these models are hard to determine because they need more historical data and will take more time. Although the models can provide relative bus arrival prediction times, it is hard for us to explain the mechanism for them. For these reasons, in this chapter we will provide a statistical approach to forecast bus arrival time.

### 10.2.2 Bus Trip Planning

Bus trip planning problems can be classified into three types:

A. *Static Network*
   In this time-independent network, the speed of each road is static, and we only consider road connectivity and length. This kind of network is suitable to describe walking or bicycling problems. Many classical algorithms can solve this problem, such as Bellman–Ford–Moore, Dijkstra, A*, Ant, Dual-network, and other improved algorithms (Zhang, Zhong, & Liu, 2009).

B. *Dynamic Network*
In this time-dependent network, the speed of each road is dynamic and affected by traffic congestion. This network exactly describes self driving and can be solved by the dynamic algorithms (Misra & Oommen, 2005).

C. *Complete Dynamic Network*
With dynamic network properties, there are also time-dependent waiting delay and interchange cost on each station. This network is suitable for bus transit and has been studied previously (Wu, Hartley, & Al-Dabass, 2005).

There are some drawbacks in the existing bus trip planning research:

1. *Data Completeness Requirement*
In a practical transport system, there are many key characters: map package provider, traffic data provider (bus trip planning), transport service provider, clients, and other transport-related departments. These different departments are often distributed at different geographical positions and communicate through the Internet, e.g., web services. In this system, the transport service provider plays a core role. It helps to integrate all data that include static GIS map data from map provider, real-time traffic data from traffic data provider (transport department), feedback from clients, and other correlative information. In order to communicate conveniently with other departments, the traffic data provider often stores the real-time data in a database system and may update it at any time. In the existing algorithms, it is necessary to download all these real-time data to a local machine before determining the shortest path. These real-time data include the arrival and departure times of each bus at each station, available seats, and so on. The data capacities that need to be downloaded from the network are shown in the gray area in Fig. 10.3(a)–(c). We call this requirement the data completeness requirement.
2. *Limited Objective*
In order to implement the algorithm efficiently, the existing algorithms usually limit their search objectives, such as shortest travel time and shortest transfer times. In existing map services, the search objective is limited to sort by time or transfer times. But users may need to search based on more properties of bus routes, for example, searching a bus route with air conditioning, or with available seats. According to many data collecting systems, for example, AVL and automatic passenger counting (APC), more and more properties related to the bus can be used. Our system enables users to customize their search objectives.
3. *Low Computational Efficiency*
Because the system will be accessed by the public, it may happen that hundreds of people access the service at the same time. In large cities, there are more than 2000 bus stations, and more than 800 bus lines. Based on such a large-scale transport network and route accessing count, existing algorithms are insufficient in their computational capability.

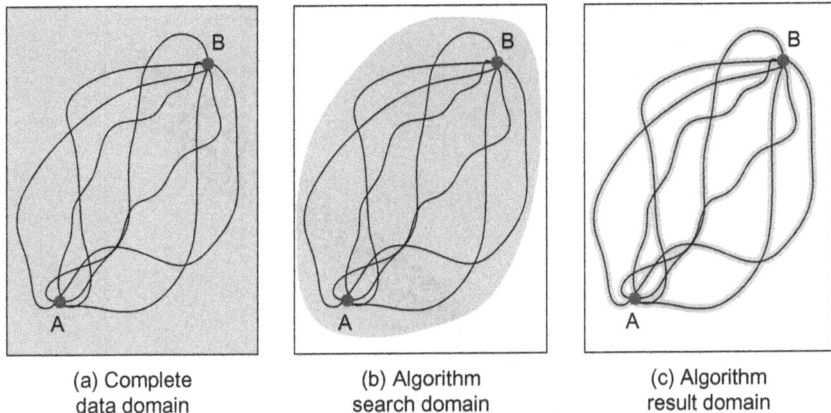


**Figure 10.3: Data domain that should be downloaded by different algorithms.**

So one purpose of this chapter is to propose a bus trip planning service system that is based on large-scale transport networks and real-time traffic data, and that enables users to customize their search objectives.

## 10.3 Bus Arrival Prediction

### 10.3.1 Factor Analysis

In a recent study of bus rapid transit, it was found that buses spend 5% of the time in traffic congestion, 20% of the service hours are stopped at intersections, 23% of the time on boarding and alighting passengers (dwell time), and the remaining 52% in traveling on the links of the urban network (Bus Rapid Transit, 2000). Obviously, the service hours spent at intersections and dwell time are the most important factors for bus delay. Therefore, in order to improve the accuracy of the bus arrival time prediction, we have to consider how to model the delay caused at bus stops and traffic signal intersections. In fact, in addition to the aforementioned factors, the departure time, bus drivers, bus location, and traffic conditions will also affect the accuracy of bus arrival time prediction.

As we know, different departure times mean different traffic conditions on the urban traffic network. For example, at midnight, the traffic condition on the network is very good, whereas in the morning peak time or afternoon peak time, there are lots of vehicles on the network and congestion occurs frequently. Meanwhile, at midnight, the number of passengers is fewer than that at peak time. In Fig. 10.4, we list the relationships between the travel time and departure time of a bus line in 1 day, where different geometric shapes mean different trips. From Fig. 10.4, we see that the travel times are different for different departure times. Therefore,

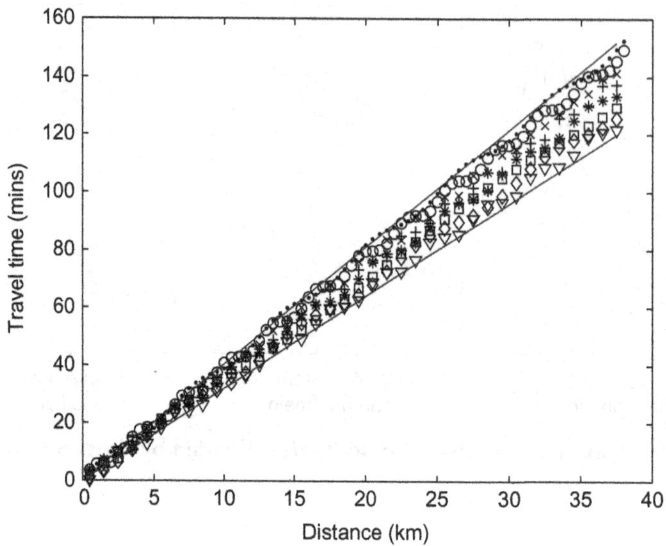


**Figure 10.4: Travel times for different departure times.**

departure time is a key factor for us to predict the bus arrival time at each bus stop, especially when we do not know the exact bus location information.

According to research by Toshiya et al., the drivers can be clustered into models of "rude," "normal," and "slow" (Hirose, Oguchi, & Sawada, 2004). For drivers in the "rude" model, they react quickly without considering potential possibility with which traffic conditions will change. For drivers in the "normal" model, they will react normally during their driving process, and for the "slow" drivers, they react slowly whilst driving. Normally, for the same driving environment, the "rude" driver will arrive earlier than "normal" and "slow" drivers, and the "slow" driver is the last to arrive. As a result, different characteristics of drivers of public buses will affect the bus arrival time at each bus stop.

Currently, most of the buses are equipped with an on-board unit (OBU) that can provide exact location information. It is easy for us to track the bus, calculate its average speed, and find its location by using this information. It will help us to capture the distance between the current location and the bus stops and other information between them, such as number of intersections. In the travel time prediction model, the location information is very helpful for us to calibrate, adjust, or modify the forecasting errors.

Intersections on the road, especially those with signals, are core parts of an urban network. Those signals have huge impact on the traffic flows. For example, the vehicles on the main street with green waves do not spend time waiting for green lights, whereas the vehicles on the crossing streets have to spend more time. These effects will lead to different service hours of public buses at intersections. Besides signals, the complex environment of intersections will also affect traffic flows. For example, at an intersection with mixed modes for passengers, jinrikishas (a small two-wheeled cart for one passenger) and vehicles, the vehicles have to wait until all passengers and jinrikishas have crossed the street. For these reasons, the conditions of intersections on bus routes are important for us to predict the arrival time at each stop.

As we mentioned in the above sections, the dwell time is another key factor that will affect accuracy of the bus arrival time prediction. At each bus stop, passengers will get on and off the public buses. If the number of passengers is relatively large, the bus has to spend more time waiting for all passengers to get on or off. Conversely, the bus will spend less time if the number of passengers is less.

From a statistical view, the buses will spend about 50% of their time traveling on the urban network for a trip from their original station to the final station. Therefore, the traffic conditions between the original station and the final station are another key factor affecting the bus travel time. They will also affect the arrival time at each bus stop. In fact, the traffic conditions include traffic speed, traffic volume, and traffic occupancy of directed links of the urban network, which results in different average speeds of the buses. By analyzing the historical data, we get the following average speed information at different times. In Fig. 10.5, we list the average speed information over a 2-week period. From the figure, we can see that the bus arrival time at each stop had strong seasonality. For different weeks, all of these trips are very similar. The trips on weekdays (from Monday to Friday) are different from the weekend (Sunday and Saturday). For weekdays, the trips are almost the same. Meanwhile, in a whole day and for different weekdays and the weekend, the peak times and off-peak times are different.

In conclusion, the main factors that affect the public bus arrival time are traffic conditions (including the departure time, weekday, and peak hour), the dwell time or sequences and number of bus stops, the number of intersections, and other factors. In this chapter, we will consider the first three key factors.

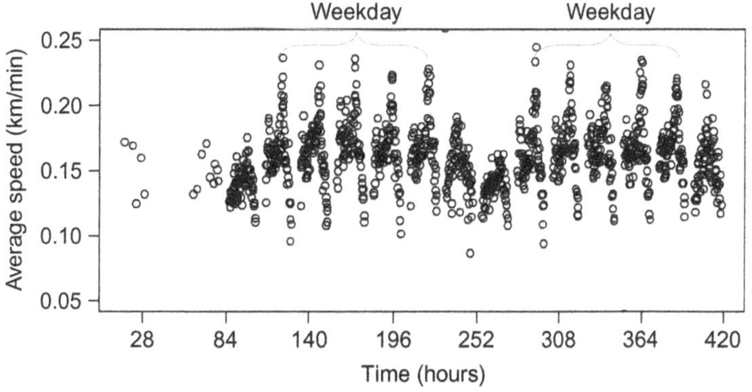


**Figure 10.5: Average speed.**

### 10.3.2 Linear Model and Parameter Calibration

In order to better understand data on public bus arrival time, we use the following figure to demonstrate the journey of a public bus. In Fig. 10.6, the bus will start from the original bus station, travel on the directed links, change direction at the intersections, stop and start at internal bus stops, and finally arrive at its destination. We call the whole traveling process from origin to destination a trip.

By analyzing the historical data, we found that the public bus arrival time consisted of two main parts: linear parts and residuals. In Fig. 10.7, we show the relationship between the bus

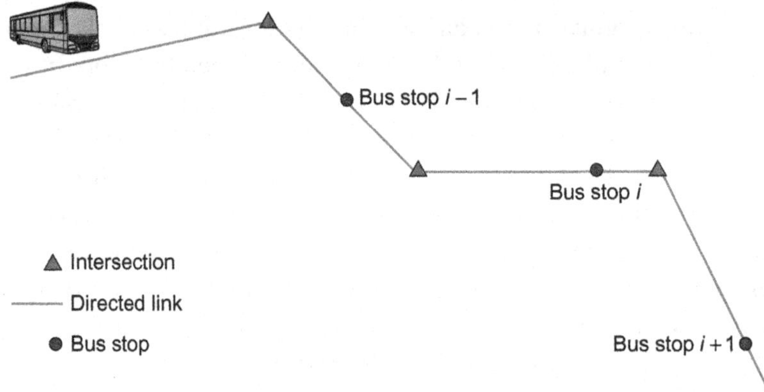


**Figure 10.6: Trips for a public bus.**

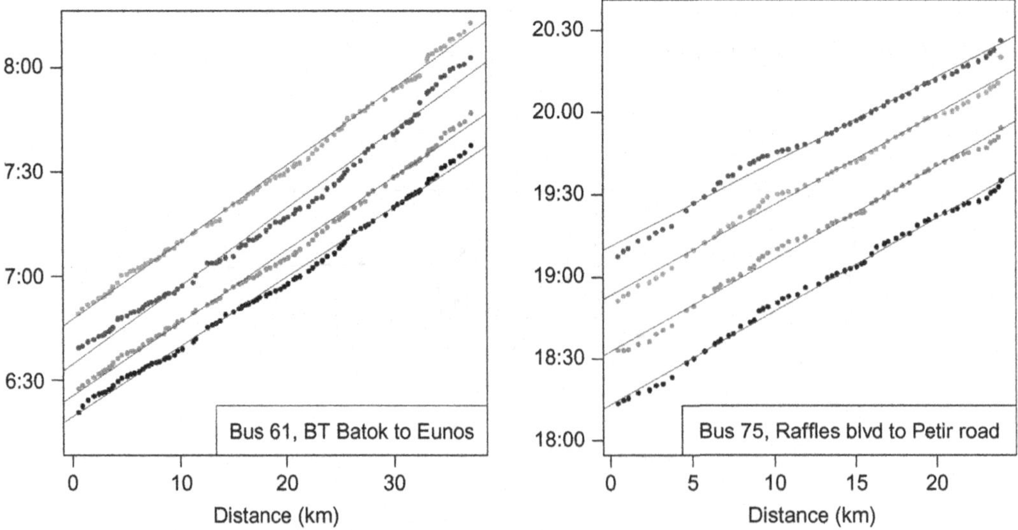


**Figure 10.7: Bus travel time versus distance.**

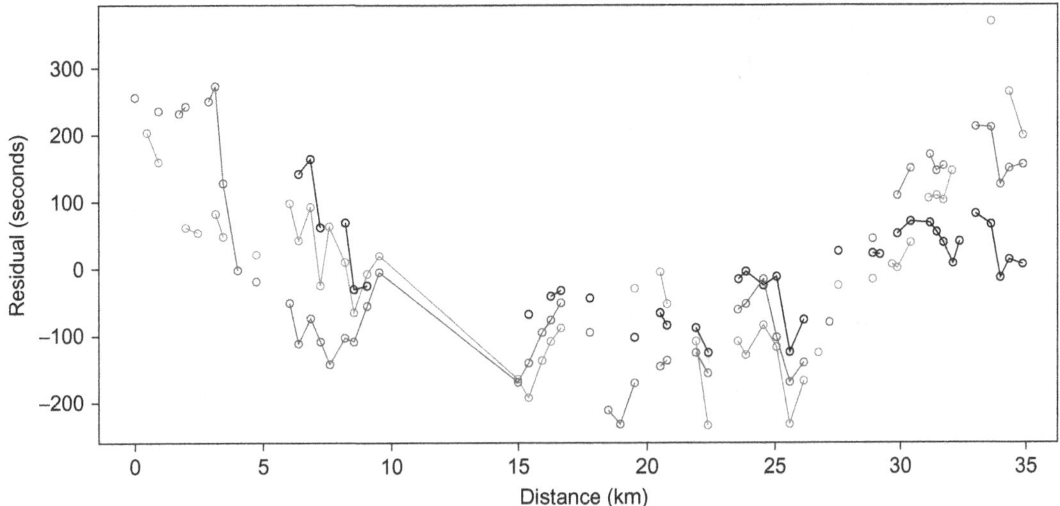


**Figure 10.8: Three consecutive days' trip residual.**

arrival time and the traveled distance. Both of these bus lines (No. 61 and No. 75) show a strong linear relationship between travel time and distance.

Besides the linear parts, Fig. 10.7 also demonstrates some residuals. Obviously, if we remove the main part (linear part), we can obtain Fig. 10.8 for those residuals.

Then the problem in forecasting the public bus arrival time is how to describe the linear part and residual part while considering the above analyzed factors such as traffic conditions, dwell time (or sequences and number of bus stops), and intersections.

From Fig. 10.6, we can easily see that the arrival time of the $i + 1$th bus stop is equal to the sum of the arrival time of the $i$th bus stop and the traveling time between those two bus stops. Naturally, we can use the following linear model to describe it:

$$\Gamma_{i+1}(t) = \Gamma_i(t) + \sum_{k=1}^{\text{No.of Links}} \alpha_k \tau_k(\mu_k(t)) + D_i(t) + \sum_{j=1}^{\text{No.of Intersections}} \beta_j \sigma_j(\eta_j(t)), \qquad (10.1)$$

where $\Gamma_i(t)$ denotes the predicted arrival time at bus stop $i$ at time $t$. *No. of Links* explains the link numbers between bus stops $i$ and $i + 1$. $D_i(t)$ denotes the dwell time at bus stop $i$ at time $t$. *No. of Intersections* shows the number of intersections between bus stops $i$ and $i + 1$. $\tau_k(t)$ represents travel time of the $k$th link from bus stop $i$ to $i + 1$ at time $t$. $\alpha_k$ represents the parameters for $\tau_k(t)$. $\sigma_j(t)$ denotes the estimated service time of the $j$th intersection from bus stop $i$ to $i + 1$ at time $t$. $\beta_j$ represents the parameters for $\sigma_j(t)$. Obviously,

$$\mu_1(t) = \Gamma_i(t),$$
$$\mu_2(t) = \mu_1(t) + \tau_1(\mu_1(t)),$$
$$\ldots,$$
$$\eta_1(t) = \Gamma_i(t) + \tau_1(\mu_1(t)),$$
$$\eta_2(t) = \mu_2(t) + \sigma_1(\eta_1(t)) + \tau_2(\mu_2(t)),$$
$$\ldots$$

In model (10.1), we have considered the factors of traffic conditions, dwell time, intersections, and departure time. As we mentioned above, the main parts of the bus traveling time are affected by them. The public bus arrival time prediction model (10.1) is a formally linear model, which has already described the main part of the bus travel time, but we have to estimate its parameters. Meanwhile, we still need to improve its accuracy by considering the other factors that we mentioned earlier.

The parameter calibration process of the linear model is very difficult. Fortunately, we can capture those factors by illustrating the historical arrival times at each stop for all bus lines. The parameter calibration logic for model (10.1) is given below:

*Parameter Calibration Logic*
*Step 1 Classification. Determine the patterns of traffic conditions by illustrating the historical data.*
*Step 2 Estimation. Determine the dwell time $D_i(t)$, traveling time $\tau_k(t)$, and service time $\sigma_j(t)$ in model (10.1) for each pattern.*
*Step 3 Evaluation. Optimize the parameters of model (10.1) by using historical data.*

In Step 1, we illustrate the historical data and try to find the subsequent patterns of traffic conditions. Figure 10.9 lists the inverse of average speed versus trip sequence order of Bus 61 in 2 weeks (from July 5, 2009 to July 18, 2009).

Figures 10.5 and 10.9 show that the traffic conditions have strong seasonality. At the same time, the pattern is strongly similar to Fig. 10.10, which demonstrates the variation for 3 weeks' traffic volumes on a directed link.

Based on these statistical analyses, we classify the models into eight clusters: Sunday peak time, Sunday off-peak time, Monday and Friday peak time, Monday and Friday off-peak time, Tuesday–Thursday peak time, Tuesday–Thursday off-peak time, Saturday peak time and Saturday off-peak time. For different clusters, the parameters are different.

Based on the classification of Step 1, we will determine the dwell time $D_i(t)$, traveling time $\tau_k(t)$, and service time $\sigma_j(t)$ in model (10.1) for each cluster. In Step 2, we first collect all the historical data, then calculate the mean and derivation of the dwell time at each bus stop, and finally determine the parameters of $D_i(t)$ by using this information. And the same method can

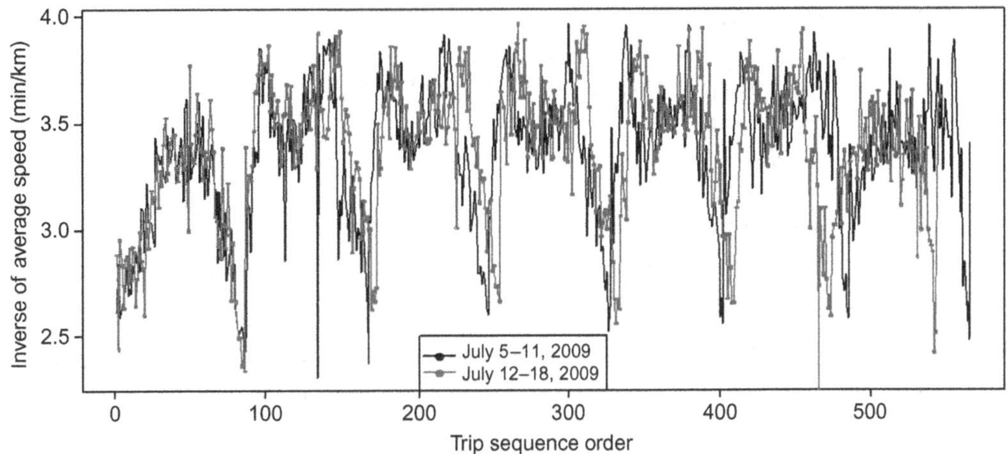


**Figure 10.9: Bus arrival patterns.**

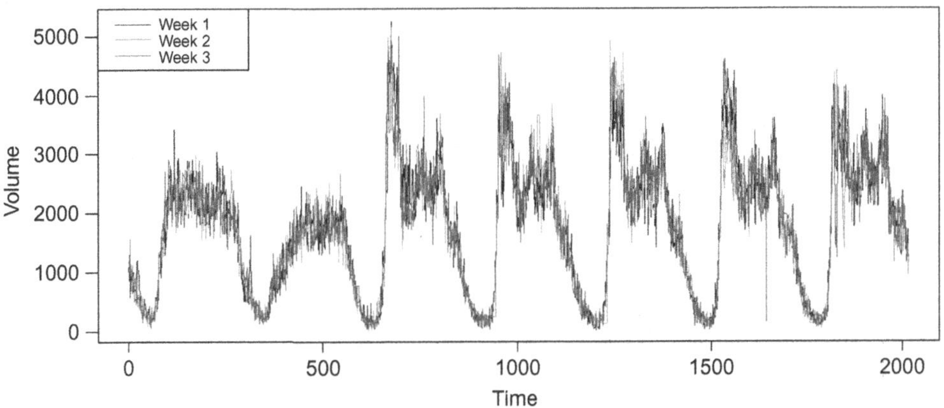


**Figure 10.10: Traffic volume patterns.**

be used for estimating the service time $\sigma_j(t)$ at each intersection. For $\tau_k(t)$ values in model (10.1), we will use a traffic information management system, IBM Traffic Prediction Tool (Min & Wynter, 2011), to determine them.

Once we determine all the main parts, we can use a least squares method to optimize the parameters in model (10.1) in Step 3.

Once we accomplish the parameter calibration process of the model parameter [(Eq. 10.1)], we can use it to forecast the public bus arrival time. During the prediction process, we will consider the error by comparing the updated GPS location of the bus and the predicted result. If the error exceeds a given threshold value, then we will feed back the latest error into the model and update it.

## 10.4 Bus Trip Planning

A system to provide bus trip planning services is given below.

### 10.4.1 System Overview

Some researchers have put much effort into the collection of real-time traffic data and prediction. These data are often stored in a remote database system and updated frequently. We will use these data to provide more bus trip planning services.

In order to respond quickly, we use semidynamic algorithms in our system. The system architecture is shown in Fig. 10.11. These algorithms use a "space-for-time" scheme and include two main steps:

1. *K-transfer*
   In this step, we generate candidate paths for each bus station pair, which have less than $K$ transfer times and $M$ walking times. Based on our research (32 questionnaires, the result

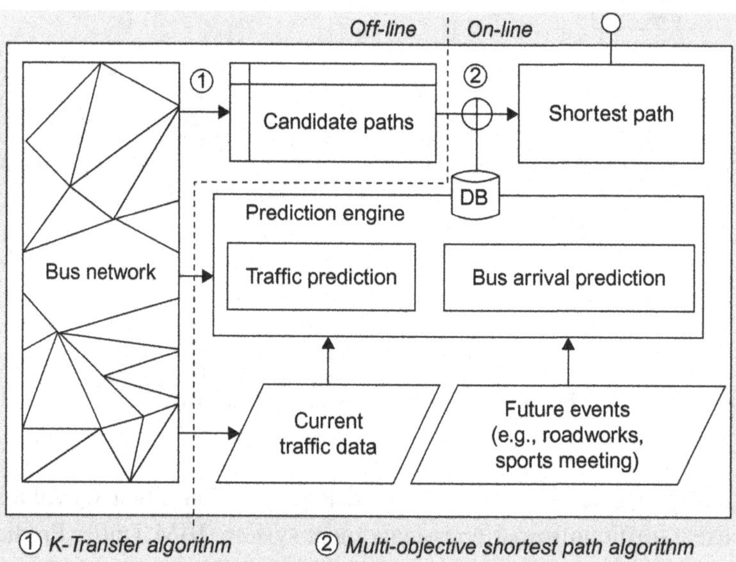


**Figure 10.11: System architecture.**

**Table 10.1: Research data on bus passengers.**

| Actual Transfer Times | 0 | 1 | 2 | 3 | 4 | 5 |
|---|---|---|---|---|---|---|
| % | 46.9 | 37.5 | 12.5 | 3.1 | 0.0 | 0.0 |
| Max interchange times | 0 | 1 | 2 | 3 | 4 | 5 |
| % | 9.4 | 50.0 | 21.9 | 18.7 | 0.0 | 0.0 |
| Max walk times | 0 | 1 | 2 | 3 | 4 | 5 |
| % | 59.4 | 34.4 | 6.2 | 0.0 | 0.0 | 0.0 |

is shown in Table 10.1), people rarely take a bus trip that has more than three transfer times ($K = 3$) and more than two walking times ($M = 2$); so later we generate paths according to these variables.

This includes a reasonable assumption that candidate paths found by the $K$-transfer algorithm will include the shortest path that people really need.

All these candidate paths are computed off-line in this step, and then stored in a database for further processing. This reprocessing step is executed only once for each city's transportation network, not for each query, so it doesn't need to be very efficient.

2. *Multiobjective shortest path*

   Different passengers may have their own definition of the "shortest path." For example, some passengers prefer the shortest waiting time and others prefer the shortest transfer times. So we enable passengers to customize their bus trip in this step. This step will be executed for each query, so it needs to be efficient.

## 10.4.2 K-Transfer

The major objective of this step is to find candidate paths for each station pair, e.g., from Station 0 to 17, as shown in Fig. 10.12. These paths should have transfer times less than $K$ and

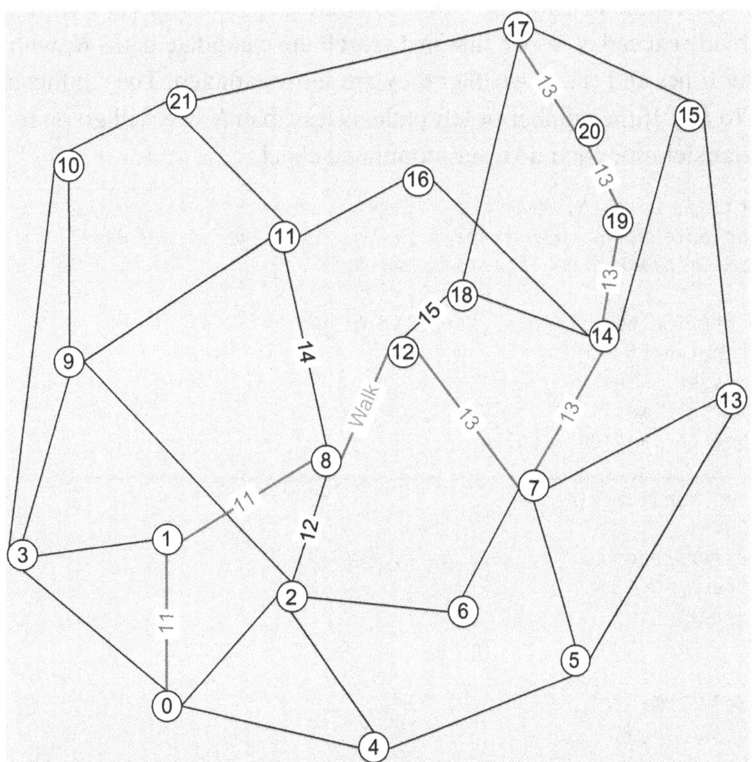


**Figure 10.12: Transportation network.**

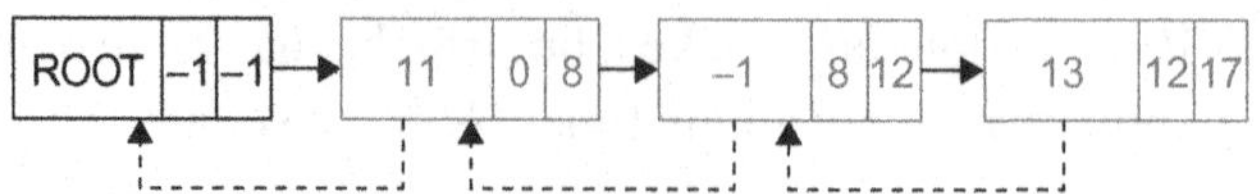


**Figure 10.13: Bus path structure.**

walking times less than $M$. The $K$-transfer algorithm (KT) is not used to solve the "$K$ shortest paths" (KSP) problem (Eppstein, 1998), but to find "$N$ paths with transfer times less than $K$."

We can use a graph to simulate a bus network, and in this graph, each path can be represented by a structured link segments as follows: {Bus ID, Start Station ID, and End Station ID}. If a segment represents walking, the part "Bus ID" uses −1. For example, there is a path that includes stations 0, 1, 8, 12, 7, 14, 19, 20, and 17, as shown in Fig. 10.12. This means taking Bus 11 from 0 to 8, then walking from 8 to 12, and finally taking Bus 13 from 12 to 17; it can be represented as link segments as shown in Fig. 10.13.

The $K$-transfer algorithm includes the following three main substeps:

a. *Find all candidate paths*
   In this substep, all candidate paths with transfer times less than $K$ and walking times less than $M$ will be found.
   The max count of candidate paths is $N$ and the count of paths with {0, 1 … $K-1$} transfer times may already exceed $N$, so we first just search the candidate paths $R_1$ with {0, 1 … $K-1$} transfer times and check whether they are self-conflicted. The conflicted paths will not be added to $R_{init}$. If the number of left paths is less than $N$, we will go on to search paths $R_2$ with {$K$} transfer times and do a self-conflicted check.

```
KTransfer(i, j, G, K, M, N)
Find N candidate paths with transfer times less than K and walk times less than
    M in the network G for station i to j
Input:
i, j: the station pairs' index for searching
G: transit network
K: max transfer times
M: max walking times
N: max count of candidate paths
Output:
R: generated candidate paths
for k:0→K-1
    KTransferBreadth(i, j, K-1, M, k, ROOT, R1)
    If size(R1) > = N
        Break
    endif
endfor
RejectSelfConflict(R1, Rinit)
if size(Rinit) < N
    for k:0→K
        KTransferBreadth(i, j, K, M, k, ROOT, R2)
```

```
    endfor
    RejectSelfConflict(R_2, R_init)
endif
SelectPaths (R_init, R, N)
returnR
```

For the segment $e$, there are two types of next-available segment that have one more transfer time than $e$: direct-transfer and walk-transfer available segments. For example, in Fig. 10.12, if $e$ is the link segment {11, 1, 8}, then the next direct-transfer available segments $T(e)$ include {12, 8, 2} and {14, 8, 11}, and the next walk-available segments $W(e)$ include {−1, 8, 12}. When we transfer one more time on $W(e)$ segments, we can get direct-transfer available segments $T(W(e))$ that include {13, 12, 7} and {15, 12, 18}. So there are four transfer available segments for $e$: {12, 8, 2}, {14, 8, 11}, {−1, 8, 12}−{13, 12, 7}, and {−1, 8, 12}−{15, 12, 18}.

*KTransferBreadth* is the major function in this substep, and it will generate candidate paths with transfer times exactly equal to $k$:

```
KTransferBreadth(i, j, K_current, M, k, e)
Input:
i, j: the station pairs' index for searching
K_current: current max transfer times
M: max walking times
k: current target transfer times
e: structured link segments
Output:
R: generated candidate paths
ke ← e.transfer + 1
if ke! = k
   if ke > K_current
      return
   else
        for each c ∈ C (e) do
          KTransferBreadth(i, j, K_current, M, k, c, R)
        endfor
      endif
endif
construct T(e) and T(W(e))
Select available segments where destination is j, add them to R
return
```

*b. Reject self-conflicted candidate paths*

After the first substep, we can get many candidate paths. But the count of these paths is too large, so we need to reject some paths in this step. For now, we will only reject two types of self-conflict: with the same bus lines or the same bus stations.

For example, there are two paths from origin station 1 to destination station 14: $P_0$ {{20, 1, 11}−{20, 11, 16}−{21, 16, 14}} and $P_1$ {{20, 1, 11}−{21, 11, 16}−{21, 16, 14}}, as

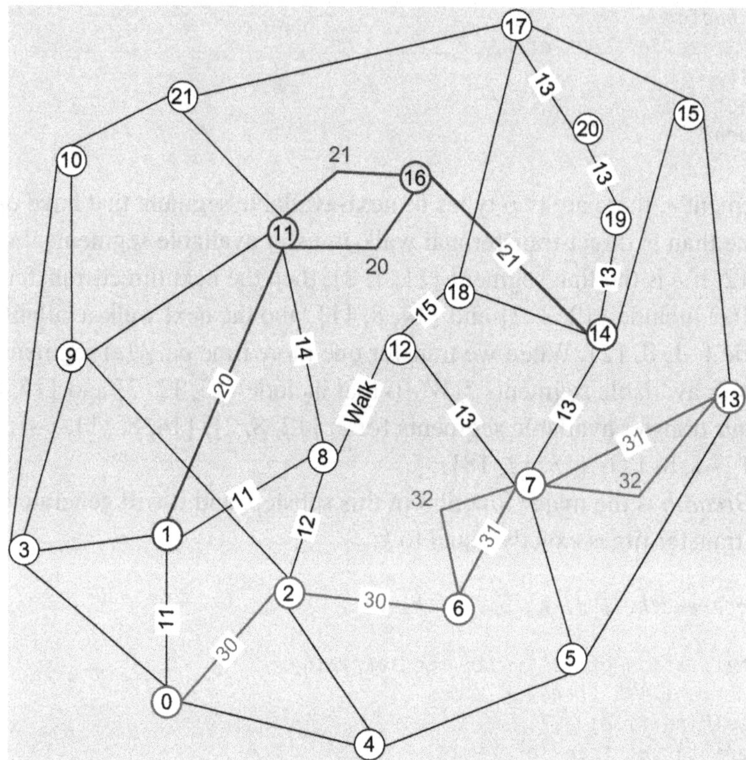


Figure 10.14: Transportation network for self-conflict.

marked in blue in Fig. 10.14. Because their bus lines are the same (20→21), we will reject one of them.

In another example, there are two paths from origin station 0 to destination station 14: $P_0$ {{30, 0, 2}–{30, 2, 6}–{31, 6, 7}–{31, 7, 13}} and $P_1$ {{30, 0, 2}–{30, 2, 6}–{32, 6, 7}–{32, 7, 13}}, as marked in red in Fig. 10.14. Because all of their bus stations are the same (0→2→6→7→13), and if one link segment is stuck in traffic congestion, both of these paths will hold up, so we will reject one of them:

```
RejectSelfConflict(R_init, R)
BL(e, f): Are e's bus lines the same with f's
BS(e, f): Are e's bus stations the same with f's
for each e ∈ R_init do
    for each f ∈ R do
        if BL(e, f) && BS(e, f)
            add e to R
            break
        endif
    endfor
endfor
return
```

c. *Select certain count of candidate paths and store in the database system*
Even after the self-conflict rejection, there will be many candidate paths, so in this substep, we will select $N$ paths according to two properties: shortest property and similar property. First, we calculate each path's shortest value $W_e$, as shown in the pseudo-code of "*Select-Paths,*" and set similarity value $S_e$ to 0. Then, we calculate this path's integrated value $V_e$ according to $W_e$ and $S_e$.
Then, we select a path that has the minimum integrated value $V_e$, and add it to the result paths. After that, we recalculate $W_e$, $S_e$, and $V_e$ for each path, and redo this operation until we have selected enough paths.
Finally, we store selected paths in the database system.

```
SelectPaths(R_init, R, N)
LC (e): bus lines count of e
SC (e): bus stations count of e
SL (e, f): similar bus lines between e and f
SS (e, f): similar bus stations between e and f
for each e ∈ R_init do
    W_e ← e.transfer* I_transfer + e.walk*I_walk + e.walkdis*I_walkdis
        +e.stationcount*I_stationcount + e.wholedis*I_wholedis + e.fee*I_fee
   S_e ← 0
    Z_e ← W_e * I_W + S_e * I_S
endfor
sort R_init by Z
for each g ∈ R_init do
    if size(R) < N
       select e ∈ R_init with min Z value
        add e to R
       delete e from R_init
       S_e ← 0
        for each f ∈ R do
           acc = (LC (e) + LC (f) + SC (e) + SC (f)) / 2
           S_e + = (SL (e, f) + SS (e, f)) / acc
        endfor
        Z_e ← W_e * I_W + S_e * I_S
        sort R_init by Z
    endif
endfor
store R into the database system
return
```

In this substep, there are eight weight values that are shown in Table 10.2.

**Table 10.2: Weight values for bus trip planning.**

| Weight | $I_{transfer}$ | $I_{walk}$ | $I_{walkdis}$ | $I_{stationcount}$ | $I_{wholedis}$ | $I_{fee}$ | $I_W$ | $I_S$ |
|---|---|---|---|---|---|---|---|---|
| Value | 20.0 | 10.0 | 1.0 | 1.0 | 1.0 | 5.0 | 1.0 | 3.0 |

### 10.4.3 Multiobjective Shortest Path

After the $K$-transfer step, all the candidate paths are stored in the database. In this step, when users send a query to the system, we will select some shortest bus paths from the database as user customized.

As summarized in Table 10.3, there are nine properties that will affect the planning result. For the column "single search," if $B_* = 1$, it means just use this property to select paths. For example, if $B_{TT} = 1$, we will select paths with the shortest transfer times, the same as the objective of Google Maps shown in Fig. 10.1. If all $B_* = 1$, we will select paths using integrated values in Eq. (10.3). "Filter value" is used to filter some paths. "Integrate weight" is used to calculate integrated value.

This multiobjective step includes three substeps:

a. *Find all the bus stations that are near the origin and destination.* In this trip planning system, the user inputs names of origin and destination stations, and then these two stations will be changed to longitude and latitude ($Pos_{\text{origin}}$ and $Pos_{\text{destination}}$). We first search stations $B_{\text{origin}}$ near $Pos_{\text{origin}}$ and $B_{\text{destination}}$ near $Pos_{\text{destination}}$:

$$\begin{aligned} B_{\text{origin}} &= \{e, \sqrt{e.Pos - Pos_{\text{origin}}} <= R_{\text{search}}, e \in E)\} \\ B_{\text{destination}} &= \{e, \sqrt{e.Pos - Pos_{\text{destination}}} <= R_{\text{search}}, e \in E)\} \\ P &= \{p, p(e, f) \in Databse, e \in B_{\text{origin}}, f \in B_{\text{destination}}\} \end{aligned} \tag{10.2}$$

b. *Fetch candidate paths from database system.*
Using the origin and destination stations as key components, we will search many candidate paths $P$ from the database.

c. *Use multiobjective evaluating function to find shortest paths (Default value of $T$ is 5).*
Combined with real-time traffic data that are read from the database, we will filter these paths $P$ to $P_{\text{filter}}$. Once the user has customized his/her objectives, we select the shortest paths $Q$ from $P_{\text{filter}}$, and send this back to the users.
For example, someone may want to find a bus path such that the whole trip time is limited to 50 minutes, and people don't need to walk during this trip with enough seats in the bus. For this integrated search, we set all single search variables $B_*$ to false and set integrate weight as shown in Table 10.3.

$$\begin{aligned} D_{\text{max}} &= \{\text{TT, WT, WAD, WHD, WTT, F, CE}\} \\ D_{\text{min}} &= \{\text{FS}\} \\ D_{\text{bool}} &= \{\text{AC, WS}\} \\ P_{\text{filter}} &= \begin{Bmatrix} p, V_i^p < \text{Max}_i, V_j^p > \text{Min}_j\, V_k^p = \text{Bool}_k \\ i \in D_{\text{max}}, j \in D_{\text{min}}, k \in D_{\text{bool}}, p \in P \end{Bmatrix} \\ Z_p &= \sum\nolimits_{i \in D} I_i^p \dot{V}_i^p \\ Q &= \{p, Z_p \in \min_T(Z), p \in P_{\text{filter}}\} \end{aligned} \tag{10.3}$$

**Table 10.3: Properties that will affect the searching planning.**

| Properties | Single Search | Filter Value | Integrate Weight | Actual Value |
|---|---|---|---|---|
| Transfer times (TT) | $B_{TT} = 0$ | $Max_{TT}$ | $I_{TT} = 0$ | $V_{TT}$ |
| Walk times (WT) | $B_{WT} = 0$ | $Max_{WT} = 0$ | $I_{WT} = 5$ | $V_{WT}$ |
| Walk distance (WAD) | $B_{WAD} = 0$ | $Max_{WAD}$ | $I_{WAD} = 0$ | $V_{WAD}$ |
| Whole distance (WHD) | $B_{WHD} = 0$ | $Max_{WHD}$ | $I_{WHD} = 0$ | $V_{WHD}$ |
| Whole trip time (WTT) | $B_{WTT} = 0$ | $Max_{WTT} = 50$ | $I_{WTT} = 1$ | $V_{WTT}$ |
| Fee (F) | $B_F = 0$ | $Max_F$ | $I_F = 0$ | $V_F$ |
| Carbon emission (CE) | $B_{CE} = 0$ | $Max_{CE}$ | $I_{CE} = 0$ | $V_{CE}$ |
| Air conditioning (AC) | $B_{AC} = 0$ | $Bool_{AC}$ | $I_{AC} = 0$ | $V_{AC}$ |
| Free seat on bus (FS) | $B_{FS} = 0$ | $Min_{FS} = 1$ | $I_{FS} = 0$ | $V_{FS}$ |
| Walk in shadow (WS) | $B_{WS} = 0$ | $Bool_{WS}$ | $I_{WS} = 0$ | $V_{WS}$ |

### 10.4.4 Path Patching

The algorithms we mentioned above are based on the research data in Table 10.1:

*"People rarely take a bus trip that has more than three transfer times ($K = 3$) and more than two walking times ($M = 2$)"*

But in poor traffic conditions, the candidate paths generated by $K$-transfer and stored in the database may not satisfy users' objectives. Then, path patching should be done at this time and more paths will be searched.

For example, there are many candidate paths for station pair (0, 17) that are stored in the database system. These paths are marked in red in Fig. 10.15. In a special case, these paths are all stuck in traffic congestion, and the minimum bus trip time still exceeds 4 hours. Then, we will select some stations as a bridge between origin and destination.

In Fig. 10.3(b), the search domain of dynamic algorithms includes any stations in the circle around the middle point of the origin and destination, and these stations can be chosen as bridge stations, as shown in Fig. 10.16. As a simple strategy, we select 11 bridge stations around the normal line between the origin and destination. In the first case, bus station 10 will be selected for the pair (0, 17), and finally combined with traffic data, we can get a shortest path {{50, 0, 3}–{51, 3, 10}–{42, 10, 21}–{41, 21, 17}}.

In normal cases, we can fetch candidate paths after a single SQL query. And for path patching, it will cost $11 \times 2 + 1$ querying time. Because SQL querying is quick enough, the time taken here has little effect on final users.

Based on this path patching, we can provide another useful service: inputs from users are origin, destination, and some middle stations they want to pass.

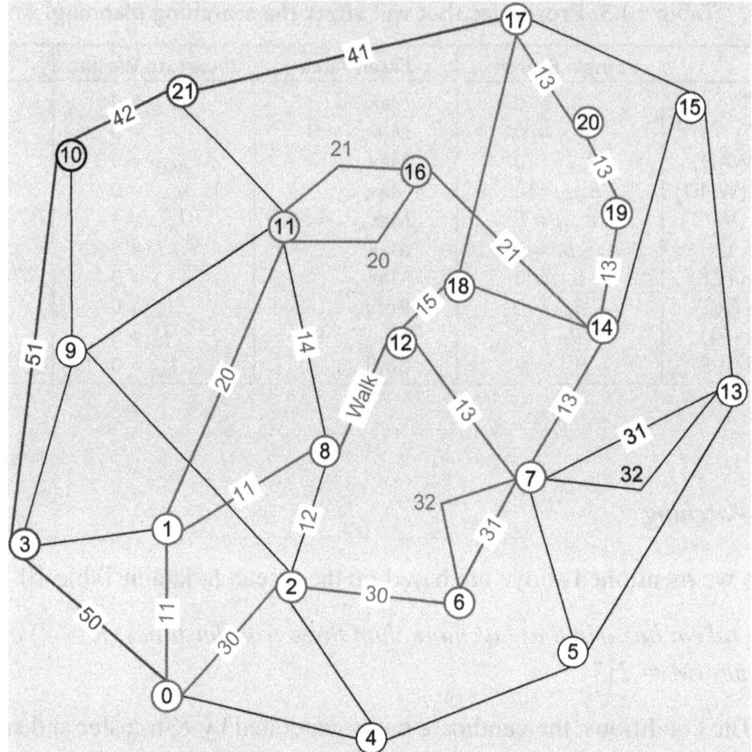


Figure 10.15: Transportation network for path patching.

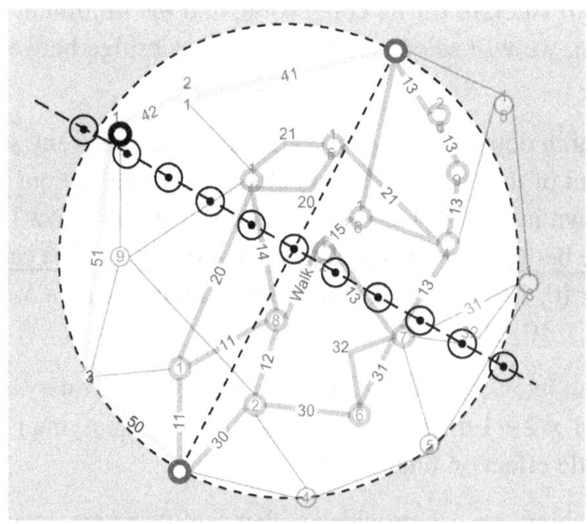


Figure 10.16: Patching circle.

## 10.5 Prototype and Experiments

In order to verify the above statistical approach, we build a web application prototype and verify the forecasting results by experiments. In Fig. 10.17, we illustrate the architecture of the traffic prediction tool and bus arrival prediction application. First of all, we collect all of the raw traffic data from the sensors such as wire loop, GPS sensor, and red-radio radar. Then we map these data into the network to generate useful traffic information such as traffic flow, speed, occupation, etc. by using data processing, analysis and fusion technology. After that, we predict the traffic conditions and bus arrival time by combining the integrated data and advanced mathematical model. Based on this useful traffic information, we can provide many high-value services such as monitoring, parking, emergency response, and other information services.

Using the proposed architecture, a web application prototype was built. The clients can use a web browser to connect the web application server and obtain useful public bus traffic information. The application server is built on a database server (DB2 Server, SQL Server, mySQL, etc.), a JSP server (Tomcat, Websphere, etc.), and Java Runtime Environment. It can run on Microsoft Windows, Linux, or Unix operating systems. The prototype is based on a subnetwork of Hong Kong city. The user interface (UI) of the prototype for bus arrival prediction and bus trip planning is shown in Figs 10.18 and 10.19, respectively. In this prototype, we can list all of the related bus stops/bus lines based on the input of the search function; for instance, when we query the bus line based on bus road, we can obtain all of the list of bus road including the input. After we have

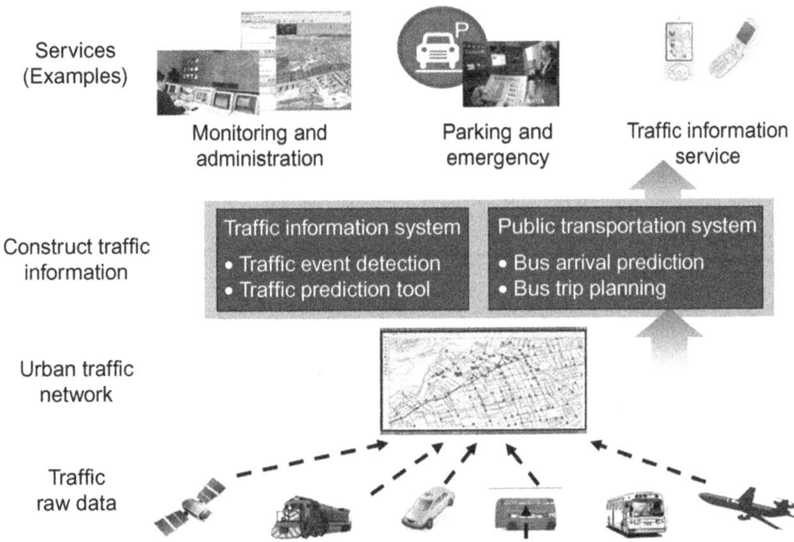


**Figure 10.17: Architecture of prototype.**

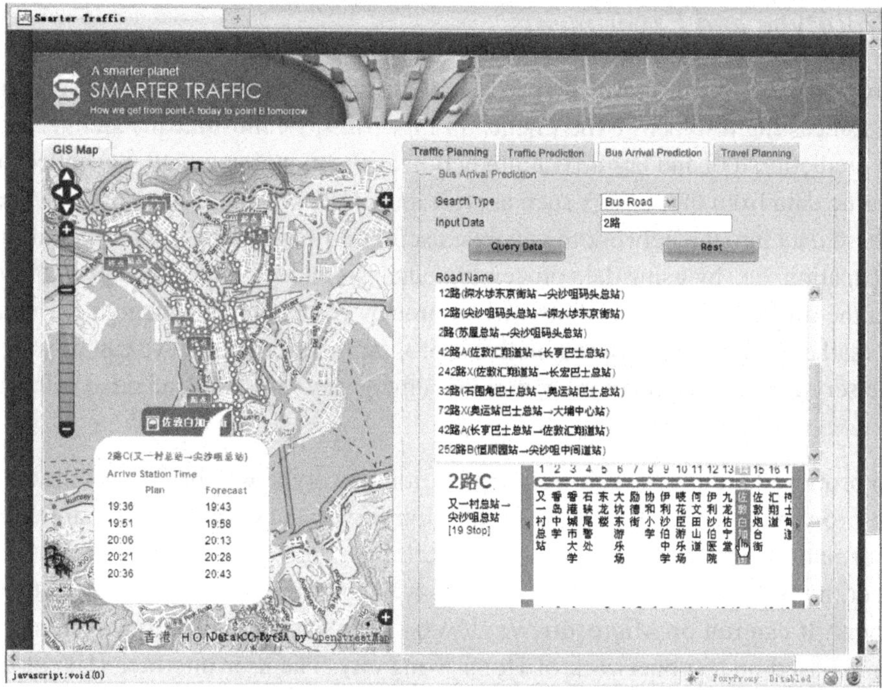


Figure 10.18: UI of bus arrival prediction.

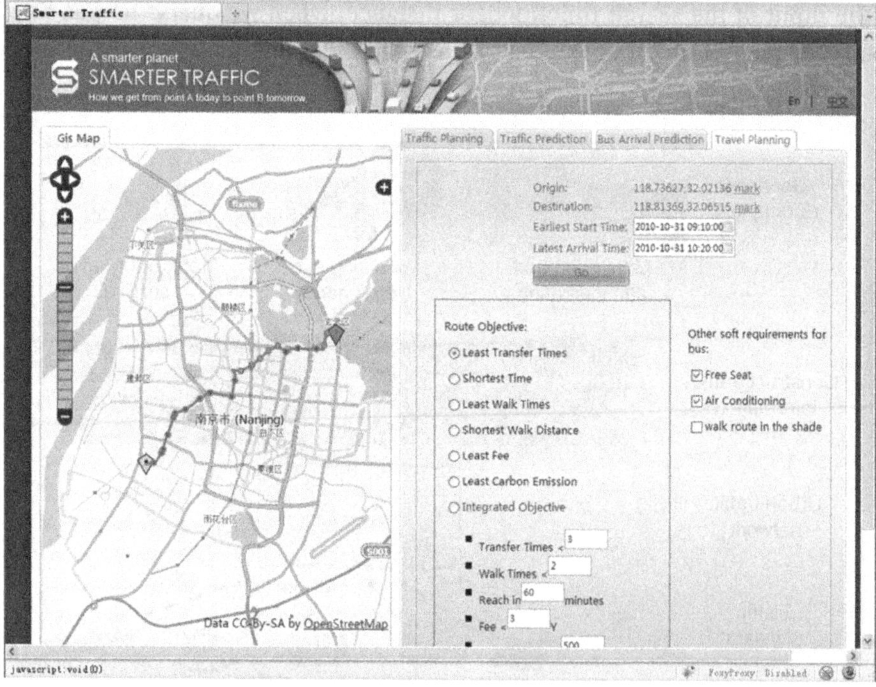


Figure 10.19: UI of bus trip planning.

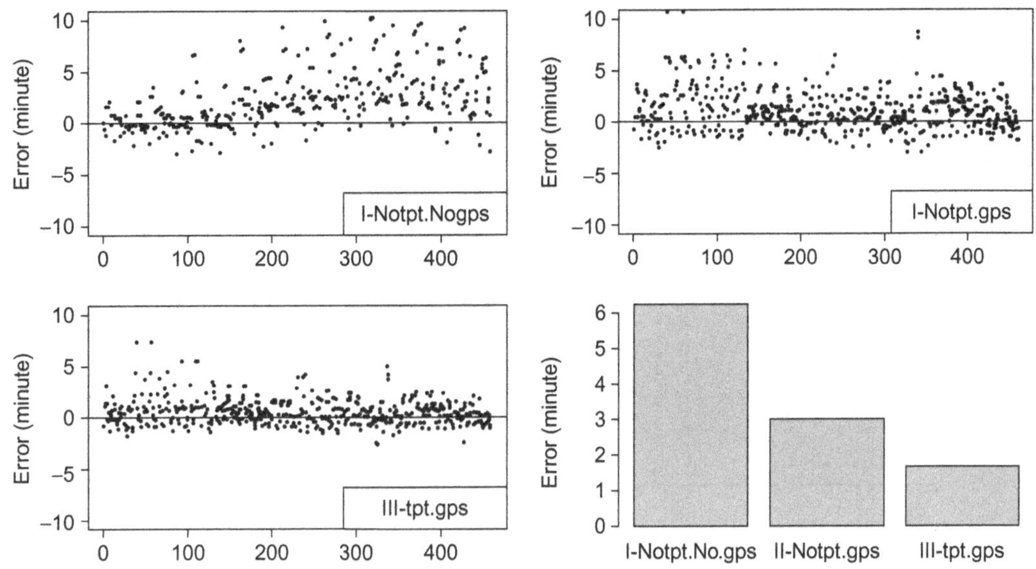


**Figure 10.20: Results for bus arrival prediction.**

obtained the bus line, then we can show it on the left map when we click on the bus line in the right window. For each stop, we will list the planned and forecasted arrival times for the following five trips in the future. When we click on the bus station, we will show all of these times on the left map.

Based on this prototype, we have done a number of experiments. Figure 10.20 shows the comparative results for the proposed statistical approach for bus-arrival prediction.

"Notpt" means we did not consider the traffic condition information provided by the traffic prediction tool in the model. "Nogps" denotes that we did not consider the bus location information in the model. "tpt" and "gps" indicate that we did consider the information mentioned above in the model.

From Fig. 10.20, it is seen that we can predict the bus-arrival time at each bus stop with an error less than 2 minutes.

Besides the above experiments on bus arrival prediction, we also carried out experiments on bus-trip planning in the real transportation network of Hong Kong and Nanjing cities of China. In the $K$-transfer step, we generate candidate paths for each bus station pair. Based on a computer with Intel Core 2 Duo CPU, E7400, 2.84 GHz, and 3.25 G memory, the $K$-transfer unit calculation for each pair takes about 258 milliseconds, as shown in Fig. 10.21. This average calculation time also indicates that direct calculation on a dynamic network takes too long.

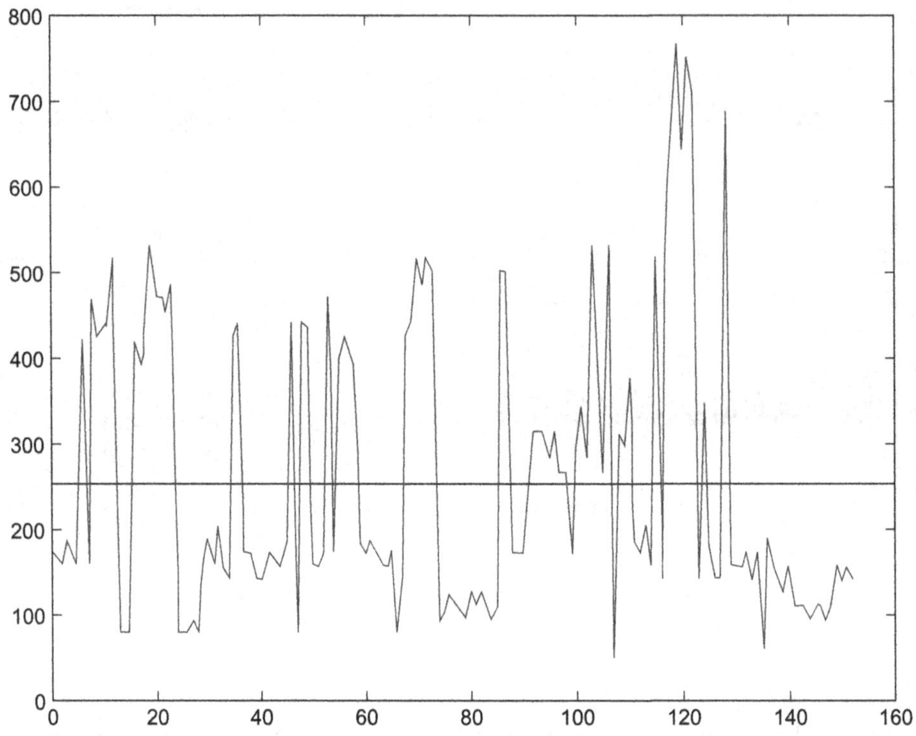


**Figure 10.21: *K*-transfer unit calculation time (150 sample station pairs).**

Candidate paths generated (*K*-transfer) for each station pair are independent of each other, so we can work on in different machines, for example, distributed terminals or personal computers. In Nanjing, there are 1345 bus stations, so this traffic network takes 1,807,680 (1345 × (1345 − 1)) times as long as *K*-transfer unit calculation. Because of the independence of the calculation, we use eight personal computers to generate candidate paths. Each computer takes about 2.3 hours and the whole procedure takes about 18 hours. These data show that we can quickly refresh all the candidate paths in the database.

When we generate candidate paths for origin station "Fenling railway station" and destination station "Wan Tau Tong Estate" in Hong Kong, the count of candidate paths generated by *KTransferBreadth* within $K - 1$ transfer times is 166 ($V^{K-1}$). After *RejectSelfConflict*, only 5 paths ($V^{K-1}_{\text{filter}}$) are left. Because $V^{K-1}_{\text{filter}}$ is less than 10 ($B^K$ is true), we will call *KTransferBreadth* within *K* transfer times, and at this time the algorithm generates another 166 paths ($V^K$). After using the *RejectSelfConflict* function again, $V^K_{\text{filter}}$ becomes 10. Statistics about the change of path counts in different steps are listed in Table 10.4. Properties of the candidate paths for Nanjing and Hong Kong are listed in Table 10.5.

**Table 10.4: Statistics on the count of candidate paths in the *K*-transfer algorithm.**

| City | Nanjing | | | Hong Kong | | |
|---|---|---|---|---|---|---|
| | Paths Count | % | Average | Paths Count | % | Average |
| $V^{K-1}$ | 0–200 | 64.53 | 27.55 | 0–200 | 20.0 | 47.87 |
| | 200–2000 | 29.52 | 743.26 | 200–2000 | 17.36 | 801.38 |
| | 2000–∞ | 05.95 | | 2000–∞ | 62.64 | |
| $V^{K-1}_{filter}$ | 0–50 | 85.29 | 8.56 | 0–50 | 43.25 | 12.70 |
| | 50–200 | 13.52 | 92.8 | 50–200 | 30.53 | 109.68 |
| | 200–∞ | 01.19 | 277.54 | 200–∞ | 26.22 | 604.72 |
| $B^K$ is TRUE | 60.52% of station pairs need check paths with *K*-transfer times | | | 25.48% of station pairs need check paths with *K*-transfer times | | |
| $V_K$ | 0–200 | 84.40 | 27.63 | 0–200 | 53.80 | 43.30 |
| | 200–2000 | 15.44 | 451.03 | 200–2000 | 40.20 | 0.40 |
| | 2000–∞ | 00.16 | | 2000–∞ | 6.00 | |
| $V^K_{filter}$ | 0–20 | 78.66 | 3.22 | 0–20 | 59.11 | 5.62 |
| | 20–50 | 21.32 | 40.63 | 20–50 | 27.37 | 31.53 |
| | 50–∞ | 00.01 | 218.90 | 50–∞ | 13.52 | 83.510 |

**Table 10.5: Properties of candidate paths.**

| City | Nanjing | Hong Kong |
|---|---|---|
| *N* (Max count of candidate paths) | 10 | 30 |
| Bus lines | 189 | 179 |
| Bus stations | 1345 | 1152 |
| Total count of candidate paths | 11,061,283 | 1,246,4315 |
| Average candidate paths | 6.11 | 9.39 |
| Average transfer times | 2.16 | 1.80 |
| Average walk times | 0.50 | 0.52 |
| Average walk distance | 0.39 | 0.39 |
| Average station count | 29.67 | 25.17 |

After the *K*-transfer step, many candidate paths are stored in the database system. Figure 10.22 shows the origin station A (marked with solid diamond) and the destination station B (marked with empty diamond). Figure 10.23 shows 10 candidate paths from A to B.

With different user objectives, the system presents different shortest paths. For example, when the user chooses shortest transfer times as trip-planning objective, the path in Fig. 10.24 will be presented, and details about this path are as follows: {walk to the station "Meijing"} – {take bus "line 13", and then get off at the station "Sansan Street"} – {walk to the station "Zhangfu Garden"} – {take bus "line 41", and then get off at the station "Yutang"} – {Walk to the destination}. Some paths based on other objectives are presented in Figs. 10.25, 10.26, and 10.27.

Figure 10.22: Origin and destination stations.

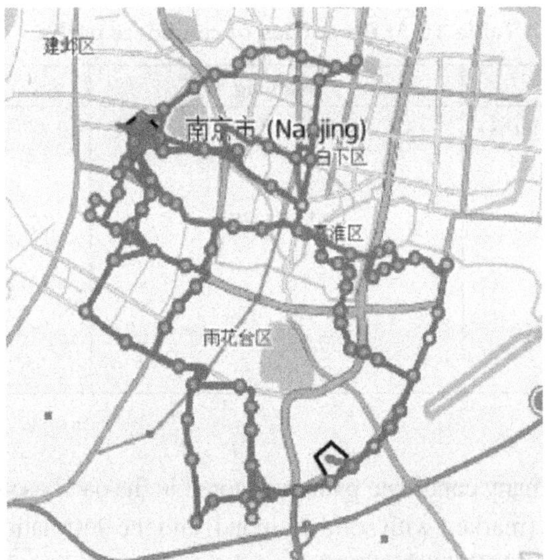


Figure 10.23: All candidate paths.

For Hong Kong, there are 43,344 station pairs (3.96%) that have no candidate paths in the database. This means that there is no suitable $K$-transfer path for these pairs. When people search paths between these stations, we may generate shortest paths in a path patching way.

**Figure 10.24: The shortest path with shortest transfer times or smallest fee.**

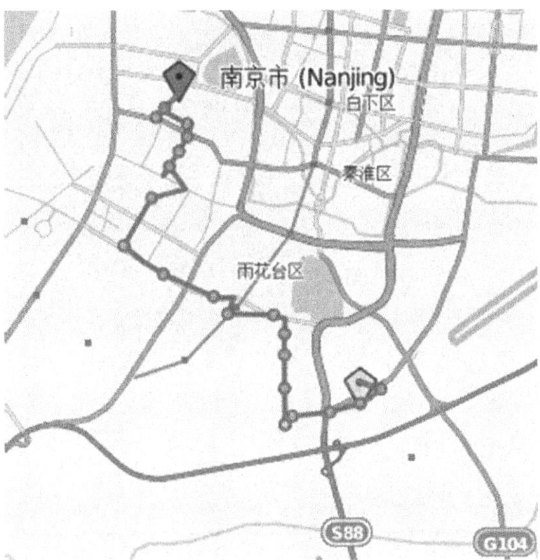


**Figure 10.25: Shortest walk distance and shortest walk times.**

The practice also shows that this system is sufficiently effective and can respond to users in real time. The average time for querying (A, from user sending a query to receiving result paths) (marked in red in Fig. 10.28) is 108.4 milliseconds and for calculating (C, from server receiving the query to sending back the result paths) is 87.4 milliseconds. C includes five

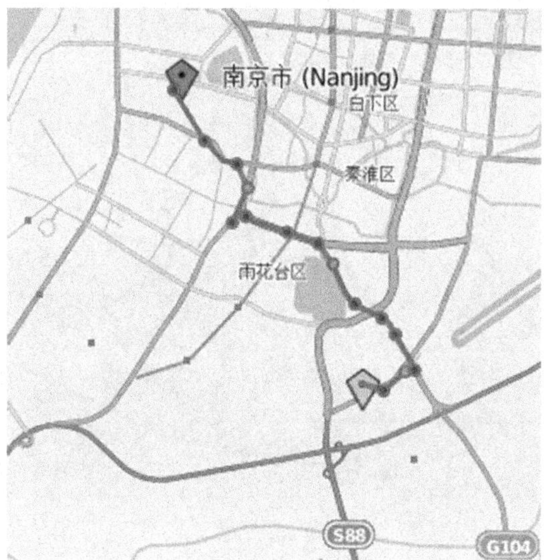


**Figure 10.26: Shortest time based on real-time traffic data.**

**Figure 10.27: Lowest carbon emission.**

parts: find neighbor stations ($F_{SQL}$) in the database, find candidate paths ($S_{SQL}$) in the database, decide and select shortest paths (D), change shortest paths to program structure (P), and construct GIS data for sending back through network (G). Two of these parts are processed by the database system, so computational time on the service server is greatly reduced, and it can respond quickly.

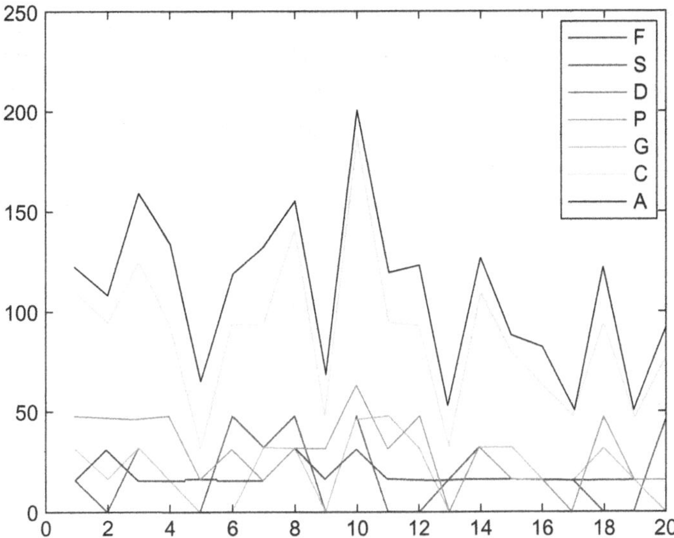

**Figure 10.28: Querying time and calculating time.**

## *10.6 Conclusions*

In this chapter, a statistical approach was presented to forecast the arrival time at each stop for public buses. Based on the assessment of all factors that will affect the bus arrival time prediction, a linear model was proposed. In the model, we have considered all the evaluated factors such as departure time, driver characteristics, dwell time, intersections, and traffic conditions. By analyzing the historical data of the bus arrival time, we classify the bus travel pattern into eight clusters. For each cluster, an algorithm was proposed to determine the parameters of the model. Based on the presented bus arrival prediction approach, a system and several algorithms for bus trip planning services were proposed. In the proposed system, we used a semi-dynamic strategy by including two steps: *K*-transfer and multiobjective step. In the first step, we generated many candidate paths for each station pair and stored them in the database system. In the second step, we just searched the shortest paths from these candidate paths according to user-customized objectives. The first step was only executed once for the whole traffic network, and the second step was executed for each user query. Because of the simplicity and efficiency of the second step, this system can respond to user queries quickly.

In order to verify the practicability and efficiency of the proposed approaches and system, a web-application prototype was built. Based on the built prototype, a number of experiments were carried out. These experiments proved that the presented bus arrival prediction approach is relatively accurate and more efficient than other methods, and the bus trip planning system is practical and efficient. In fact, there are three key points that have a great effect on the proposed bus trip planning system: (1) quality of candidate paths in the $K$-transfer step; (2) querying efficiency of the database system in the multiobjective step; and (3) bridge stations in the path patching step. For future work, we need to give more attention to the following problems: (1) improve $K$-transfer's calculating efficiency of candidate paths; (2) how to refresh the database content when the bus network changes, for example, add new bus lines; (3) how to interact with other systems more efficiently; and (4) how to process clients' feedback, for example, when many people choose the same bus line, there may not be a free seat.

## References


Abdelfattah, A. & Khan, A. (1998). Models for predicting bus delays. *Transportation Research Record: Journal of the Transportation Research Board, 1623*, 8–15.

American Public Transportation Association (2010). *Public transportation: Moving America forward.* Retrieved from http://www.publictransportation.org/

Bus Rapid Transit (2000). *Metro rapid. Urban transport of China.* Retrived from http://brt.volpe.dot.gov/projects/losangeles.html

Eppstein, D. (1998). Finding the k shortest paths. *SIAM Journal Computing*, *28*(2), 652–673.

Hirose, T., Oguchi, Y., & Sawada, T. (2004) Framework of tailormade driving support systems and neural network driver model. *Journal of International Association of Traffic and Safety Sciences*, *28*(1), 108–114.

Horizon Research Consultancy Group (Horizon Group) (2009). Foton Chinese Index for Mobility (research report on Chinese citizens lives' mobility index), Beijing, China, December; http://www.cctv.com/english/special/news/20091227/101335.shtml

Jeong, R. & Rilett, L. (2004). *The prediction of bus arrival time using AVL data.* Paper presented at the 83rd Annual Meeting of the Transportation Research Board, Washington, DC, January.

Li, F., Lin, H., Yuan, Y., & Min, W. (2011). *Public bus arrival time prediction based on traffic information management system.* Paper presented at the 2011 IEEE International Conference on Service Operations and Logistics, and Informatics, Beijing, China, July.

Lin, W. & Zeng, J. (1999). Experimental study of real-time bus arrival time prediction with GPS data. *Transportation Research Record: Journal of the Transportation Research Board, 1666*, 101–109.

Min, W. & Wynter, L. (2011). Real-time road traffic prediction with spatio-temporal correlations. *Transportation Research, Part C, 19*(4), 606–616.

Misra, S. & Oommen, B. (2005). Dynamic algorithms for the shortest path routing problem: Learning automata-based solutions. *IEEE Transactions on Systems, Man, and Cybernetics, Part B*, *35*(6), 1179–1192.

Ojili, S. (1999). A prototype of a bus arrival time notification system using automatic vehicle location data (Master's thesis, Texas A&M University, College Station).

Ramakrishna, Y., Ramakrishna, P., Lakshmanan, V., & Sivanandan, R. (2006). Bus travel time prediction using GPS data. *Proceedings of Map India*, New Delhi, India, January; http://www.gisdevelopment.net/proceedings/mapindia/2006/student%20oral/mi06stu_84.htm

Seema, S. & Alex, S. (2010). Dynamic bus arrival time prediction using GPS data. Paper presented at 10th National Conference on Technological Trends (NCTT09), Kerala, India, November 2009.

Shalaby, A. & Farhan, A. (2003). Bus travel time prediction model for dynamic operations control and passenger information systems. Paper prepared for presentation at the 82nd Annual Meeting of the Transportation Research Board, Washington, DC, January.

Steven, I., Ding, Y., & Wei, C. (2002). Dynamic bus arrival time prediction with artificial neural networks. *Journal of Transportation Engineering*, *128*(5), 429–438.

Sun, S., Zhang, C., & Yu, G. (2006). A bayesian network approach to traffic flow forecasting. *IEEE Transactions on Intelligent Transportation Systems*, *7*(1), 124–132.

Suwardo, S., Madzlan, N., Ibrahim, K., & Oyas, W. (2009). *Bus travel time prediction in mixed traffic by using statical neural network.* Workshop dan Simposium XII Forum Studi Transportasi antar Perguruan Tinggi (FSTPT), Universitas Kristen Petra.

Urban Mobility Report and Appendices (2010). *What does congestion cost us?* Retrieved from http://mobility.tamu.edu/ums/report/congestion_cost.pdf

Wall, Z. & Dailey, D. (1999). *An algorithm for predicting the arrival time of mass transit vehicles using automatic vehicle location data.* Paper presented at the Transportation Research Board 78th Annual Meeting, Washington, DC, January.

World Urbanization Prospects (2007). *The 2007 revision population database.* Retrieved from http://esa.un.org/unup/

Wu, Q., Hartley, J., & Al-Dabass, D. (2005). Time-dependent stochastic shortest path(s) algorithms for a scheduled transportation network. *International Journal of Simulation Systems, Science & Technology, 6*(7–8), 53–60.

Zhang, L., Zhong, G., & Liu, X. (2009). *Dual-network model and its application in algorithms for solving the designated vertex shortest path problem.* Paper presented at the 2009 MASS'09 International Conference on Management and Service Science, Wuhan, China, September, pp. 1–4.

CHAPTER 11

# Mass Customization Manufacturing and Its Application for Mobile Phone Production

**Gang Xiong**

*State Key Laboratory of Management and Control for Complex Systems, Institute of Automation, Chinese Academy of Sciences, Beijing, China*

## 11.1 Introduction

Prior to the 1980s, manufacturers applied production processes of either customized and crafted products or standardized mass-produced products. The timeline summarizes the changing dominance of competitive priorities: price until the mid 1980s, quality until the early 1990s, flexibility until the mid 1990s, and agility or responsiveness thereafter. However, nowadays customers want a product with the highest quality, fastest delivery, and highest level of product customization, and yet he/she is unwilling to pay the commensurate price (Kumar, 2007). The marketing environment in new economic times urgently requires a new mode of production.

In 1970, Alvin Toffler, an American sociologist and futurist, gave his creative suggestion in his book *Future Shock*: "Provide customized products and services to meet customer's special requirements with the cost and speed of standard mass production." In 1987, Stan Davis envisioned that through an efficient deployment of flexible and agile processes, a one-of-a-kind product could be manufactured to customer specification without sacrificing scale economies. In this way, customers are able to purchase a customized product at the cost of a mass-produced item. He named this type of production mode mass customization (MC) in his book *Future Perfect*. In 1993, Pine and Davis wrote their book *Mass Customization: The New Frontier in Business Competition*. The goal of mass customization is to provide enough varieties in products and services so that almost everyone finds exactly what they want at a reasonable price.

Mass customization becomes possible due to a convergence of several factors: market pressure, developments in information technology (IT) capabilities enabling customer integration into product co-design, evolution of customer relationship management (CRM) as a strategy, improvements in enterprise resource planning (ERP) software consistent with

Service Science, Management, and Engineering.


DOI: 10.1016/B978-0-12-397037-4.00011-9

personalization needs, and mass customization tools such as modularity and delayed differentiation, which help reduce manufacturing costs and cycle times (Kumar, 2007). The advances in manufacturing technology, information technology, and novel management methods have made mass customization a standard business practice, where producers realize product customization at low cost and customers get the benefits of customized products and quick service with relatively low prices simultaneously. The latest IT technologies are Web 2.0, Bite and Torrent (BT), Peer to Peer (P2P), Business to Consumer (B2C), Business to Business (B2B), search and recommendation engines, and so on.

### 11.1.1 Definitions and Analysis

Mass customization has many different definitions that can be given either broadly or narrowly (Giovani, Borenstein, & Fogliatto, 2001). The broad, visionary concept was given by Davis (1987): "Mass Customization as the ability to provide individually designed products and services to every customer through high process agility, flexibility and integration, thus Mass Customization systems may reach customers as in the mass market economy but treat them individually as in the pre-industrial economies."

I. P. McCarthy (2004) gives the definition of mass customization as: "The capability to manufacture a relatively high volume of product options for a relatively large market (or collection of niche markets) that demands customization, without tradeoffs in cost, delivery and quality."

At an operational level, mass customization will often mean different things to different groups of firms because factors such as the product volume and variety ratio, the complexity and value of product complexity, the point of customer involvement, the degree of customer involvement, the type of product modularity offered, and the nature of the customized offering and the perceived value will vary for different firms.

Many researchers propose similar but narrower, more practical concepts: "Mass Customization system uses information technology, flexible processes, and organizational structures to deliver a wide range of products and services that meet specific needs of individual customers, at a cost near that of mass produced items. Mass Customization is seen as a systemic idea involving all aspects of product sale, development, production, and delivery, full-circle from the customer option up to receiving the furnished product" (Giovani et al., 2001).

MacCarthy, Brabazon, and Bramham (2003) analyzed previous work in the literature and summarized the fundamental operation modes of mass customization as shown in Table 11.1.

All the benefits of mass customization come with important challenges in the areas of product design, manufacturing, and information systems support (Smith & Sánchez, 2008). Giovani et al. (2001) summarized the success factors of mass customization as shown in Table 11.2. However, research in enabling technologies for mass customization in practice is scarce.

**Table 11.1: Fundamental operation modes of mass customization.**

| | NBIC | Motorola | European Bicycle | Computer | Commercial vehicle |
|---|---|---|---|---|---|
| Lampel | TC + CS | CS | TC + CS | CS | TS + PS |
| Ross | CoreC | CoreC | CoreC | CoreC | CoreC + PPC |
| Alford | Optional | Optional | Optional | Optional | Core |
| Duray | Involver | Assembler | Assembler | Assembler | Fabricator |
| Giovani DS | Fabrication + Assembly | Assembly | Assembly + Fabrication | Assembly | Design |
| Gilmore and Pine | Collaborative | Collaborative | Collaborative | Collaborative | Collaborative |

TC, tailored customization; CS, customized standardization; PS, pure standardization; CoreC, core customization; PPC, post-product customization.

**Table 11.2: Success factors of mass customization.**

| MC Enablers and Related Success Factors | Enabler-Related Success Factors (Organization Based) |
|---|---|
| Processes and methodologies: | |
| Agile manufacturing | Knowledge |
| Supply chain management | Value chain |
| Customer-driven design and manufacturing | Customizable products |
| Lean manufacturing | Value chain |
| Enabling technologies: | |
| Advanced manufacturing technologies | Technology, customizable products |
| Communication and networks | Technology, knowledge |

On the other hand, mass customization has some limitations:

1. It requires a highly flexible production technology that is too expensive and time consuming to apply for most small and medium enterprises (SME).
2. It requires an elaborate system for eliciting customers' wants and needs, which is more difficult to achieve than it appears.
3. It requires a strong direct-to-customer logistics system. Fulfillment is the weak link of e-commerce and mass customization.
4. People are not willing to pay more to enjoy the customized products or services.

### *11.1.2 Literature Review*

Since 1987, scientists of academic institutes have done much research on mass customization. A detailed review of the mass customization literature, including some mentioned here, is provided by Giovani et al. They generate eight levels of mass customization ranging from pure customization to pure standardization. Kumar, Gattoufi, and Reisman (2007) studied and analyzed the trends and directions of the research published in 1124 MC papers that have appeared in journals and magazines since 1987. The publication outlet data conform to an S-shaped curve, establishing maturity of the mass customization field. The publication data show that the mass customization field has passed through four stages of growth: incubation or slow (1987–1992), exponential (1993–2003), stable, and mature (2003–2005) (Kumar et al., 2007). The current state of mass customization can be characterized as healthy and robust both in terms of research volume as well as applications in the business world.

The first world conference on mass customization and personalization was held in 2001 and has been held once every 2 years (Chase, Jacobs, & Aquilano, 2006). The Workshop MCP 2005 was held in China. The 2009 world conference on mass customization and personalization took place in Helsinki, Finland. In 2004, a Chinese scholar, Dan Bin (2004), published his book *Mass Customization: To build up the core competence of your enterprise*. Elizabeth and Attahiru (2009) suggested a queuing model of delayed product differentiation. Cao, Wang, Law, Zhang, and Li (2006) proposed an interactive service customization model. Finnish scholar, Jari, described how to implement mass customization in the electronics industry (Partanen & Haapasalo, 2004). More literature reviews can be found in the study by Giovani et al. (2001).

In 1998, with the support of China's 863 High-Technology Development Foundation, the author Xiong and Nyberg (2000) started the research on "Intelligent production and optimized scheduling system based on push/pull production mode used in process industry." "Mass production" is a push production mode; what is made will be sold. "Customization" is a pull production mode; what will be sold is made. Pull production mode is mainly used and push production mode compensates it, and their advantages are combined to meet the requirements of "mass customization." These research achievements were applied in industrial practice in a global mobile phone manufacturer.

Two decades after its inception, mass customization is now showing unmistakable signs of moving in the mass personalization direction, making a market of reality, at least in selected industries. Thus, at the lower end of the personalization spectrum are manufacturing companies engaged in producing hard, configurable products, whereas at the high end of the spectrum are service companies whose products can be totally configured and delivered electronically (Kumar, 2007).

### 11.1.3 Industrial Practices

A manufacturer competes with competitors on price, quality, flexibility, delivery, and service. Mass customization can potentially satisfy two competing priorities, price and customization, simultaneously, so it is applied by more and more companies. Many large companies, such as computer manufacturer Dell, computer and mobile phone manufacturer Lenovo, and mobile phone manufacturer Nokia, have successfully implemented mass customization.

According to Feitzinger and Lee (1997), "the key to mass-customizing at Hewlett-Packard effectively is postponing the task of differentiating a product for a specific customer until the latest possible point" in the supply chain; three major organizational principles are modular product design, modular process design, and agile supply networks.

In 1998, the market share of Dell computers increased about 54%, and its profit increased about 62%. In 1999, 50% of Dell computers were sold by using e-business and mass customization. In 2001, Dell continuously broke its market and profit records when its competitors were losing their competences and market shares. In fact, it was a "mass customization" strategy that contributed to Dell's core competences and success. By using mass customization, Dell's cash conversion cycle, the time when a company pays its suppliers to the time payment is received from customers, is –37 days, whereas most companies have a weeks or months long cash-conversion cycle. Due to the modularity of PCs, the customer pays for the custom-configured computer even before the PC is assembled. Many of Dell's superior performance parameters such as customer service and short delivery lead times are ascribed to its successful mass customization strategy.

A completely customized garment will be ready in the TC2 company within hours and will be delivered to the customer in 3–5 days, to earn an extra USD 15 for a pair of customized jeans. Within the automobile industry, Ford, GM, Chrysler, and BMW have made significant efforts and resources to institute mass customization practices. Duray (2002) has demonstrated based on a sample of 126 companies that mass customizers perform financially better than non-mass customizers. Lenovo, Nokia, and many industrial manufacturers have similar success stories that are described in many media.

It is predicted that 30% of products will be customized in 2015. Some of them are transferred from mass production, such as automobile industry and household appliance industry. Some of them are transferred from normal customization, such as ship and industrial steam turbines. Furthermore, some others are transferred from batch mode, such as the machine tool industry and aviation industry. Currently, mass customization research and practice are developing very quickly. In fact, mass customization has become the main production mode of the 21st century.

The structure of this chapter is as follows. The production processes and several production modes of mobile phone components are described in Section 11.1. The mass customization manufacturing solution of the four main phases (marketing, R&D, production, and purchasing), especially customized order processing, process quality assurance, statistic process control, and architecture, are described in Section 11.2. The solution's practice results on production cost, customer satisfaction, product and process quality, and the overall benefits are given to prove the effectiveness of mass customization in Section 11.3. Finally, some conclusions are drawn in Section 11.4.

## 11.2 Mobile Phone Production Process Description

Every customer wants to buy unique electronic products, such as mobile phones, which are composed of hardware platforms, software platforms, user interfaces, and accessories. Hardware platforms include different types of components such as Wireless Local Area Network (WLAN), Blue Tooth (BT), Global System for Mobile Communications (GSM), General Packet Radio Service (GPRS), Code Division Multiple Access 2000 (CDMA 2000), Wideband Code Division Multiple Access (WCDMA), Time Division-Synchronous Code Division Multiple Access (TD-SCDMA), and Worldwide Interoperability for Microwave Access (WiMax); software platforms contain different operating systems such as BlackBerry 4.3, Palm 5.4, Symbian 9.2, Windows Mobile 6; and different operators include China Mobile, Vodafone, O2, Orange, T-Mobile, Verizon Wireless, AT&T Mobility, and TeliaSonera. User interfaces have different languages such as English, Chinese, Spanish, German, Finnish, and accessories consist of charger, battery, memory cards, data cables, hand-free ear phone, user manual and so on.

To meet these customer requirements, and pursue sufficient production efficiency, mobile phone production processes of large manufacturers are normally arranged in several phases as shown in Fig. 11.1, mainly composed of an engine operations (ENO) part and a supply operations (SO) part.

### 11.2.1 Mobile Phone Production Processes

In the engine operations (ENO) part, the Barcode (BARC) labeling station phase, panel flash phase, flash/alignment phase, final user interface phase, and bulk packing phase are mandatory, whereas other phases are optional. In the supply operations (SO) part, the labeling phase, insert assembly phase, master cartoning phase, palletizing phase, and shipping phase are mandatory, whereas other phases are optional. To realize mass customization, four levels of production order are assigned to four phases, that is, the Barcode (BARC) labeling station phase, panel flash phase, bulk packing phase, and sales cartoning phase. Detailed descriptions of these main phases are given below.

The Barcode (BARC) labeling station phase is the first phase in the line and the first variation point for the module production. Here, the PWB (printed wire board) enters the line. The modules are attached to the first work order level (WO_1) for the first variation containing module code, and the quantity to be produced fetched from the manufacturing execution system (MES) database. Module code and hardware ID defines the surface mounted device (SMD) components that are placed on the PWB. Operators can enter information about PWB, such as supplier and various information for tracking purposes. A barcode label is printed and attached to every PWB on the panel. One panel usually contains four, six, or eight PWBs. The PSN is printed. The panel ID and a unique production serial number (PSN) are generated for every module from the local parameter files.

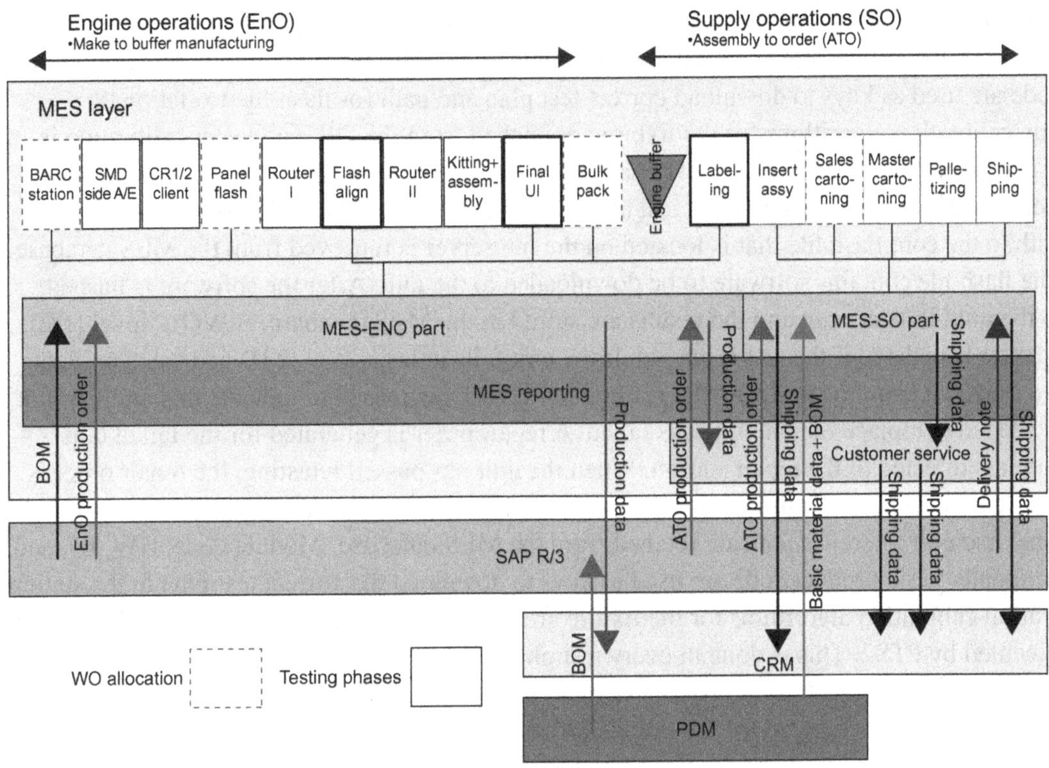


**Figure 11.1: Mobile phone production process.**

The PSN consists of factory code, line code, and running sequence number. The PSN label is printed on the barcode label and attached to the unit. Basic information about the module is created and saved in the MES database. The PSN label is used by barcode readers to identify and control the unit or panel in the next phases. Client shows the current status of the work order. When the work order is completed, the client asks the operator to accept the new work order.

In the SMD A/B phase, components are inserted to side A (top) and side B (down) of the printed wire boards by using the SMD machines. This phase is not connected to the MES database.

The change request (CR) 1/2 client phase reads the barcode label with a barcode reader and acts as a monitor. If modules have a CR or agamid alarm, the unit will be stopped at the first test phase.

The panel flash phase is a basic product code (BPC) attachment client, which is the second variation point for the unit (basic product). BPC, which comes from the work order, is given to the unit, and is attached to the second work order level (WO_2). BPC defines, for example, the color of the cover.

In the flash/alignment phase, the test phase, the PSN is read from a label and unit reference data are fetched from the MES database. Module code, HW_ID, and optionally basic product code are used as keys to download correct test plan and path for flash file to station. Station calibration algorithms for the fixture are fetched from the DB, and tester calibration is executed. This is done in every test phase. The test plan consists of test steps, and each test step has an algorithm to carry out test and parameters for valid limits of tested features. The path to the core flash file that is located on the file server is retrieved from the MES database. The flash file contains software to be downloaded to the unit. After the software is flashed to the unit, tests are run and the results are stored in the MES database. A WO_2 level (cf2) counter is updated if the test is passed. If the unit fails in tests, WO_2 ID and basic product are removed from the unit, and the cf1 counter is then decreased to indicate that another unit is needed to replace the one that has failed. A repair order is generated for the failed unit and it is directed to the repair station. When the unit has passed retesting, the repair order is closed and the passed unit is moved to the next phase. At alignment, the PSN is read from a label and unit reference data are fetched from the MES database. Module code, HW_ID, and optionally basic product code are used as keys to download the correct test plan to the station. Station calibration algorithms for the fixture are fetched from the DB, and tester calibration is executed by PTS2. This is done in every test phase. The test plan consists of test steps; each test step has an algorithm to carry out test and parameters for valid limits of tested features. A repair order is generated for the failed unit and it is directed to the repair station. When the unit has passed retesting, the repair order is closed and the passed unit is moved to the next phase.

In the kitting client phase, the PSN and the basic product code are read from escort memory. Work instructions are presented in pictorial form, and assembly is done manually by an operator according to the part list. The client gives an alarm to the operator when the basic product code changes.

In the final user interface phase, the user interface-related functions are tested. PSN is read and the correct test plan is called from the MES database using module code, HW_ID, and basic product code as keys. No software is programmed in the unit. The test plan contains information regarding test steps, used algorithms, and limit values for the tested features. Test results are stored in the MES database. In case the unit fails in tests, the repair order will be opened and the unit directed to the repair station.

The labeling phase is also a test phase; a language flash file (its name is PPM) is flashed into the unit. The PSN is read, and unit reference data are fetched from the MES database. Data are used to get an extended flash file that contains, for example, the language software that defines the language version for the unit. Module code, product code, and SW version are used as keys to find the correct language flash file. The test plan is searched with module code, HW_ID, and product code (can be searched with other keys also). Product code is allocated to the unit and attached to the work order level 3. This is the third variation point for the unit. An electronic serial number (ESN) is created from a file server. The final tests are run. If the unit passes tests, the unit is completed. The WO_3 level (cl2) counter is updated if the tests are passed. Type label is printed. If the unit fails in the tests, the WO_3 ID and product code are removed from the unit and the cl1 counter is decreased. A repair order is generated for a failed unit and it is directed to the repair station. When the unit has passed the retesting, the repair order is closed.

The units are sent to and packed in a distribution center (DC) either in bulk or if the line is equipped with inline packing, they are packed at the end of the line. If the units are not packed at the end of the line, they are put into bulk for later packing in the distribution center. The bulk is automatically opened for each label tester and has a unique bulk master ID that is attached to the bulks. A new bulk is opened when the previous one is full. If integrated packing is in use, units are packed (no bulk is used) at the end of the line or stored for later packing in integrated packing station(s).

When supply operations (SO) lines are separate from engine operations (ENO) lines, the engines will be packed and sent to an engine buffer.

Bulk functions are called to handle attachment of units to the bulk. After the bulk is ready, it is sent to the distribution center for final packing and delivery. Each label tester (station_id) fills its own bulks. Physical bulk can be handled only by the station/tester with which it has been created. Bulk master ID (BMI) is line and station specific. If a unit is reworked, MC or DC rework bulk is used. A client can close and reopen bulk.

After the engines are opened again on another SO line, its ENO-related data are retrieved from the databases of the MES. When every check is passed, the engine will be assembled into phone product, be flashed UI software, and then tested.

In the sales cartoning phase, every phone is finally packed into its own sales carton according to the customer's requirements. Fixed numbers (e.g., 12) of sales cartons are packed again into one master carton. Many master cartons are packed into one pallet. And then, those pallets are shipped out of production lines for transportation.

### 11.2.2 Mobile Phone Production Modes

There are six different production modes to realize the customized manufacturing of mobile phones (Fig. 11.2), that is, buy to order (BTO), make to order (MTO), assemble to order (ATO), pack to order (PTO), ship to order (STO), and customer to order (CTO).

In ATO mass production mode, the whole mobile phone production process is composed of an engine operations (ENO) part and a supply operations (SO) part. The MES-ENO (Fig. 11.3) part mainly manages engine operations phases; the MES-SO (Fig. 11.4) part

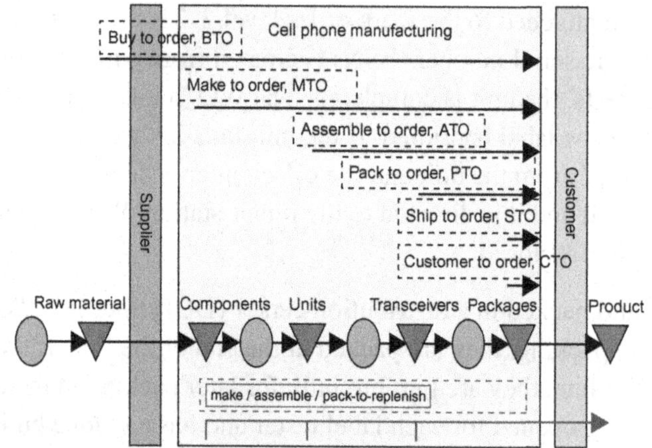


**Figure 11.2: Production modes of mobile phones.**

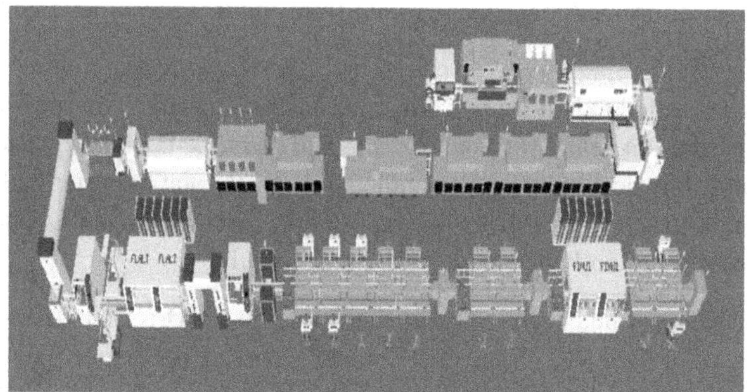

**Figure 11.3: MES-ENO line.**

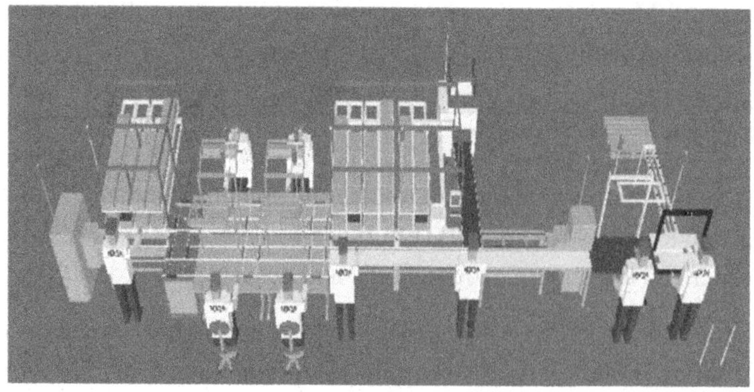

**Figure 11.4: MES-SO line.**

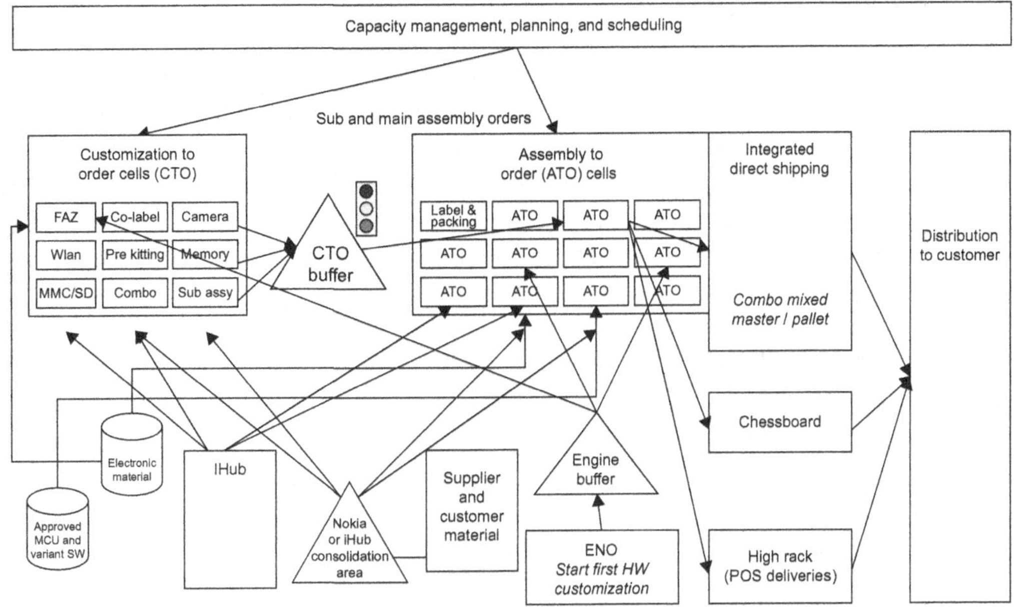


**Figure 11.5: ATO and CTO of mobile phone manufacturing.**

mainly manages supply operations phases. MES is applied to collect, analyze, and report the production data of ENO and SO about product material, order, quality, and status. Its details are as follows:

ATO and CTO are the latest modes, mainly to realize mass customization. The detailed descriptions of ATO and CTO are shown in Fig. 11.5.

## 11.3 Mass Customization Manufacturing Solution

The mass customization manufacturing solution for mobile phone production is described below. First, the main manufacturing phases of mass customization are introduced. Then, the production order processing of mass customization is described in detail. Finally, quality control, technical architecture, and practice results of the solution are given.

### 11.3.1 Main Manufacturing Phases of Mass Customization

There are normally four main phases involved in the mass customization in a mobile phone enterprise, that is, the marketing department executes customized product order function with customer relationship management (CRM), the R&D department executes customized product design function with product data management (PDM), the production department executes customized production function with ERP+MES, and the purchasing department executes customized product material function with supply chain management (SCM).

#### 11.3.1.1 Customized Product Order

To meet the mass customization requirements from the marketing department, existing CRM is modified, where the product specification, order, and requirements are recorded, transferred to the R&D department for product design, production department for product making, and purchasing department for material preparation. At the same time, the customer can create his/her own product order and track every phase of its processing, until receiving the product. The extensible function of CRM is to realize the whole product lifecycle management.

#### 11.3.1.2 Customized Product Design

To meet the mass customization requirements from the product R&D department, existing PDM is modified, where the customized product is designed according to the customized product order and requirements. After enough hardware components and software components are made and accumulated, their combination can create as many variations as needed; the customized product design can easily be executed.

#### 11.3.1.3 Customized Product Material

To meet the mass customization requirements from the material department, existing supply chain management (SCM) is modified, where the correct product material is purchased at the correct time, quantity, quality, and price to meet the production requirement and reduce the storage. To realize full lifecycle management of material quantity and quality from every supplier, and to transfer material data to the production department, marketing department, and customer department, corresponding feedback is necessary to improve the purchasing quality and efficiency.

#### 11.3.1.4 Customized Production

Customized production is a production mode where the customized product is made by assembling the existing parts and components only after the customized product order is received. To meet the mass customization requirements from the production department, existing ERP and MES are modified to a mass customization production management system as shown in Fig. 11.6, and the system is used to fulfill the targets given below:

1. Solution can meet the requirements of product order as small as only one, or as big as millions.
2. Production profits can be made even when product order quantity is as small as only one.

### 11.3.2 Order Processing of Mass Customization

The MES, together with SAP R/3 ERP, CRM, and PDM, consists of the IT systems that realize mass customization. The four layers of data management of ERP and MES are

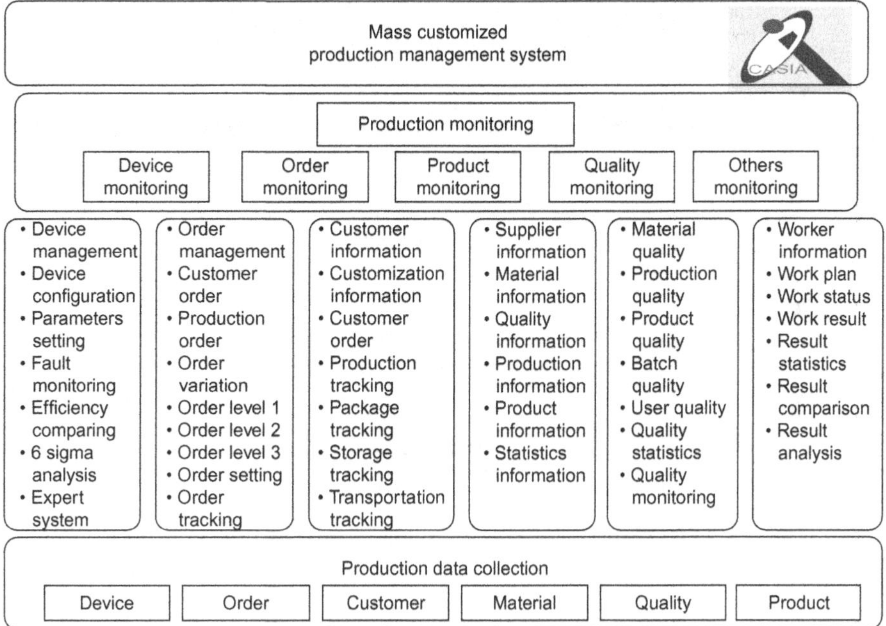


**Figure 11.6: Main functions of mass customization production management system.**

described in Fig. 11.7. First, almost all related data, such as material resource plan (MRP), part lists, sales orders, production order, scheduling, releasing, and delivery, are collected to assure the second step. Then, mass customization production is mainly executed by strict and accurate order and version management. Finally, data are used for analyzing, reporting, and archiving. Almost any kind of data can be collected and managed for different purposes.

The order processing (Fig. 11.8) is the main tool to realize mass customization. An example of customized production is particularized. In the MES of the mobile phone manufacturer, current work order structure has four work order levels (Fig. 11.9) and one level means one variation point:

1. Work order level 1 (WO_1) is for manufacturing modules. They number between 10 and 50.
2. Work order level 2 (WO_2) is for manufacturing basic product codes. They number between 20 and 100.
3. Work order level 3 (WO_3) is for manufacturing product codes. They number between 50 and 100.
4. Work order level 4 (WO_4) is for sales package. They number between 20 and several hundred. A unit is attached to a work order at every variation point. The work order

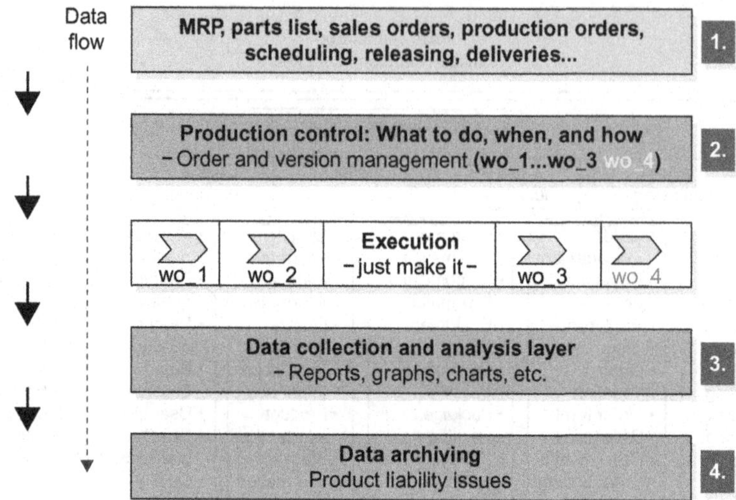


Figure 11.7: Four layers of data management of ERP and MES.

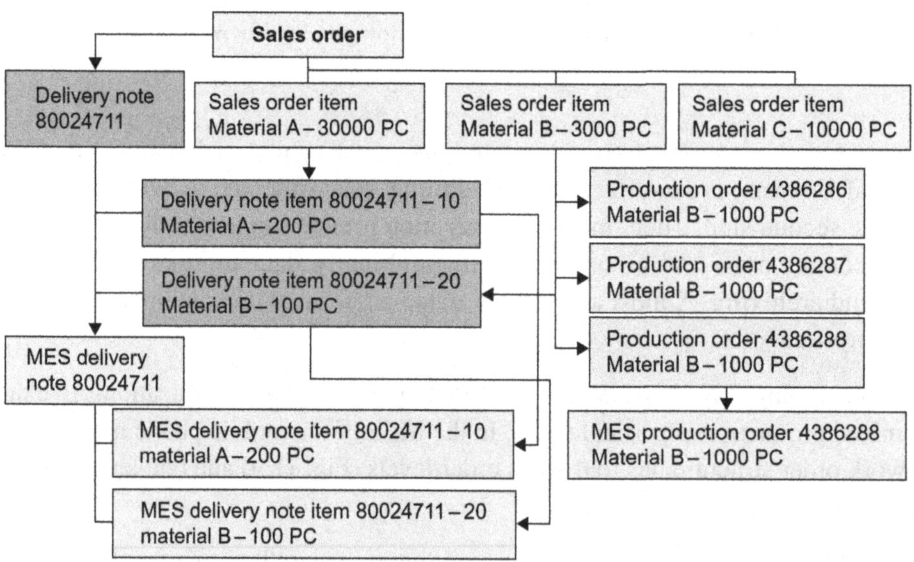


Figure 11.8: The order processing workflow.

attachment routine is dynamic and follows the last work order priority sequence at each variation point. If a unit fails at a certain test phase, it is removed from the work order and the next unit in line will dynamically replace it.

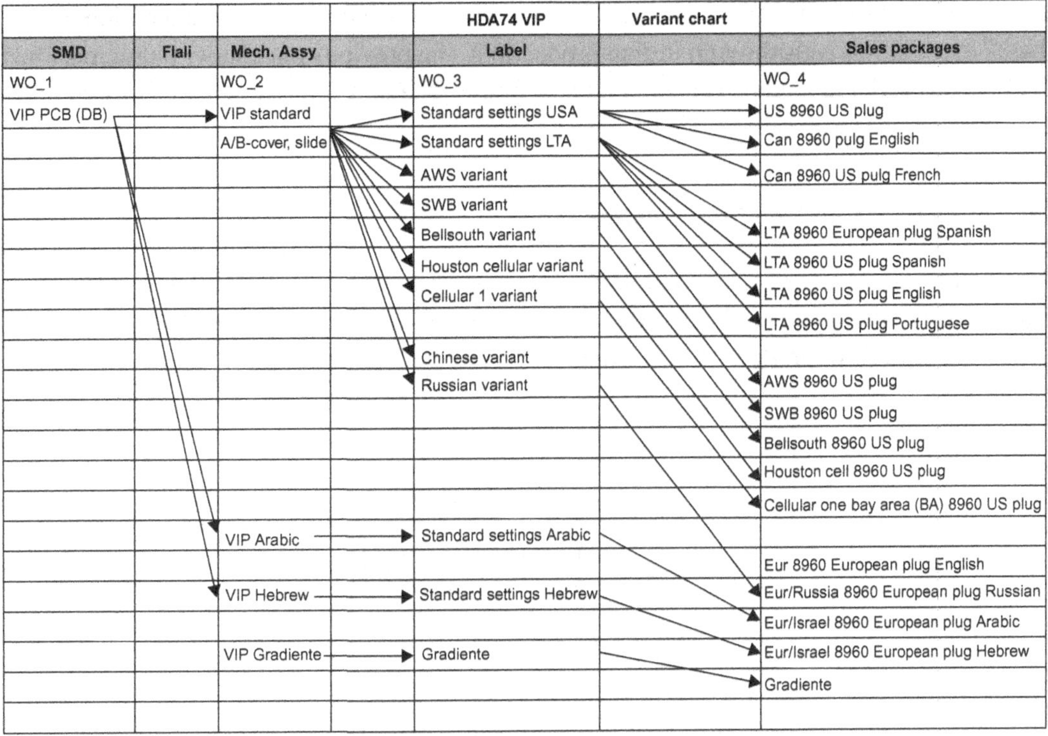


**Figure 11.9: The order variation of mass customization.**

Let us assume:

*WO_1 variation number = 20*
*WO_2 variation number = 50*
*WO_3 variation number = 50*
*WO_4 variation number = 200*

Then the work order variation of the four levels can create a number of products as large as $20 \times 50 \times 50 \times 200 = 1{,}000{,}000 = 10$ million!!

The work order has the following status codes at each work order level:

5. PENDING (in queue) – work order pending in queue means it is waiting to be processed.
6. ACTIVE – work order is in progress.
7. CLOSED – work order is completed and closed.
8. LOCKED (in edit) – work order is being modified.
9. ATTACHMENT STOPPED – unit attachment to work order is stopped.
10. HOLD – work order is put on hold (paused) status.

A module is linked to a phase order. The phase orders define all mandatory and optional phases that a unit travels through during production. The previous mandatory phase must be completed before the module can enter its next phase. The unit can return to any passed phase in order to rerun the phase.

### 11.3.3 Quality Control of the Solution

The quality assurance measures include statistical process control (SPC) and mobile phone quality testing, as shown in Fig. 11.10. An SPC system (Fig. 11.11) is applied for production line control and semi-product phase testing, such as sensitivity analysis, testing the effect of a

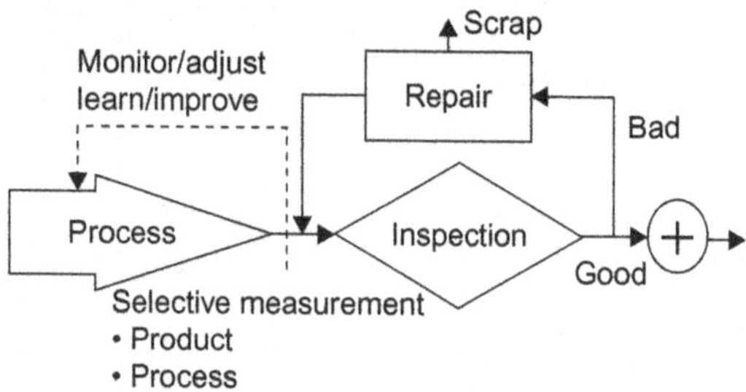


**Figure 11.10: The process and product quality assurance concept.**

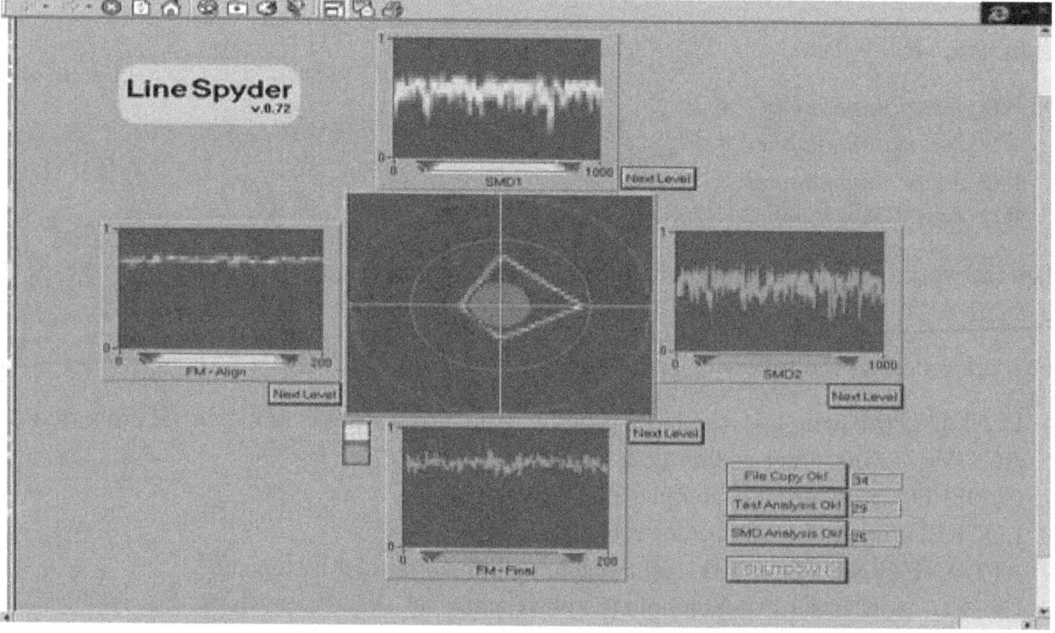


**Figure 11.11: The statistical process control of an SMD machine.**

single factor treatment, and so on. The overall process status online can be represented by just a few indicators. One indicator can represent the status of one process phase, and the closer to zero one indicator is, the better situation the process phase has.

Let us take an example, the calculation of an SMD indicator: "fuzzy SMD process indicator for SpyderMap" is the average of "fuzzy paste print machine indicator," "fuzzy SMD machine indicator," and "fuzzy reflow oven indicator." The fuzzy paste print machine indicator is the "XR-chart of test pads" of the paste print machine. The fuzzy SMD machine indicator is calculated with "picking error of nozzles," "picking error of feeders," and "XR-chart of cycle time." The fuzzy reflow oven indicator is "XR-chart of temperature."

Mobile phone quality is verified by testing plan execution with testers on different phases. Here is a simple test plan example of DHA14:

*Step 1. First time tests;*
*Step 2. Read HW and SW versions;*
*Step 3. Audio tests;*
*Step 4. Calibrate charge current;*
*Step 5. Keypad test;*
*Step 6. Set base station and call test;*
*Step 7. Set call tests to cellular tester;*
*Step 8. Check keypad test result;*
*Step 9. LCD Test A all ON;*
*Step 10. Call tests part 1;*
*Step 11. LCD Test A all OFF;*
*Step 12. Call tests part 2;*
*Step 13. LCD Test A BORDER;*
*Step 14. Call tests part 3.*

Calculation of test indicator:

$$\mathrm{FQI} = \mathrm{mean}\,(\mathrm{PQI}_1 \ldots \mathrm{PQI}_\mathrm{N})$$

where $\mathrm{PQI}_j = \mathrm{mean}\,(\mathrm{MQI}_{j1} \ldots \mathrm{MQI}_{jN})$, $j = 1, \ldots, N$, and $\mathrm{MQI}_{ji} = F(m_{ji})$, $i = 1, \ldots, M$.

FQI is the fuzzy quality indicator, PQI the product quality indicators, MQI the measurement quality indicators, *m* the different parameters of the measurement device, and *F* the function.

Besides the online test, there are extreme environment tests and statics tests executed manually with a sale package. Together they comprise a complete quality assurance system.

### 11.3.4 Technical Architecture of the Solution

The technical architecture of the solution can be described as shown in Fig. 11.12. Its service-oriented architecture (SOA) three-tier structure includes client, middleware, functions,

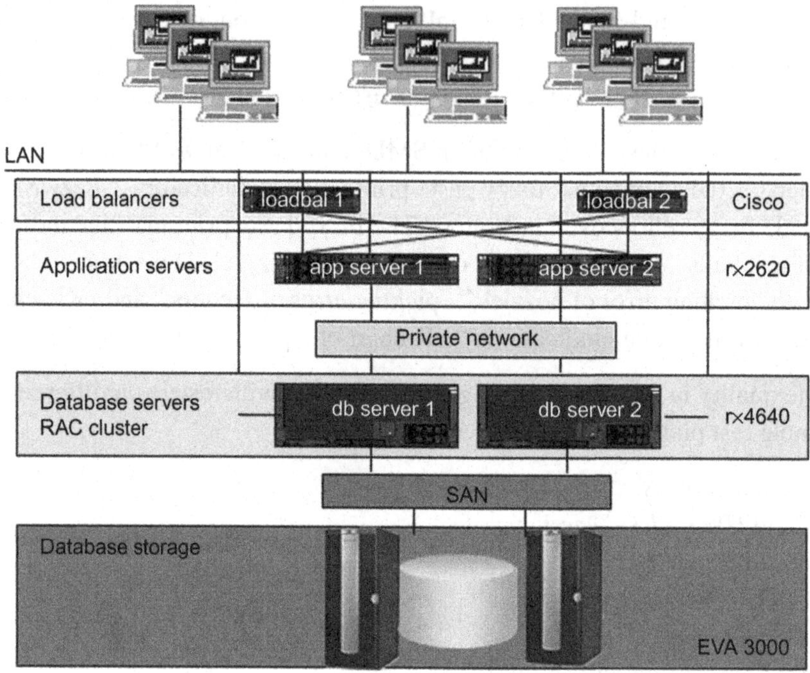


**Figure 11.12: The technical architecture of the solution.**

and database (DB). When the product data and production data are collected and saved in the database, the system can provide different reporting functions. System intelligence (business logic) is built into individual functions or services. Functions communicate with the Oracle DB. Tuxedo middleware knows where the services (functions) are, how to call them, how to deal with parameter interfaces, and how to arrange different calls to different function servers. Clients on the production line call Tuxedo middleware with the service name and parameters, to perform the wanted function(s). The clients can be machine clients (like testers) or user clients.

The main components on the server side are:

*HP servers, UX 11.23 DB servers use 8 GB memory, 2 × 1.3 GHz Itanium CPU*
*Load balancers, Cisco*
*Oracle 9.2.0.6*
*Tuxedo 8.1, WebLogic 8.1*

There are two separate servers for the application; application servers use 4 GB memory, 2 × 1.6 GHz Itanium CPU. Application server 1 is the primary server for all clients, on which all production processes are running for one Tuxedo instance, and background processes can be started manually if application server 2 is down for a long time. Application server 2 is a backup server for all clients. Background processes are also running there for another Tuxedo instance. All production processes are running there but not accessed by clients during the normal process.

Clients use Weblogic that is running in application servers 1 and 2. Connections from the clients go first to one of the two load balancers. The load balancers divide the service requests between the two application servers, which were mentioned earlier.

There are two servers in a cluster for the database. Each server has its own database instance. A real application cluster (RAC) controls the system. Connection to the database is done using cluster features. Oracle Net connects randomly to one of the listeners on server 1 or 2 that redirects based on load information. The listener selects an instance in the following order: least-loaded node (1-minute node load average) and least-loaded instance (number of connections to each instance). If the instance fails, RAC automatically moves the connection to the other instance. If a client is executing a query, the client will not see any interruption. All transactions are rolled back.

### 11.3.5 Practice Results of the Solution

The solution described above is one of the main success factors of mobile phone manufacturers. The practice results of the solution are very positive. In fact, large-volume mobile phone manufacturers have many partners such as Sanyo supplies battery, Excel provides for logistic services, Ibiden supplies PWB, Verso supplies accessories, Elcoteq provides PWB and engine, and Foxconn provides for low-end phones. This perfect supply chain can make 400–500 million mobile phones every year, and its manufacturing costs occupy only 6–10% of production costs, lower than its competitors. Every mobile phone is made against its customer order to meet the specific requirements of each specific user. The reliability of the production line is improved with statistical process control (SPC). The quality of semiproduct and final product are assured by different types of strict testing.

The screen shown in Fig. 11.13 mainly provides work order online monitoring function of customized production of mobile phones, where all active orders are listed. For every work order, its production line, product code, and module code are shown and its total quantity, pass quantity, and fail quantity are summarized.

Figure 11.14 shows the production quality online monitoring function of customized production, where all failed PWBs, modules, units, engines, and phones are tracked and analyzed to find the reason for failure nearly real time. The statistical process control (SPC) analysis mainly uses the six sigma method and tool.

The root reason may be the "material" quality from suppliers. If so, the whole batch of material will no longer be used and will be recalled. Negotiation will start with the supplier. The material can be battery, PWB, monitor, or chips.

The root reason of failure may be from the production line. For example, some machine's parameters are not adjusted or calibrated correctly. If so, production will be stopped to repair the machine in trouble, and all products made by the machine will be recalled.

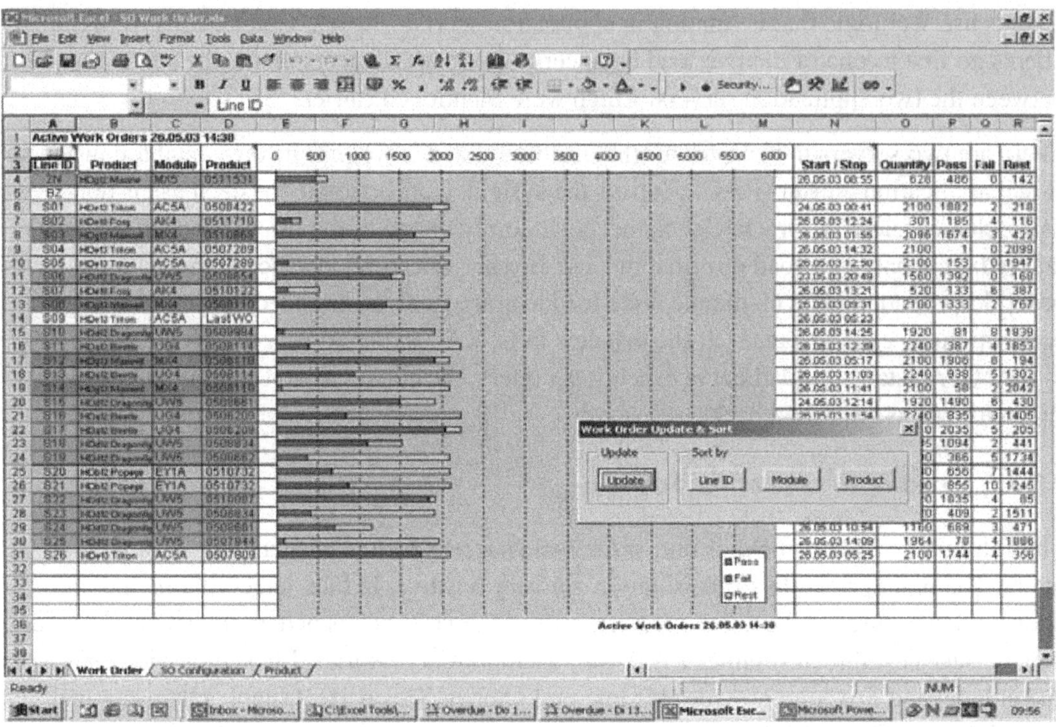


Figure 11.13: Work order monitoring of customized production.

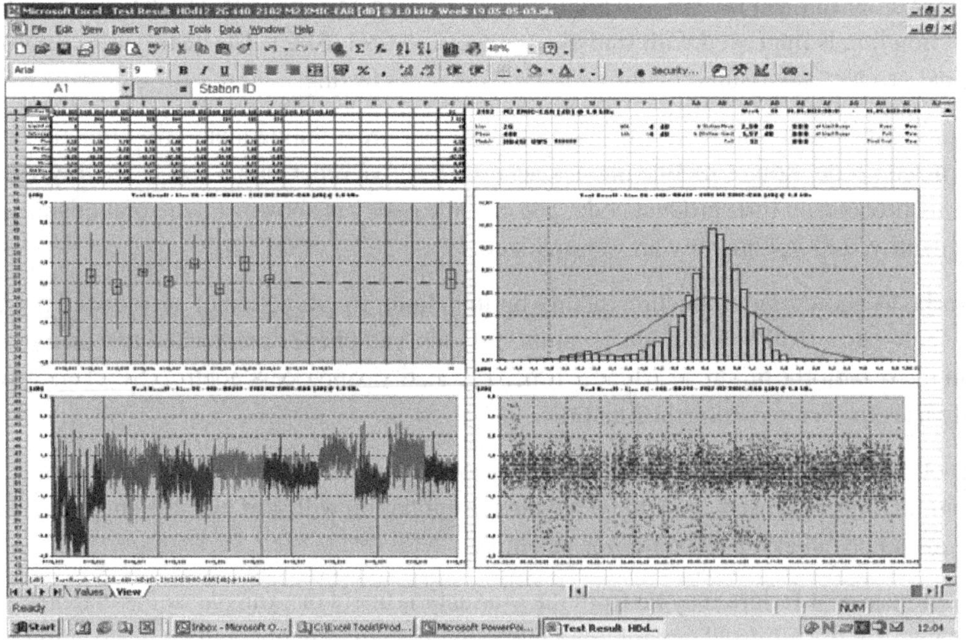


Figure 11.14: Production quality monitoring.

The root reason may also be from a production worker. For example, if a worker is careless or tired, then mistakes are made. If so, the worker's work will be stopped and recalled.

All kinds of quality, quantity, and status data can be compared between different times, different production lines, different materials, different workers, and different factories to find possible mistakes, exceptions, and problems. And then, the production quality and efficiency can be improved continuously.

The solution can support production rates as high as 7 seconds per phone in a factory. The solution itself is also of high quality. For example, its service availability can be as high as 99.97%, which means less than 3 hours unplanned breaks occur every year. The overall efficiency of the solution can also be seen on the key metrics scorecard (Fig. 11.15) of operational excellence.

For end-to-end manufacturing costs, more emphasis will be on analysis of total end-to-end manufacturing costs, including inventory carrying cost (ICC), sub-assembly manufacturing value added (MVA), and compound product cost (CPC). Follow-up of E2E manufacturing costs is very critical to avoid suboptimization of cost, for example, when deciding on what is manufactured in-house (CPC) and what is outsourced (MVA). For design for manufacturing

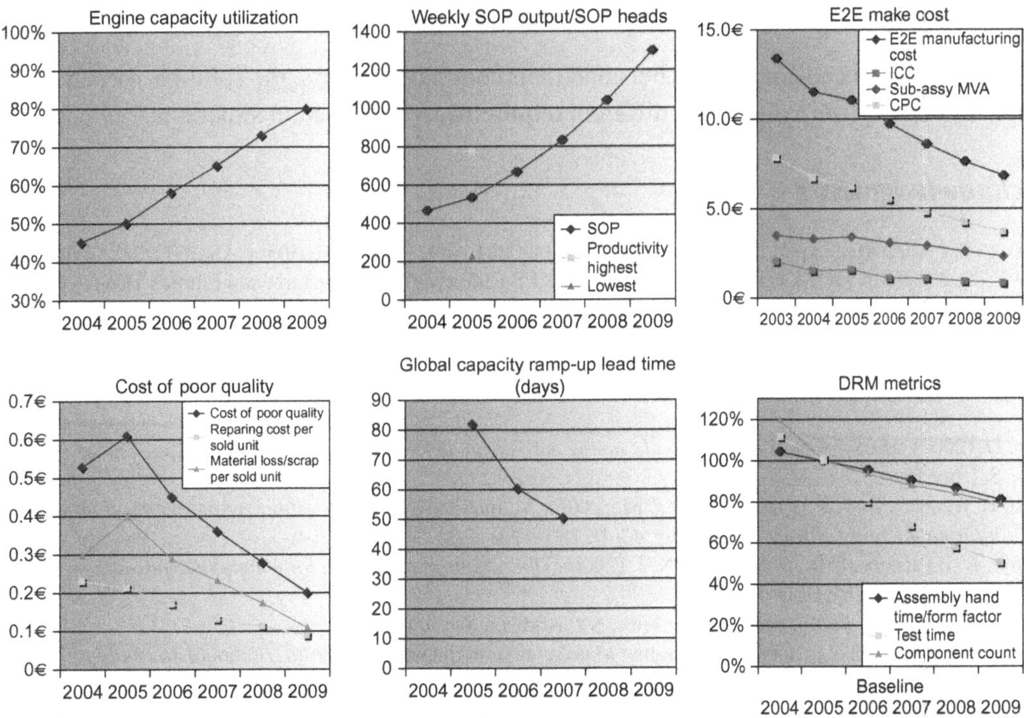


**Figure 11.15: Key metrics scorecard of MC solution.**

(DFM) metrics, key performance indicators (KPI) for "product manufacturability" are as follows: assembly time per form factor, test time, and component count.

Now, the story below becomes fact. Every morning, customer orders are collected and transferred to factories and the cargo transports all necessary "materials" from suppliers' storages to the mobile phone factory. ERP and MES help the factory arrange suitable workers, production lines, and the time for making the phone to fulfill the order according to the customer's requirements. In the afternoon, all phone products are made and shipped to the cargo waiting outside. In the evening, all these are transported to different customers all over the world. The response time from the creation of the customer's order to receipt of the phone can be as short as several days. The customer order can be as small as only one. The storage of material and product can be as small as "ZERO." Mass customization is fully realized.

## 11.4 Conclusions

In this chapter, a type of mass customization manufacturing solution is given for mobile phone production, which has been successfully applied to a large-volume mobile phone manufacturer. The solution can easily be applied to other mobile phone manufacturers, transformed and applied to other manufacturers of electrical goods such as laptops, and can be referenced by other mass customization industries such as the furniture industry and the clothing industry.

With the help of academic researchers and their continuing efforts, the solution will be improved and optimized to fulfill different requirements for different situations.

## Acknowledgments

This work is supported in part by NSFC 61174172, 70890084, 60921061, 90920305; CAS 2F09N05, 2F09N06, 2F10E08, 2F10E10; MOST I11B200070; Finnish TEKES Cloud Software Program; and Finnish TEKLA R&D Fund.

## References

Bin, D. (2004). *Mass customization: To build up the core competence of your enterprise*. Beijing, China: Science Press.

Cao, J., Wang, J., Law, K., Zhang, S., & Li, M. (2006). An interactive service customization model. *Journal of Information and Software Technology, 48*(4), 280–296.

Chase, R. B., Jacobs, F. R., & Aquilano, N. J. (2006). *Operations management for competitive advantage*. New York, NY: McGraw-Hill/Irwin.

Davis, S. M. (1987). *Future perfect*. New York, NY: Addison-Wesley.

Duray, R. (2002). Mass customization origins: Mass or custom manufacturing? *International Journal of Operations and Productions Management, 22*(3), 314–328.

Elizabeth, M. J. & Attahiru, S. A. (2009). A queueing model of delayed product differentiation. *European Journal of Operational Research, 199*(3), 734–743.

Feitzinger, E. & Lee, H. L. (1997). Mass customization at Hewlett-Packard: The power of postponement. *Harvard Business Review, 75*, 116–121.

Giovani, D. S., Borenstein, D., & Fogliatto, F. S. (2001). Mass customization: Literature review and research directions. *International Journal of Production Economics, 72*, 1–13.

Kumar, A. (2007). From mass customization to mass personalization: A strategic transformation. *International Journal of Flexible Manufacturing System, 19*, 533–547.

Kumar, A., Gattoufi, S., & Reisman, A. (2007). Mass customization research: Trends, directions, diffusion intensity, and taxonomic frameworks. *International Journal of Flexible Manufacturing System, 19*, 637–665.

MacCarthy, B., Brabazon, P. G., & Bramham, J. (2003). Fundamental modes of operation for mass customization. *International Journal of Production Economics, 85*(3): 289–304.

McCarthy, I. P. (2004). Special issue editorial: The what, why and how of mass customization. *Production Planning & Control, 15*(4), 347–351.

Partanen, J. & Haapasalo, H. (2004). Fast production for order fulfillment: Implementing mass customization in electronics industry. *International Journal of Production Economics, 90*(2), 213–222.

Pine, B. J. & Davis, S. (1993). *Mass customization: The new frontier in business competition*. Boston, MA: Harvard Business School Press.

Smith, N. R. & Sánchez, J. M. (2008). Editorial: Special issue on intelligent systems for mass customization. *Journal of Intelligent Manufacturing 19*, 505.

Toffler, A. (1970). *Future shock*. New York, NY: Bantam Books.

Xiong, G. & Nyberg, T. R. (2000). Push/pull production plan and schedule used in modern refinery CIMS. *Robotics and Computer-Integrated Manufacturing, 16*(6), 397–410.

CHAPTER 12

# ACP Approach-Based Plant Human–Machine Interaction Evaluation

**Xi-Wei Liu*, Xiao-Wei Shen*, Dong Fan*, and Masaru Noda†**

*State Key Laboratory of Management and Control for Complex Systems, Institute of Automation, Chinese Academy of Sciences, Beijing, China

†Graduate School of Information Science, Nara Institute of Science and Technology, Nara, Japan

## 12.1 Introduction

This section introduces the background and related research pertaining to this chapter.

### 12.1.1 Background

Recently, to improve the usability and quality of various software systems, designs of human–machine interaction (HMI) have gained much attention in the academic and industrial domains. In particular, the HMI running in the central control room of a chemical plant has high requirements of safety and reliability, and these prototype systems need rigorous evaluation and testing before application. There are many evaluation criteria, experience-based guidelines, and heuristic evaluation methods to support the design of the HMI, but it is difficult to integrally predict and analyze the dynamic characteristics of the system by using these methods. Moreover, it is impossible to analyze the capabilities and performance levels of the plant operator.

Meanwhile, with the rapid development of information and control technologies, the efficiency and the automation levels of the safety-critical chemical process system are improved. Therefore, plant operators have to face more complex, large-scale, highly integrated production systems, which results in new risks and failure modes. Statistics (Aas, 2009) show that 60–85% (the statistical data comes from different industrial domains) of industrial accidents originate from human errors. A deficiency of traditional HMI design is that they only considered rationality from an engineering point of view, but ignored the operational habits and cognitive capabilities of the users. We believe that it is a positive research trend to mimic human information processing from the viewpoint of cognitive psychology and then evaluate the prototype of HMI for a human-centered design.

Service Science, Management, and Engineering.


DOI: 10.1016/B978-0-12-397037-4.00012-0

HMI is a complex system including human operators, plant process control systems, and management of people and equipment. Traditional design and evaluation of HMI mainly contains a checklist about its visual effect, such as color, graphic expression, and layout. However, it is emphasized by the researchers of complex systems that effective conclusions should be based on investigation of the objective system as a whole. It is considered as a deficiency in complex system research that only the engineering complexity is involved but the social and psychological complexity is omitted. To cope with this problem, Wang proposed a new methodology called the ACP approach (2004a, 2004b, 2004c, 2004d, 2006), which includes three parts: Artificial systems, Computational experiments, and Parallel execution. In this chapter, we propose a research road map applying the ACP approach to explore the usability of the complex HMI in a chemical plant, and try to cover all related factors, such as user panel design, human cognition, and memories during the course of fault detection and isolation (FDI). Based on the ACP approach, an artificial system is firstly constructed with various agents and models for the objective plant and then computational experiments are carried out to evaluate the usability of the objective plant HMI. By the integration of the real and artificial systems, usability and operability of the real HMI and efficiency of the artificial system can be improved.

### 12.1.2 Related Research

Under a certain hardware environment, the usability of an HMI system is decided by two aspects: human and user interaction. Correspondingly, studies on human errors and usability evaluation are involved. Various human models are built for these two domains. Several of these models were investigated in this research.

#### 12.1.2.1 Model Human Processor

As shown in Fig. 12.1, the model human processor (MHP) proposed by Card, Moran, and Newell (1983) is used to explain and predict how a human responds to a stimulus. The MHP is composed of memories and processors. The memories are characterized by storage capacity,

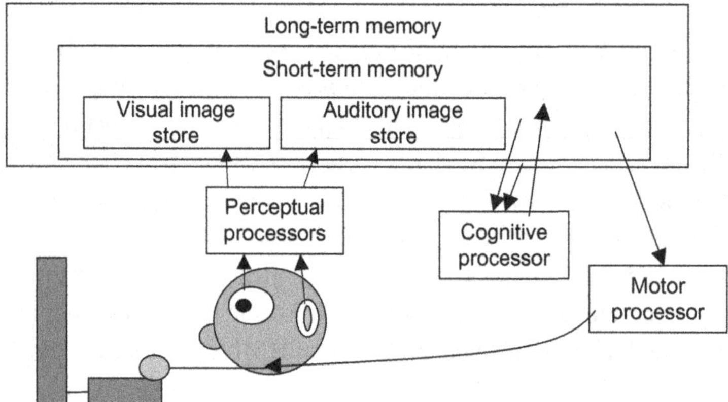


Figure 12.1: Model human processor.

decay time, and type of coding, and the processors by cycle time. The values of these attributes were determined by empirical studies.

Three subsystems are included in the MHP: perceptual, cognitive, and motor subsystems. The perceptual subsystem consists of sensors and associated buffer memories. A visual image store and an auditory image store hold the output of the sensory system after the output is symbolically coded. The cognitive subsystem receives symbolically coded information from the sensory image stores in the short-term memory (STM) and uses previously stored information in long-term memory (LTM) to make decisions about responses. The motor subsystem carries out the responses.

The MHP is a conceptual model, which provides only a research framework. It can be employed in various applications after embedding information-processing procedures and related data.

#### *12.1.2.2 Operator Model for Nuclear Power Plant*

Takano et al. at the Central Research Institute of the Electric Power Industry (CRIEPI) proposed a simulation system for the behavior of an operating group (SYBORG) to simulate and analyze the cognitive process of the operators and the behavior of operation teams (Takano, Sawayanagi, & Kabetani, 1994; Takano, Sasou, Yoshimura, Iwai, & Sekimoto, 1994). SYBORG simulates behavior of three operators—one is the leader of the team and the others are followers with different roles. It is assumed that the leader does not observe or touch the control panel. The leader model collects information of the plant via communication.

The operator model consists of attention, thinking, action, and utterance micromodels. The thinking micromodel introduces the "mental model mechanism," which describes and illustrates how the operators predict plant behavior and make decisions to prevent the deterioration of its conditions. It was developed based on cognitive science, group dynamics, and also on interviews with nuclear power plant operators. Each operator model has some knowledge bases (KBs). They store knowledge pertaining to the relations between (1) events and parameters, (2) events and causes, (3) change of parameters and interlock, (4) change of parameters and carrying out countermeasures, and so on.

The above-mentioned operator model is not enough for simulating the complete behavior of the operator. So, some other characteristics relating to team behavior have been incorporated. Thus, the authors introduced the human–human interface model that comprises the task assignment, disagreement, and utterance management micromodels. Human personality, credibility, position, etc. are considered in the model.

The SYBORG considers a large number of KBs and is applied to simulate some particular situations. However, its application is restricted to several cases. Complex structure is helpful to simulate details of human behavior, but it introduces many uncertain factors as well. In this research, its mental model mechanism is referred to and built into a KB.

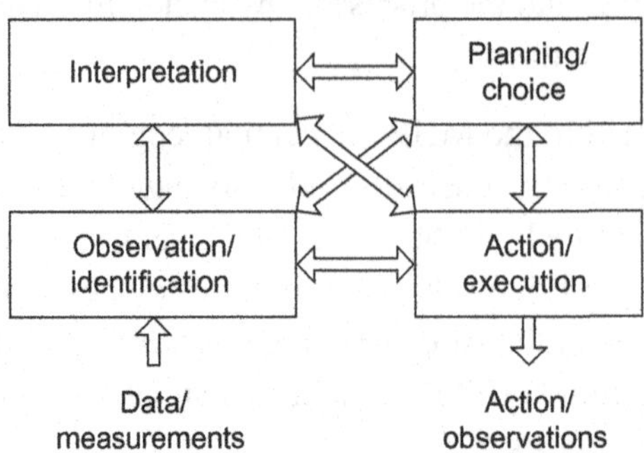


**Figure 12.2: Simple model of cognition.**

### *12.1.2.3 Simple Model of Cognition*

Hollnagel proposed a simplified model of cognition (SmoC) (1998) as shown in Fig. 12.2. SmoC illustrates four types of human behaviors. Based on this model, a person observes and identifies a visual or auditory signal, interprets the signal, plans and decides what operation has to be done, and finally initiates and executes an action.

In the case of plant operation, the first step of the SmoC information-processing model addresses the observation and identification of graphic items on user panels. The second step describes how an operator interprets and organizes the information into a memory unit. The third step addresses the planning and decision-making processes involved. The fourth step refers to execution of the planned actions.

Corresponding to MHP, these four types of human behaviors are executed by three processors. The cognitive processor performs interpretation, plan, and choice. The perceptual processor concerns observation and identification. The motor processor involves action and execution. SmoC provides a sequence of these human behaviors, and some intermediate steps may be bypassed during the information processing.

## 12.2 Artificial Human–Machine System

To study a human–machine system (HMS) from the viewpoint of the ACP approach, we built an artificial HMS for a chemical plant, as shown in Fig. 12.3, which includes a chemical plant agent, an HMI agent, and an operator agent.

The MHP is a conceptual framework, modeling a human operator as an information-processing system. In this study, based on the MHP, a cognitive information-processing agent workable on a PC is developed as a virtual subject to detect and isolate causes of failure. The cognitive processor sends commands to a motor processor to move gaze point or to push a button to confirm the status of the associated variables. The motor processor executes the

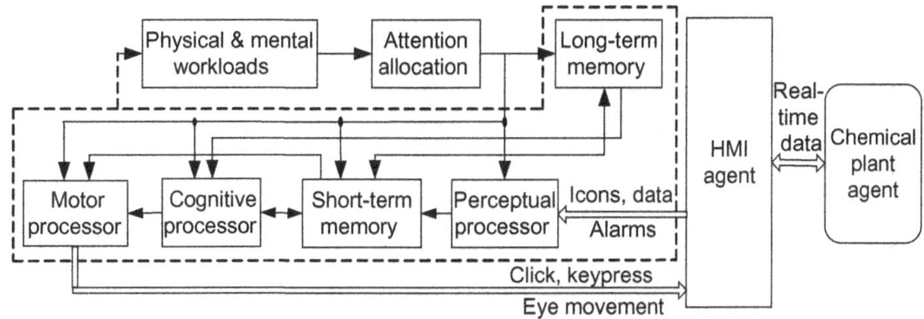


**Figure 12.3: Framework of artificial human–machine system.**

commands from the cognitive processor. Knowledge bases for variable information (VI-KB), failure–symptom relations (FS-KB), and alarm management (AM-KB), as well as a normal-state-monitoring (NSM) procedure and an abnormal-state-supervising procedure (ASSP), are constructed in the LTM of the agent.

The VI-KB includes color, position, and the normal range of every process or control variable on the user panels. The operator agent uses the VI-KB to find the position of a relevant graphic item on the HMI. Once a malfunction occurs, some process variables (PVs) change outside of their normal ranges and even violate alarm limits. These changes are considered symptoms of malfunctions. The FS-KB is built based on a general cause–effect analysis and contains all the known failures with these symptoms. The AM-KB includes responses to these alarm messages and has rules to convert an alarm status of the plant monitoring system to a symptom.

To simplify HMI evaluation, we assume that the operator agent never errs due to lack of knowledge. The KBs and procedures are created based on process risk analyses, operational experiences, and expert reviews. In the LTM, VI-KB, FS-KB, and AM-KB are implemented by a table, a matrix, and if–then rules, respectively. When plant status is normal, human perception is in passive mode without definite intention. In an emergency, however, the operators mainly concentrate on certain items or areas. We assume that active perception directly transfers detected items into the STM.

In this research, the chemical plant agent is an independent boiler plant simulator, which is equipped with communication function and all variables can be accessed freely in real time by a software interface; the HMI agent is created based on a set of existing user interfaces including their graphic (static) and response (dynamic) characters.

## 12.3 Experimental Environment

This section includes three parts introducing the chemical plant agent with possible malfunctions for experimental scenarios and the configuration of its alarm system.

### 12.3.1 Overview of Boiler Plant Simulator

A boiler plant simulator for training is installed in a distributed control system (DCS), which includes a field control station (FCS) and several PC-based information command stations (PICSes). The DCS is made by Yokogawa Electric Corporation and can be equipped as an actual control system in a chemical plant. All real-time data in the plant simulator are accessed by using the object linking and embedding for process control (OPC) technology.

Figure 12.4 shows a sketch of the system's data communication. The boiler plant simulator is used as a plant model. Through the special field device network, FCS is connected with PC1 on which an OPC DA (data access) server is installed. Another computer, PC2, is connected to PC1 via the local area network. According to the OPC DA standard, the user interface model in PC2 can access the plant model in real time.

A process flow diagram of the boiler plant simulator is shown in Fig. 12.5. Tags of continuous PVs in the simulator are listed in Table 12.1. As a target state, the simulated boiler plant produces 80 t/h of superheated steam at 485 °C. The whole plant control system includes four subsystems of control: feeding water, steam temperature, combustion, and furnace pressure. In normal situations, the demand load of the simulated boiler plant randomly changes from 77.9 to 82.4 t/h, which determines the normal ranges of all variables in the plant. Table 12.2 shows the normal fluctuation of PV and manipulated variable (MV) values. The user interface system of the boiler plant includes the following panels: overview, operational, engineering, trend, and alarm summary. Figure 12.6 shows the overview panel as an example.

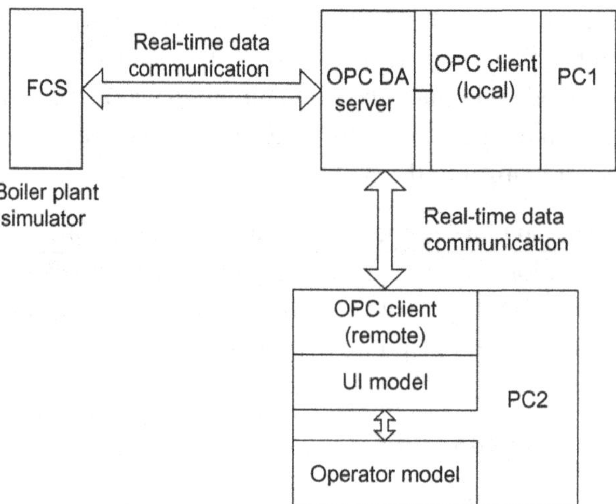


**Figure 12.4: Sketch of data communication.**

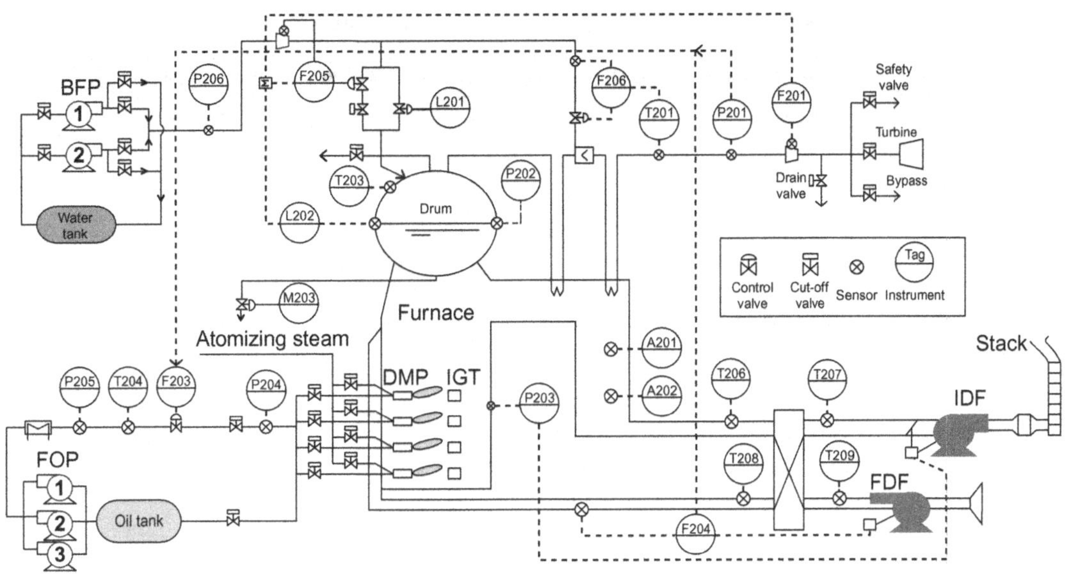


Figure 12.5: Process flow diagram of a boiler plant simulator.

Table 12.1 Tags in boiler plant simulator.

| Tag | Description |
|---|---|
| A201 | Oxygen concentration |
| A202 | CO concentration |
| C208 | $O_2$ and CO analyzer selector |
| F201 | Main steam flow rate |
| F202 | Fuel flow rate |
| F203 | Fuel auto selector |
| F204 | Air flow rate |
| F205 | Drum feedwater flow rate |
| F206 | Desuperheater spray flow rate |
| L201 | Drum water level control (for a small valve) |
| L202 | Drum water level control (for a large valve) |
| M203 | Drum blow valve control |
| P201 | Main steam pressure |
| P202 | Drum pressure indicator |
| P203 | Furnace pressure |
| P204 | Burner-head pressure |
| P205 | Fuel pump outlet pressure |
| P206 | BFP outlet pressure |
| R080[a] | Wind speed |
| R034[a] | Fuel oil viscosity |
| T201 | Main steam temperature (for control) |
| T202 | Main steam temperature (for measurement) |
| T203 | Drum water temperature indicator |
| T204 | Fuel temperature |

[a]R080 and R034 are not measurement points.

Table 12.2: Normal fluctuation of PV and MV values.

| Tag | Low PV Value | High PV Value | Low MV Value | High MV Value |
|---|---|---|---|---|
| A201 | 2.61% | 2.78% | 49.1% | 49.9% |
| A202 | 23.5 ppm | 25.7 ppm | 0% | 0% |
| C208 | – | – | 49.1% | 49.9% |
| F201 | 77.9 t/h | 82.4 t/h | – | – |
| F202 | 6.78 t/h | 7.23 t/h | 51.1% | 53.7% |
| F203 | 51.1% | 53.7% | 51.1% | 53.7% |
| F204 | 67.9% | 71.9% | 47.7% | 55.5% |
| F205 | 75.59 t/h | 81.11 t/h | 42.9% | 44.7% |
| F206 | 1.68 t/h | 1.88 t/h | 40.61% | 43.49% |
| L201 | −1.63 mm | 1.12 mm | 0% | 0% |
| L202 | −1.63 mm | 1.12 mm | 75.9 t/h | 81.4 t/h |
| P201 | 79.3 kg/cm$^2$ | 80.6 kg/cm$^2$ | 67.9% | 72.3% |
| P202 | 83.3 kg/cm$^2$ | 84.6 kg/cm$^2$ | – | – |
| P203 | −14.9 mm$H_2O$ | −5.8 mm$H_2O$ | 66.1% | 72.9% |
| P204 | 3.75 kg/cm$^2$ | 4.1 kg/cm$^2$ | 0% | 0% |
| P205 | 12.6 kg/cm$^2$ | 12.9 kg/cm$^2$ | – | – |
| P206 | 96.7 kg/cm$^2$ | 97.2 kg/cm$^2$ | – | – |
| T201 | 480.4 °C | 489.8 °C | 1.67 t/h | 1.88 t/h |
| T202 | 480.4 °C | 489.8 °C | – | – |
| T203 | 297.9 °C | 298.8 °C | – | – |
| T204 | 89.1 °C | 90.0 °C | – | – |

- Unavailable item.

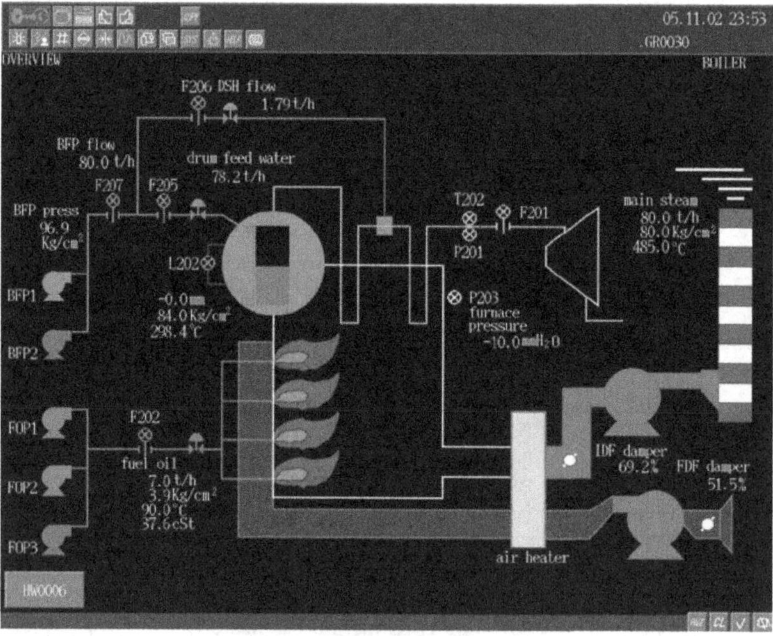


Figure 12.6: Overview panel.

### 12.3.2 Assumed Malfunctions

We assume the following 11 malfunctions in the boiler plant:

Mal-1: Indicated by FOP1 failure. A fuel pump (FOP1) failure decreases fuel oil flow rate (F202) and burner-head pressure (P204). The FOP1 icon flashes in red, so this malfunction is easy to identify.

Mal-2: Indicated by burner extinction. Extinction of all burners decreases the pressure (P201) and flow rate (F201) of the main steam. After this malfunction, the icons of the burner's fire disappear.

Mal-3: Indicated by forced draft fan (FDF) degradation. An FDF degrades, which reduces air intake (F204) and pressure in the furnace. Air–fuel ratio control correspondingly decreases the fuel oil flow rate. The FDF icon flashes after this malfunction.

Mal-4: Indicated by induced draft fan (IDF) trip. An IDF trip reduces air exhaust and increases furnace pressure. The IDF icon flashes in this case.

Mal-5: Indicated by oil heater failure. This causes a drop of oil temperature (T204), and then the oil flow rate decreases due to viscosity (R034) increase.

Mal-6: Indicated by P204 sensor failure. Burner-head pressure sensor (P204.PV) failure forces the measured variable to remain at a low value. This fully opens a control valve of the fuel flow rate. Then the fuel oil flow rate increases out of control.

Mal-7: Indicated by a fuel leak. It actually decreases the oil flow rate to burners and causes a state where the heat is insufficient to produce the desired steam flow rate.

Mal-8: Indicated by BFP1 trip. A water-feeding pump (BFP1) trip interrupts the water supply to the drum and the desuperheater, which may explode the water tube. This malfunction can be detected by BFP1's flashing icon.

Mal-9: Indicated by a water leak. Water tube leak increases furnace pressure (P203). Water flow rate (F205) slightly decreases.

Mal-10: Indicated by $O_2$ sensor failure. Oxygen sensor (A201.PV) indicates a small value that causes an increase of the air–fuel ratio.

Mal-11: Indicated by turbine trip. This drastic malfunction causes sharp changes of many variables.

### 12.3.3 Alarm System

In the monitoring and supervising software of the DCS, function blocks are defined as the basic unit for performing control and calculations. Continuous control, sequence control (sequence tables and logic charts), and calculations are performed by function blocks. These blocks are interconnected in a manner similar to the conventional instrument flow diagrams and combined to design the control function. Each function block, denoted by a tag name, represents the smallest unit of control; for example, the tag name F201 is used to indicate a proportional-integral-derivative (PID) function block for a flow rate controller. A tag in a plant control system has several items that are defined for various alarm limits. Table 12.3 shows an example of a PID function block.

In this research, two upper, two lower, and the rate-of-change limits of a PV value are considered for alarm settings. These limits are denoted by the symbols HH, PH, PL, LL, and VL, respectively. Fault detection and identification also considers two alarm limits of an MV value, which is denoted as MH for MV upper limit and ML for MV lower limit. When a variable value exceeds one of its upper limits or becomes less than its lower limits, an alarm will appear with sound and flashing marks.

Alarm status shows the status information of a function block or tag. Table 12.4 lists all of the alarm statuses in the DCS.

If the PV value of a tag is in its normal range, its alarm status is NR, and if its PV value exceeds its high alarm limit PH, its alarm status is HI. Here, PH is a symbol used to define the high alarm limit of a tag, such as F201.PH = 85.0 t/h, and HI is used to show the alarm status of a tag, which means its PH alarm limit is violated.

Process alarms indicate the abnormality of a process system, whereas system alarms reflect the hardware malfunctions of ICS and FCS. In this study, we consider only process alarms.

**Table 12.3: Data items of a PID function block related to alarm management.**

| Symbol (Data Item) | Description | Modifiable | Range |
|---|---|---|---|
| ALRM | Alarm status | NO | - |
| AFLS | Alarm flashing | NO | - |
| AF | Alarm check | NO | - |
| AOFS | Alarm suppression | OK | - |
| SH | Process variable (PV) scale high limit | NO | - |
| SL | PV scale low limit | NO | - |
| HH | High-high alarm (limit) | OK | SL-SH |
| LL | Low-low alarm (limit) | OK | SL-SH |
| PH | High alarm | OK | SL-SH |
| PL | Low alarm | OK | SL-SH |
| VL | Velocity (rate-of-change) alarm | OK | ±(SL-SH) |
| DL | Deviation alarm | OK | ±(SL-SH) |
| MSH | Manipulated variable (MV) scale high limit | NO | - |
| MSL | MV scale low limit | NO | - |
| MH | MV output high alarm | OK | MSL-MSH |
| ML | MV output low alarm | OK | MSL-MSH |
| SVH | Setpoint high limit | OK | SL-SH |
| SVL | Setpoint low limit | OK | SL-SH |

**Table 12.4: Alarm statuses.**

| Symbol | Description | Symbol | Description |
|---|---|---|---|
| NR | Normal | VEL+ | Velocity alarm + |
| HH | High-high alarm (status) | VEL– | Velocity alarm – |
| HI | High alarm | MHI | Output high alarm |
| LO | Low alarm | MLO | Output low alarm |
| LL | Low-low alarm (status) | | |

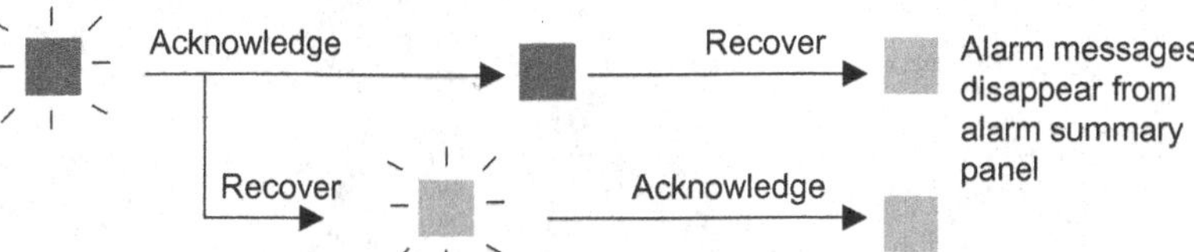


**Figure 12.7: Alarm acknowledgment.**

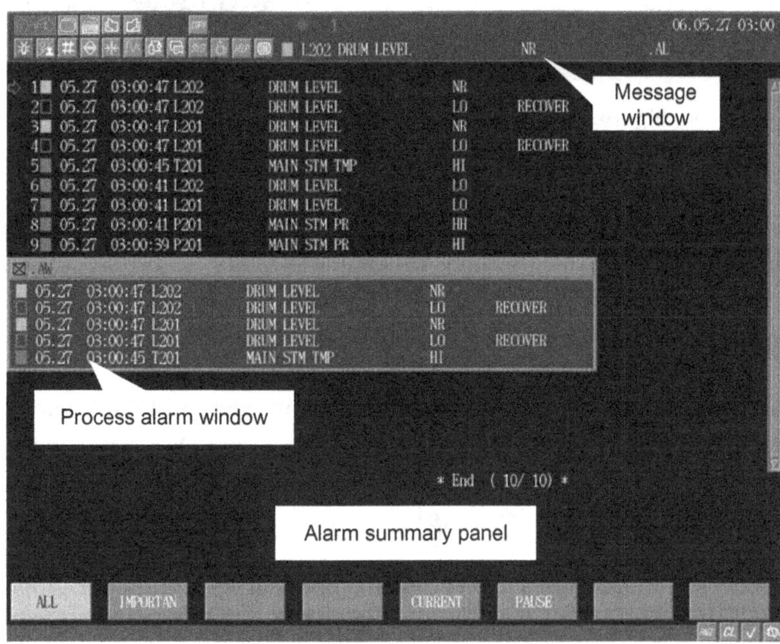


**Figure 12.8: User interfaces for alarm message.**

Sound and flashing marks with an alarm require the operator to acknowledge the alarm information. After acknowledgment, the alarm sound will be eliminated and flashing will be paused. Figure 12.7 shows the sequences of an alarm acknowledgment.

Alarm messages can be displayed on several user interfaces, as shown in Fig. 12.8. A message window, which is always on top of the monitoring screen, can show the latest

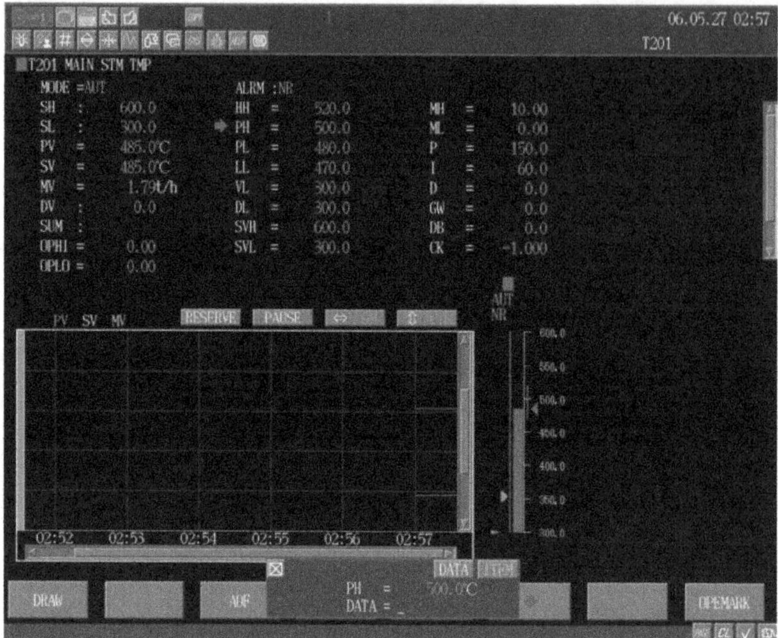


**Figure 12.9: Tuning panel to modify alarm limits.**

alarm. The five latest alarms are listed in the process alarm window, which can be summoned from the message window. On the alarm summary panel, we can check the 200 latest alarms. Figure 12.9 shows the tuning panel through which alarm limits can be directly modified.

## 12.4 Design of Computational Experiments

This section involves the construction of failure–symptom relations knowledge and the definition of ASSP for FDI behavior and the details about how FDI is performed.

### 12.4.1 Failure–Symptom Bipartite Graph

The relations between failure causes and symptoms after sufficient time for propagation for all assumed malfunctions can be simply represented in matrix form based on the results of cause–effect analysis. In the matrix, a row corresponds to a symptom and a column corresponds to a cause of failure. For instance, Table 12.5 shows a matrix, where $F_m$ is the $m$th failure cause and $S_n$ is the $n$th symptom; $FL_m$ is the number of all symptoms for the $m$th cause of failure, and $SL_n$ is the number of all causes related to the $n$th symptom. FL and SL values reflect the complexity of cause and effect, respectively. If a cause of failure $F_m$ can cause a symptom $S_n$, the element in the $n$th row and the $m$th column is set to 1. Obviously, the total value in the $n$th row is $SL_n$, and the total value in the $m$th column is $FL_m$. Even when a set of assumed malfunctions is added, the matrix is easily modified.

Table 12.5: Matrix of cause–effect relation.

| | $F_1$ | $F_2$ | ... | $F_m$ | ... | $F_M$ | SL Value |
|---|---|---|---|---|---|---|---|
| $S_1$ | 0 | 1 | ... | 1 | ... | 0 | $SL_1$ |
| $S_2$ | 1 | 1 | ⋱ | 0 | ⋱ | 0 | $SL_2$ |
| ⋮ | ⋮ | ⋮ | | ⋮ | | ⋮ | ⋮ |
| $S_n$ | 1 | 0 | ... | 1 | ... | 1 | $SL_n$ |
| ⋮ | ⋮ | ⋮ | ⋱ | ⋮ | ⋱ | ⋮ | ⋮ |
| $S_N$ | 0 | 1 | ... | 1 | ... | 1 | $SL_N$ |
| FL value | $FL_1$ | $FL_2$ | ... | $FL_m$ | ... | $FL_M$ | |

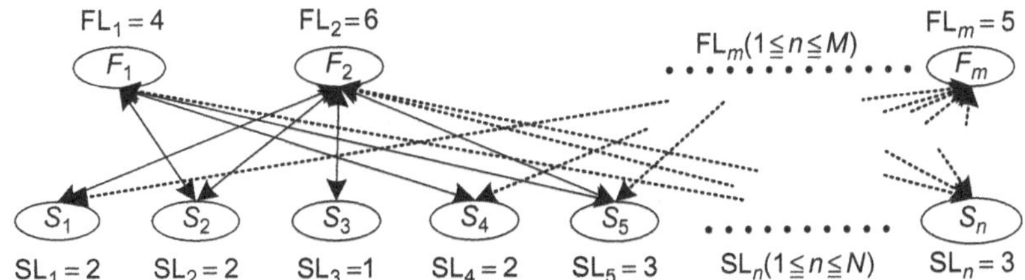


Figure 12.10: Failure–symptom links.

The matrix is also illustrated by a bipartite graph shown in Fig. 12.10. The graph has two layers. The upper layer shows all causes of failure and the lower one shows all symptoms. An element 1 of the matrix in Column 1 is shown by a solid line between related cause of failure and symptom in Fig. 12.10. Dotted lines in Fig. 12.10 also indicate these connections, but the other ends of the dotted lines are omitted due to space limitation. $FL_m$ is the number of links connected with $F_m$ and $SL_n$ is the number of links connected with $S_n$. We define the association strength $AS_{m,n}$ of an FS link between $F_m$ and $S_n$ in a systematic way as follows: for any $(m, n)$

$$AS_{m,n} = \frac{\dfrac{w_{m,n}}{SL_n}}{\sum_{S_i \in A_m} \dfrac{w_{m,i}}{SL_i}} \tag{12.1}$$

where $w_{m,n}$ indicates the weight of the $F_m$–$S_n$ link, and $A_m$ is a set of indices of all symptoms connected with $F_m$. In other words, $AS_{m,n}$ is a contribution ratio of $S_n$ to cause of failure $F_m$. For any cause of failure, the total AS of all links in set $A_m$ is normalized to 1.0, i.e., the AS value becomes 1.0 after complete propagation.

### 12.4.2 Abnormal-State-Supervising Procedure

A human operator can employ various tactics to identify the cause of failure in an emergency. We assume that the following procedure is activated after detecting an alarm. The outline of the procedure is shown in Fig. 12.11.

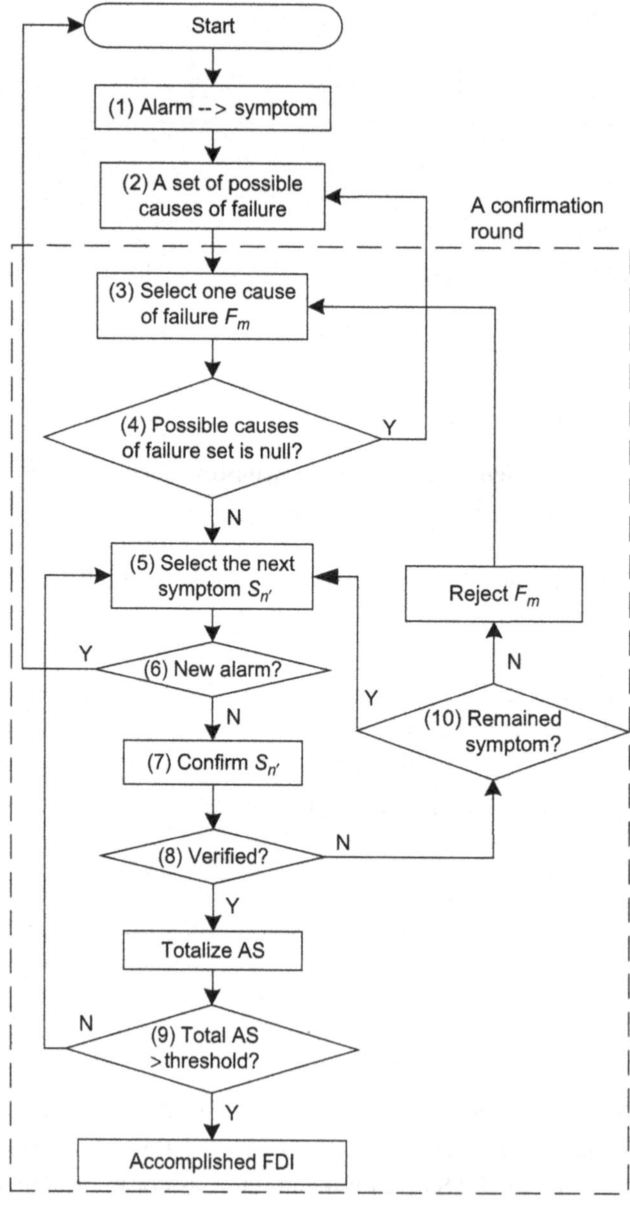


**Figure 12.11: Abnormal-state-supervising procedure.**

1. Based on AM-KB, interpret the newly detected alarm as the *n*th symptom $S_n$ and acknowledge the alarm.
2. Based on FS-KB, assume that the causes of failure that connect to all alarmed symptoms are a set of possible ones, and reject others without a connection to $S_n$.
3. Select one cause of failure $F_m$ whose AS value $AS_{m,n}$ is the largest among those of the possible causes.
4. If all possible causes of failure are rejected, return to step (2) to start a new round of confirmation.
5. Select the next symptom $S_{n'}$ whose AS value is the largest among those of all unconfirmed symptoms that connect to $F_m$.
6. If a new alarm is detected, restart the procedure.
7. Confirm $S_{n'}$ by checking the trend data of its corresponding PV $Tag_{n'}$ on a graphic panel.
8. If the value of $Tag_{n'}$ changes outside of its normal range and accords with $S_{n'}$, add the $AS_{m,n'}$ of the link between $S_{n'}$ and $F_m$ to the total AS value; otherwise, go to step (10).
9. When the total AS value becomes larger than the specified threshold, the FDI process is accomplished.
10. If a symptom that connects to $F_m$ remains, go to step (5). Otherwise, reject $F_m$ from the set of possible causes of failure and return to step (3).

This ASSP can cope with multiple alarms. When a new alarm is issued, the ASSP is restarted and the set of possible causes of failure is modified by taking into account the corresponding symptom of the new alarm.

Symptoms converted from the alarms remove the irrelevant causes of failure from the set of possible causes of failure and then these removed causes will not be checked again during the entire FDI process. After checking all symptoms for a cause of failure, if its total AS value is less than the threshold, the cause of failure is temporarily rejected in that confirmation round, but it is not removed from the set of possible causes of failure and will be considered again in the next round. If failed confirmations temporarily reject all possible causes of failure, the ASSP is restarted and the set of possible causes is updated as mentioned above. The restarting ASSP generates a number of confirmation rounds. The dashed line area in Fig. 12.11 shows a confirmation round.

### 12.4.3 Task Decomposition and Workload Estimation

Human behavior in the FDI process is classified into physical and mental subtasks. The latter includes perception, cognition, STM, and LTM activities, respective examples of which are reading an alarm message, remembering a previous alarm, searching for a symptom in the KB, and rejecting a cause of failure.

Physical subtasks include gaze point movement and finger movement. The physical workload for these subtasks is assessed by the magnitude of motion: for example, keyboard input workload is assessed by the number of key presses. An FDI track is an information flow diagram composed of these subtasks. In this study, the FDI track from detecting an alarm to successfully identifying a cause of failure is generated automatically based on the proposed ASSP of the operator agent. On the other hand, the mental workloads for perception, cognition, STM, and LTM subtasks are scaled to several levels based on the required information of the subtasks, human subject experiments, and expert questionnaires by referring to the VACP (visual, auditory, cognitive, and psychomotor) workload method (McCracken & Aldrich, 1984). The VACP method is a simple, diagnostic workload measurement tool based on task analysis in four dimensions; however, in this research, we define the dimensions as perceptual, cognitive, motor, and mental (LTM and STM) and physical respectively.

## 12.5 Expected Effects

This section introduces examples of FDI track and workload estimation, based on which the plant HMI can be evaluated and improved.

### 12.5.1 Example of FDI Track

Even a simple operation may include many subtasks. This makes human behavior analysis very troublesome. However, based on the structure of the operator agent shown in Fig. 12.3, we can define a part of the subtask sequence as an operational stage. Every operational stage has at least one STM subtask, which may follow after a perception, cognition, or LTM subtask. Therefore, an operational stage is processed in the order of perception, cognition, LTM, and physical subtask, but it may not include all types of subtasks.

As an example, we suppose a malfunction causes a low alarm of a temperature variable T201.PV. Figure 12.12 shows the generation of an FDI track after the alarm (T201.LO). The first subtask is a perception subtask, through which the virtual subject captures the alarm message. Then, the STM subtask is performed to store the alarm information. Through AM-KB, the alarm information is converted to a symptom T201.PV.Low and stored in the STM. The following physical subtasks are performed to acknowledge the alarm, and the first operational stage ends at the vertical bar with a sequence number. Sequentially, the cognitive processor searches FS-KB in the LTM for a set of possible causes of failure. The cause of failure with the maximum AS value, and its corresponding symptom, P203.PV.High, are selected and stored in the STM. This is the second stage. The cognitive processor searches VI-KB in the LTM for the information of P203.PV and then stores it in the STM. According to the position information of P203.PV, the virtual subject switches to the corresponding user panel from the alarm summary panel. The third operational stage is accomplished here. From the user panel, P203.PV is captured by the perceptual processor and then stored in the STM. The cognitive processor fails to verify P203.PV.High. After the fourth operational stage, the FDI process continues until the total AS value is larger than the specified threshold.

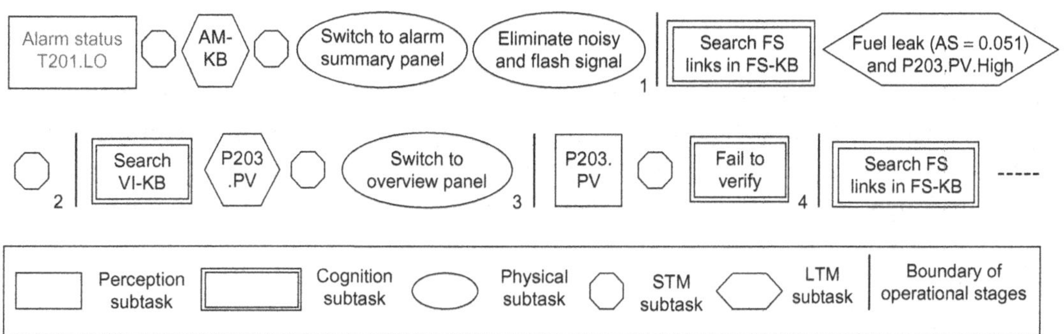


**Figure 12.12: FDI track generation after an alarm.**

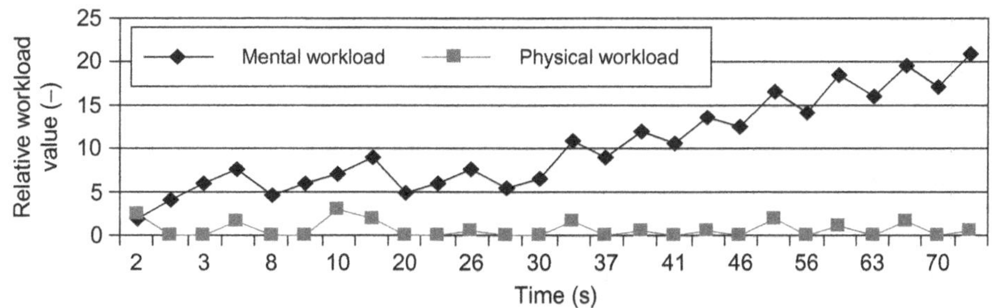


**Figure 12.13: Workload graph during FDI.**

## 12.5.2 Example of Workload Estimation

Figure 12.13 is an example of a workload graph during FDI in the first 80 seconds after the first alarm is captured. Mental workload is a weighted sum of the estimated workload values of perceptual, cognitive, motor, STM, and LTM subtasks.

## 12.5.3 HMI Evaluation and Improvement

We propose the following procedure to evaluate and improve the real HMI:

1. Build an HMI agent that includes all of the graphic and interactive information.
2. Build VI-KB, FS-KB, AM-KB, and ASSP based on the target process, expert reviews, operational experiences, and process risk analyses.
3. Through computational experiments, we can obtain the FDI tracks with changes in physical and mental workloads, and the required time to isolate a cause of failure. These FDI tracks and workload changes show the FDI performance of a design scheme of the HMI.
4. Redesign the HMI including its user panels and the alarm settings.

Repeat steps 1–4 until an acceptable result is obtained.

## 12.6 Conclusions

We proposed an HMI evaluation method based on the ACP approach and showed the expected effects, through which the FDI performance can be improved by comparing the evaluation results of different HMI design schemes. Obviously, the method can be used as support for the plant HMI design. In the future, we will apply the methodology for an ethylene plant.

## References

Aas, A. L. (2009). Probing human error as causal factor in incidents with major accident potential. In *Third international conference on digital society,* Cancun, Mexico.

Card, S. K., Moran, T. P., & Newell, A. (1983). *The psychology of human-computer interaction.* London, England: Lawrence Erlbaum Associates.

Hollnagel, E. (1998). *Cognitive reliability and error analysis method.* Oxford, UK: Elsevier Science.

McCracken, J. H. & Aldrich, T. B. (1984). *Analyses of selected LHX mission functions: Implications for operator workload and system automation goals* (Technical Note ASI479-024-84). Fort Rucker, AL: Army Research Institute Aviation Research and Development Activity.

Takano, K., Sasou, K., Yoshimura, S., Iwai, S., & Sekimoto, Y. (1994). *Behavior simulation of operation team in nuclear power plant-development of an individual operator model* (Research Report S93001). Tokyo: Central Research Institute of the Electric Power Industry.

Takano, K., Sawayanagi, K., & Kabetani, T. (1994). System for analyzing and evaluating human-related nuclear power plant incidents: Development of a remedy-oriented analysis and evaluation procedure. *Journal of Nuclear Science and Technology, 31*(9), 894–913.

Wang, F. Y. (2004a). Artificial societies, computational experiments, and parallel systems: An investigation on computational theory of complex social-economic systems. *Complex Systems and Complexity Science, 1*(4), 25–35.

Wang, F. Y. (2004b). Computational experiments for behavior analysis and decision evaluation of complex systems. *Journal of System Simulation, 16*(5), 893–897.

Wang, F. Y. (2004c). Computational theory and methods for complex systems. *China Basic Science, 6*(41), 3–10.

Wang, F. Y. (2004d). Parallel system methods for management and control of complex systems. *Control and Decision, 19*(5), 485–489.

Wang, F. Y. (2006). On the modeling, analysis, control and management of complex systems. *Complex Systems and Complexity Science, 3*, 26–34.

CHAPTER 13

# Cloud of Health for Connected Patients

**Timo Nyberg[*], Gang Xiong[†], and Jani Luostarinen[‡]**

*[*]School of Science, Aalto University, Espoo, Finland*

*[†]Institute of Automation, Chinese Academy of Sciences, Beijing, China*

*[‡]Mobile Health Services, International Health Promotion Association (IHPA), Helsinki, Finland*

## 13.1 Introduction

Cloud of Health is an important part of today's information systems, and is mainly introduced in this context. The Cloud of Health is understood broadly; physically, it is an extended information system including many types of data transfer from hard-wired broadband network resources to wireless and mobile connection destinations, and including different types of information, information processing, information sources, and possibly actuating devices connected to the rather complex system. These information sources and actuating devices may be personal health measurement devices, remotely managed artificial organs, camera surveillance systems, health cards, mobile tracking devices, or of course traditional laptops, pad computers, smartphones, etc. It must be emphasized that legal and medical aspects play a central role in the realization of health-related systems. Naturally, the software and information plays a vital role in the total Cloud of Health.

The Cloud of Health is a complex ecosystem comprising a multitude of stakeholders. Stakeholders like insurance institutions, pharmaceutical companies, medical equipment companies, medical education and research organizations, healthcare providers, patient organizations, citizens, etc., are all more and more networked with each other through magnanimous information about health.

Here, the concept of Cloud of Health is used to describe the new and emerging application of the Internet, mobile, and wireless technologies to:

i. "connect" the patient to expert advice and information knowledge databases;
ii. "connect" patients to each other in self-help groups;
iii. "connect" the patient to monitoring devices for self-diagnosis; and
iv. "connect" the patient physiological measurement data to the clinician.

**Service Science, Management, and Engineering.**


DOI: 10.1016/B978-0-12-397037-4.00013-2

We also summarize the health problems of twenty-first-century lifestyles and examine the attributes of Short Messages System (SMS) to health promotion with reference to relevant innovative applications of SMS. SMS messaging is still, and will be for many years to come, the most common information channel in the world.

## 13.2 Drivers for Change in Health Care

Three drivers that change the healthcare landscape and the mindsets of patients, public, and clinicians can be identified.

First, a world with increasing wealth fare and improving medical technologies results in longer lives of citizens, but at the cost of an increased incidence of chronic conditions (European Commission ICT, 2007). Societies everywhere in the world face the challenge of delivering affordable health care with high quality to all their citizens. Some policy makers have realized that an affordable way to deal with chronic diseases is delaying the onset, or preventing diseases, through increased awareness that lifestyle plays an important role in these chronic diseases.

Second, there is an evolution attributed to the Cloud of Health, which is moving today's patients from those that are "informed" to those that are "expert." In the early days of the Internet, a patient walked into a doctor's surgery or consultant's office with a printout from a medical website, which was to some clinicians a source of amusement and to others was an irritation. Nowadays, the clinical practitioner will often refer the patient to websites for information. In many cases the care is managed through an Internet care management system, which includes communication processes, medical data collection, and intelligent decision support systems. Anyone who is motivated can research the Internet to become knowledgeable about most subjects, and patients are often very motivated to find out more about their particular condition from the Internet. Self-help patients are an increasing group using the Internet.

> *...Patients no longer want information provided only by medical professionals; people find that exchange of experiences with other patients and ex-patients is the most reassuring and efficient way of getting support...*
>
> ***Source: Department of Health (UK) Self Care Support summary of work in progress (2005–2007)***

Third, there is such a market penetration of mobile and smartphone technology that it is unusual for those under 45 not to have a mobile or smartphone in most developed countries.[1] The prevalence of mobile and smartphone technology is such that it provides an infrastructure and momentum for new innovative medical applications. At the same time, many wireless technologies have become widespread. For example, WiFi and BlueTooth are commonly used for broadband connectivity at home, office, and in public buildings. These technologies

[1] Wikipedia. www.wikipedia.org/wiki/Mobile_phone.

are constantly improving their functionality, and merging into devices such as the iPhone/iPad, Android phones/pads, and Windows Mobile devices. Potentially, there are different Personal Health Assistant (PHA) devices. The advantages of using mobile or smartphones are that they are familiar devices to many people and they are very cost-effective. Mobile health care systems can be easy and quick to install, and systems can be updated without needing to return the device to a central location. One day in the future, wideband data connection and hand-held supercomputers will become available.[2] Health programs employing just traditional websites have reported that the need to have a computer always on is a problem, and users were not using the sites for more than short periods of time.

One of the observable issues taking place across the world is the struggle to develop and implement interoperable patient databases fed by primary, secondary, and community care systems. These are ambitious and challenging programs requiring an enormous investment of resources over many years. So far the results are scarce. At the same time, technological developments are being harnessed in small but innovative ways to bring quick beneficial results improving care processes and the lives of patients. These require typically small amounts of resources and increasingly employ the equipment of the current consumer, the existing skills and knowledge of the patients. These devices will help shift part of the diagnosis, responsibility, and control from the medical specialist at the hospital to the patients at home.

A few of these innovations are described below, some of which are already in place and in use, while others are on the near horizon. It is likely that some will remain in niche markets but others will become absorbed into mainstream life.

## 13.3 Applications

### 13.3.1 Expert Knowledge—Websites

There are an enormous number of health-related websites from trusted sources offering expert help and advice. In the United Kingdom, NHS Choices[3] is a service that aims to put the individual at the center of their health care. It has been designed to help visitors make choices about their health through the practical aspects of finding and using different health-related services, for example assistance with smoking cessation and reduction in the consumption of alcohol while increasing exercise. The extract below shows the Map of Medicine page which one can select to view expert medical advice for a particular disease.

Map of Medicine produces health guides (see Figs. 13.1 and 13.2) to help individuals prepare for discussions with clinical practitioners or for anyone who wishes to find out more information about a health issue.

---

[2] Hand-held supercomputers "on way"—BBC News October 22, 2007. www.news.bbc.co.uk.

[3] NHS Choices site—About Us/www.nhs.uk/Pages/homepage.aspx.

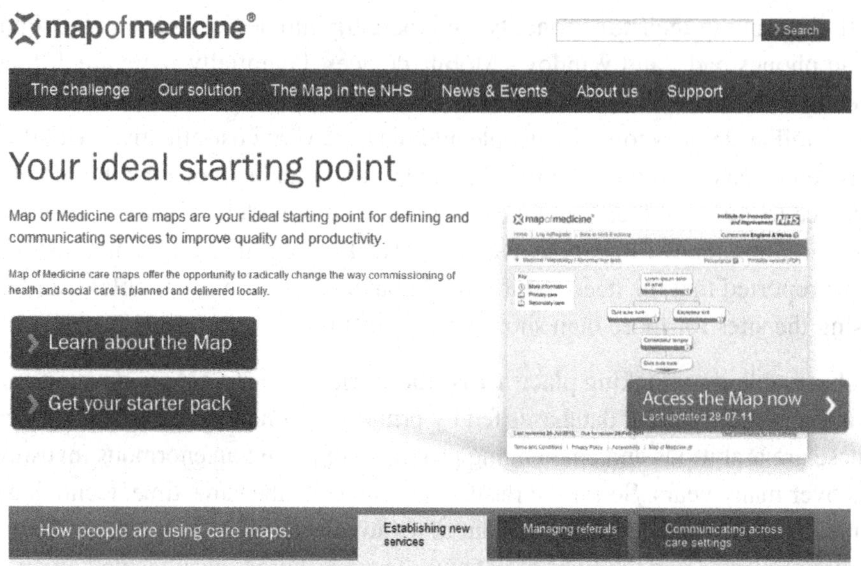


**Figure 13.1: Starting page of www.mapofmedicine.com.**[4]

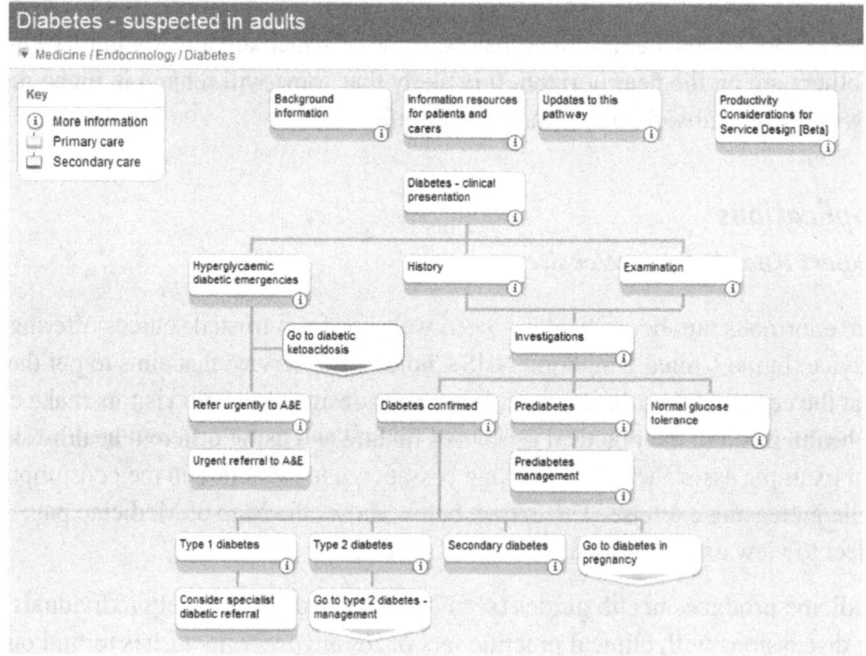


**Figure 13.2: Health guides.**

[4] Map of Medicine. www.healthguides.mapofmedicine.com.

Health guides have been developed by the Map of Medicine to give a clear understanding of the steps to better manage individual's health. It gives the same trusted information available to healthcare professionals.

### 13.3.2 Self-Help

In addition to the websites run by national healthcare organizations are those websites run by charities and others, which provide advice, support and discussion forums. One recent Internet phenomenon is the growth of social networking websites, such as "Facebook." Since approximately 30% of people suffer from a long-term medical condition, inevitably social interaction on the Internet will include sharing of health issues and experiences.

Websites such as those of Diabetes UK report thousands of visitors a day. Increasingly, new and updated information can be acquired by subscribing to a web feed such as RSS. In the typical scenario of using web feeds, a site providing information will publish a link to which the user can register within "reader" software on their own device. RSS on health information sites will enable patients to keep informed of the latest information relevant to their condition in an automated manner that is easier than checking sites manually. Receiving information like this has advantages over emails that require rules to sort incoming data into separate folders, and it is easy to remove the RSS link if the information is no longer wanted. St Johns Ambulance in the United Kingdom offers the facility to subscribe to medical advice feeds and download first-aid procedures to mobile devices.

Originally developed in the United States in the 1970s as a self-help program to aid people with arthritis, the Expert Patient Programme has since been developed for a range of conditions across a number of countries to help patients with chronic problems to take more control over their health by understanding and managing their conditions. This has had great success in empowering people with chronic conditions, giving them control, a greater sense of independence, and leading to an improved quality of life. Expert patients are said to have more effective conversations with health professionals, visit the doctor less, have less time off work, and are less likely to suffer acute episodes resulting in admission to hospital (Lorig et al., 1999a). Participants report life-changing experiences acquiring the skills, knowledge, and confidence to deal with the mental and emotional barriers to self-help.

It is easy to see how the patient can be connected with self-monitoring devices linked to trusted medical databases described in this chapter to help with initiatives such as the expert patient.

### 13.3.3 SMS

There are a growing number of applications utilizing the fact that more and more people possess mobile phones. SMS messaging is showing an expanded use in various health applications.

#### 13.3.3.1 Reminders for Appointments

Missed appointments are a significant waste of scarce resources. They are not only a waste of the clinician's time but also a waste of all the administrative processes that surround making and preparing for that appointment. Although this varies from specialty to specialty, an average of 10% of all patients in the United Kingdom miss their hospital outpatient appointment. The number of missed appointments in 2005–2006 was calculated to be £6.8 million,[5] whereas the estimated costs vary from £300 million to over £600 million per year; any reduction in the percentage of failure to show up for an appointment would save significant resources.

The main reason that patients miss appointments is simply that they forget. Applications are now being used to send a timely reminder SMS message to the patient's mobile phone stating the time and date of their appointment. The advocates of these systems report great success in reducing the failure to attend rate and resulting efficiencies in throughput in busy clinics and surgeries.

Another reason for failure to attend is the difficulty in canceling or changing an appointment time. SMS can help again because it is technically possible to cancel or confirm an appointment using reply text.

#### 13.3.3.2 Sexual Health Service

One medical area where the characteristics of the mobile phone can be usefully exploited is in the treatment and prevention of sexual diseases because young people are an important risk group and they are most likely to have a mobile phone constantly available. Diseases such as Chlamydia are a high disease burden and the United Kingdom has initiated a National Chlamydia Screening Programme to control the disease by early detection and treatment. One of the challenges to overcome is the long waiting times in genitourinary medicine clinics.

Newham General Hospital in London has been using an SMS messaging service for patients waiting for sexual health test results to reduce unnecessary hospital visits and waits. Instead of returning for a second lengthy visit to the clinic, eligible patients are given a negotiated time and date for receiving their results by an SMS message on their mobile phone. By preventing delays and inconvenience for patients, it also frees up appointment time in the clinic, allowing services to be offered to more patients. The hospital reports that the user-friendly service news has spread among the community and among young people and that as a result more people are coming forward for screening.[6]

---

[5] Missed hospital appointments "up"—BBC News May 08, 2006. www.news.bbc.co.uk.

[6] Newham University Hospital Annual Report 2004–2005.

At the same time, a text messaging service was piloted in hospital to allow patients to send blood sugar readings via SMS. These were read by a diabetes specialist nurse, who texted or called to give appropriate advice. This project was aimed at improving care for women with antenatal diabetes by reducing the frequency of outpatient attendances and providing aggressive management of blood sugar levels by the diabetes nurse. The service was very popular among the patients.

#### *13.3.3.3 Medication Non-Compliance*

Failure by patients to take medication as prescribed by clinicians is another significant problem. Although there are many reasons for this, undoubtedly forgetfulness is one cause. The results of avoiding the medication can be very serious for some patients. Companies offer to help patients and carers by sending SMS reminders direct to their mobile phones. Developments in this area can include the ability to send a message to a central system recording when a pill container was opened as the best indication that the patient has taken the medication. The central server can be programmed to send an alert to a carer or health worker if no compliance message has been received.

### 13.3.4 Measuring Device Linked by Personal Health Assistant to Expert Systems

These applications can be used for a wide range of conditions. One important group is those patients suffering from long-term chronic diseases such as diabetes, asthma, and arthritis. To this list we need to add obesity, which is increasingly recognized as a major health problem in developed countries.

People with Long-Term Conditions (LTC) are the most intensive users of health services. In the United Kingdom, patients with an LTC account for 80% of all GP consultations and they use approximately a third of all bed days, whereas in the United States, 78% of expenditure[7] is on patients with chronic diseases. Technology that automates chronic care processes can improve the life of the individual and release monies to be reallocated within health care.

An example of such automation is the use of measuring devices linked by a hand-held PHA to monitor chronically ill patients as they go about their normal daily routines. Daily monitoring can help patients with chronic diseases maintain appropriate drug, diet, and exercise regimes. It is worth mentioning that the PHA can also be used for managing the legal and safety requirements. Most smartphones have the necessary processing power and other features required in a PHA, but they are not medical devices.

[7] Source: October 10, 2003, Jane Horvath Reining in the Cost of Chronic Illness referring to Medical Expenditure Panel Survey, 1998.

Figure 13.3 shows an application of a PHA device, which can give instant feedback to the patient or send the recorded data to be stored and analyzed by either a software decision support system or a clinician treating the patient. The software decision support may be useful for either a diagnosed patient actively managing their condition or for a nondiagnosed individual who wants to prevent illness occurring.

The diagram shows a blood pressure device but equally this could be a blood glucose monitor, bathroom scales fitted with a wireless module, a wireless ECG heart monitor, a lung function spirometer, pulse oximeter, etc. All of these and more can be used in conjunction with a PHA operating as a mobile health monitor, which transmits data to a central server where the results are analyzed, and the individual is given expert advice from decision support software or from connected clinicians.

An increasing number of people are opting for self-diagnosis. In Germany, for example, 1.2 million self-measurement devices are sold annually (Eckert, Gleichmann, Zagorski, & Klapp, 1997). Figure 13.4 illustrates automated interaction between an electronic patient record and a decision support system accessing directly trusted evidence-based medical databases.

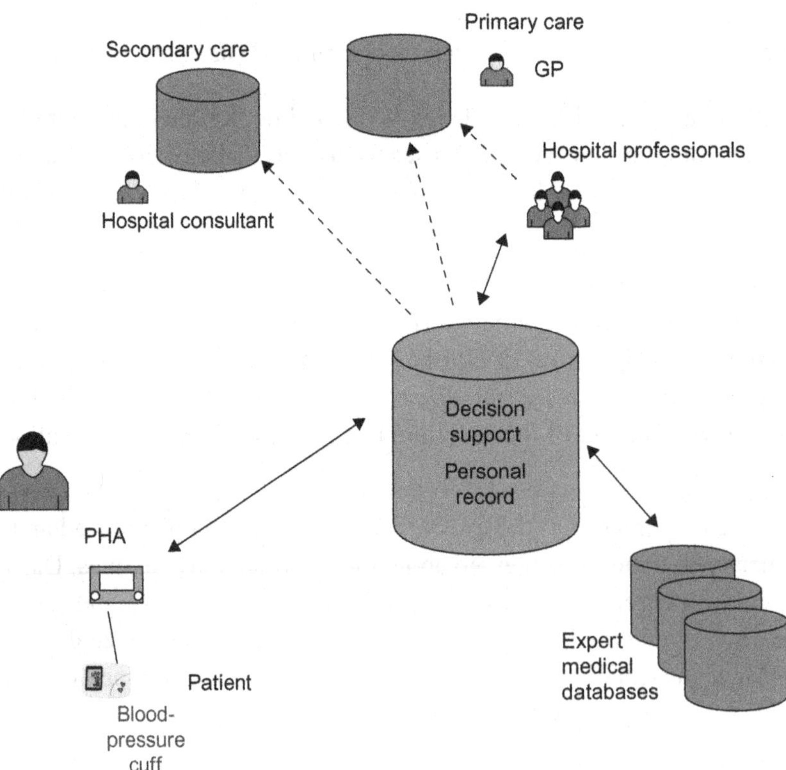


**Figure 13.3: A Personal Health Assistant (PHA) device is a stand-alone feedback system and an interface of an individual to the chronic disease management system.**

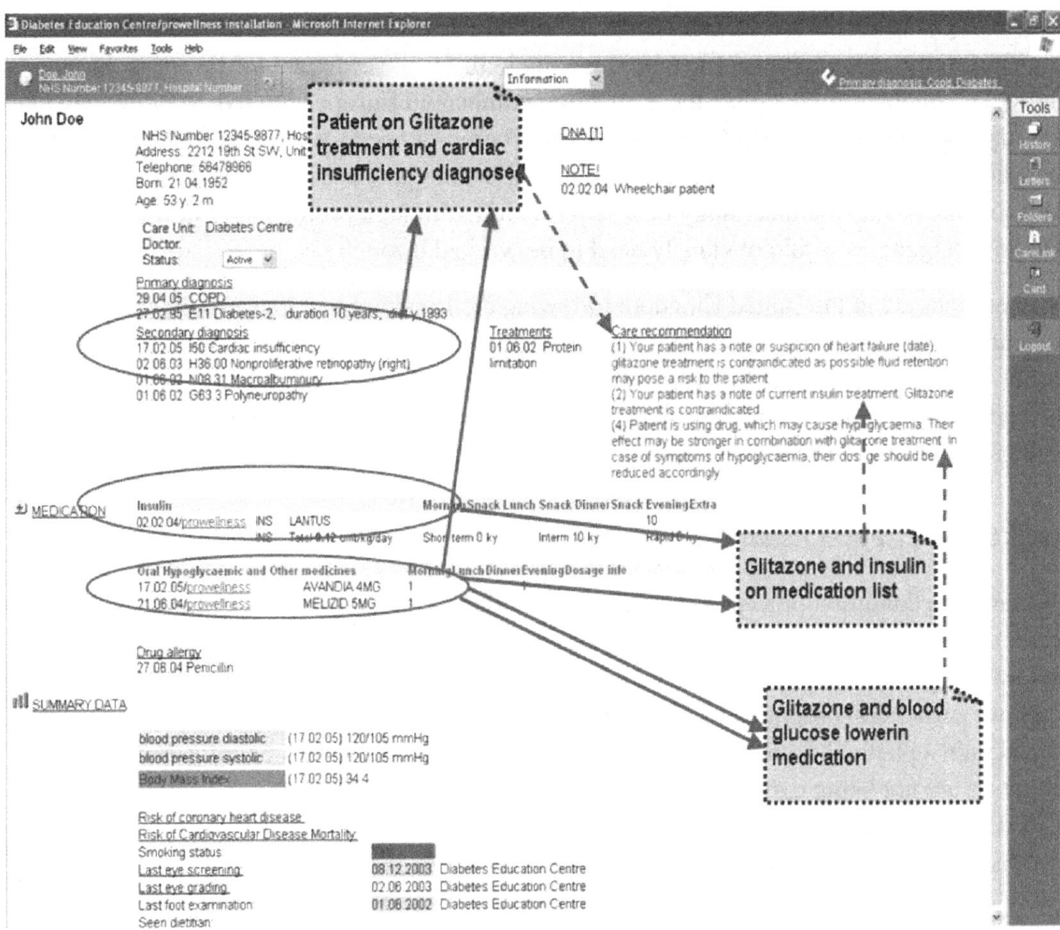


**Figure 13.4: A scenario is depicted of the possible automated interaction between an electronic patient record and a decision support system accessing directly or indirectly trusted evidence-based medical databases such as EBM Guidelines and The Cochrane Collaboration. The solid arrows depict input to the decision support and the dotted arrows indicate the output of recommendations. It is now possible to fit this decision support system onto hand-held devices like the PHA with the necessary security protection.**[8]

## 13.3.5 Medical Data on a SIM Card

Issues such as privacy and cost are delaying the introduction of electronic ID cards. However, mobile phone users already carry a form of ID card called a SIM (Subscriber Identification Module) that is embedded in their handset. Modern SIM cards have enough spare capacity to hold basic medical data applications including patient identification, and the storage of basic medical data such as drug allergies and blood group. This would also provide an important step toward the patient taking ownership of their own electronic medical records.

[8] The Finnish Doctors Association Duodecim, 2007.

Although governments are concentrating on building up national databases within the EU, there are significant numbers of people traveling both for pleasure and for business. In particular, among retired people there are large numbers in Europe who live in more than one country each year for extended periods of time. There is a need for cross-border health care for when someone falls sick in another country. A mobile phone with a SIM card containing basic medical data would certainly be a step forward, although protocols for accessing that data in emergencies would obviously need to be worked through.

A scheme started in the United Kingdom following the London bombings in July 2005 to standardize an emergency contact name in the mobile phone's address list, which continues to gain support. The contact ICE is stored in the phone's address book. It is the number of the person that the phone owner would want to be contacted "In Case of Emergency." In an emergency situation, police and medical staff can then quickly find out who they need to phone.

### 13.3.6 Exercise and Rehabilitation

To manage a condition or to prevent illness, individuals need to exercise. Limited medical resources mean that every patient can visit a physiotherapist only occasionally. Wireless and mobile devices like smartphones can be used to record, monitor, and provide feedback to phone owners, in order to advise and motivate them as individual patients and help them to maintain a planned exercise routine. Once again, a care worker can be informed if the exercises are not being carried out, so that the reason for noncompliance is explored, and the plan is adjusted or further advice is given if required. The use of such devices to encourage exercise could alleviate depression and reduce lethargy.

### 13.3.7 Links to Nutrition Information Expert Knowledge Systems

The importance of diet to health is well known, but information about eating in recent years has often left people confused. Some of the blame for this confusion lies in the hands of the media, who will often pick up issues based partly on their entertainment value. Good expert advice is available, but it needs to be packaged and delivered in a meaningful manner for the public.

Existing technologies can be applied in a number of ways. Smartphones could capture the bar code, and send it to a central database that could provide individual-based advice, for example if the person has an allergy that may be aggravated by the food product or requires a low-salt diet.

The individual could subscribe to a service that allows them to record their food consumption and receive personalized coaching on whether it is matching a target plan.

Just as in Japan, where mobile phone companies provide immediate notification of earthquakes and other natural disasters to their customers, in cases when there has been a contamination of a food product, an alert message could be sent by a central service to mobile phones.

### 13.3.8 Epilepsy Alert

In the Western world, approximately 5 million individuals are affected by epilepsy and many of them are children. Of the approximately 400,000 new cases diagnosed each year, nearly one-third are children. Of great concern to parents is the thought that the child will have a seizure when they are on their own, for example outside the house or in bed asleep. A prolonged seizure can lead to brain damage and even result in death.

An Israeli company has developed a watch-like sensor unit worn by the patient on the hand or the foot, which detects and processes the specific vibrations of epileptic seizures and transmits an alarm to carers.

### 13.3.9 Body-Worn Sensors

Wired sensor devices restrict the movement of the patient being monitored. As a result, some suppliers have developed the ability to capture information wirelessly from sensors embedded in clothes or within a "plaster." This enables the patient to live a normal life, moving about their home or work environment freely (Figure 13.5). These sensors are able to provide continuous monitoring and, for example, detect early life-threatening signs of imminent cardiac problems. With appropriate software, the data can be transmitted to a central remote server for analysis by using GSM or GPRS, and an alert is sent to the patient and clinical support when necessary.

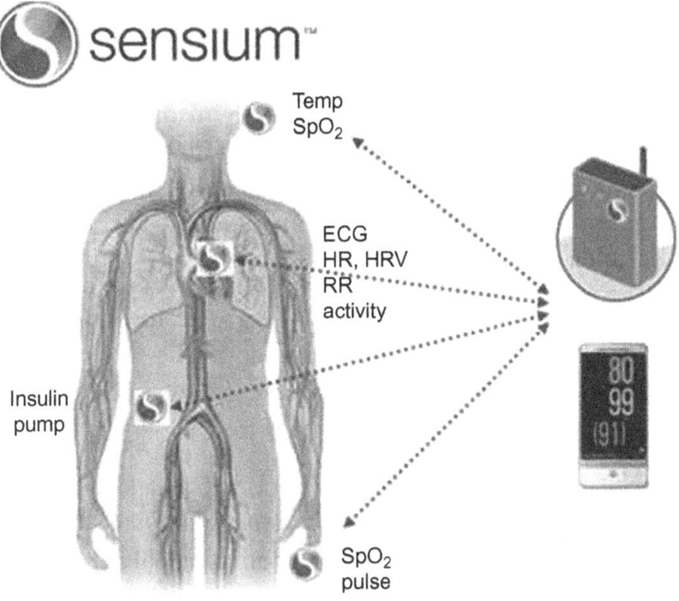


**Figure 13.5: Body-worn sensors have been tested widely.**[9]

[9] www.toumaz.com.

## 13.4 Conclusions of Cloud of Health

David Colin-Thomé, National Director for Primary Care of the NHS, has written:

> *When the NHS was conceived it was logical to bring the patient to the specialist; now, advances in technology and medicine present us with the opportunity to bring the specialist to the patient. Bringing care closer to home. Making it personal rather than impersonal.*

Over the last 50 years, we in the West have concentrated on providing excellent services for the diagnosed patient. We have done this very well but the resulting increase in people's life span now needs to trigger a new phase that will develop strategies to provide both the diagnosed patient and the healthy individual with the information and tools to self-monitor and manage their lifestyle to stay as well as possible.

These strategies can build on health innovations such as those described in this chapter. The traditional boundaries of health care need to be expanded to take in the currently healthy individual, employing the Internet and personal hand-held devices to provide the information and feedback to enable all people to make their right lifestyle choices to prevent or delay the onset of chronic diseases. By employing familiar consumer technologies and actively encouraging self-care by using the largest collaborative resources available, the public can be motivated and mobilized to help meet the challenges we face and improve health outcomes and patient satisfaction within current funding levels.

## 13.5 Case Study of SMS Messaging in Health Promotion

In the Western world, increasing affluence and improving medical technologies mean that people are living longer and, as a result, more are living with chronic conditions (European Commission ICT, 2007). Societies face the challenge of delivering health care with high quality to all their citizens, at affordable cost, and increasingly policy makers need to prevent diseases through education and awareness.

One of the greatest challenges facing the NHS today is the promotion of "healthy living." Major health problems such as obesity, smoking, alcohol abuse, and sexually transmitted diseases are the consequence of life choices that people make on a day-to-day basis.[10] The cost of poor lifestyle choices is significant.

The most common sexually transmitted infection (STI) in England is Chlamydia, and the number of diagnoses has increased steadily since the mid-1990s. The annual cost of Chlamydia and its consequences in the United Kingdom is calculated to be more than £100 million.[11]

[10] Partnerships for Better Health, Small change, big difference: healthier choices for life. UK Department of Health.

[11] www.dh.gov.uk/en/Policyandguidance/Healthandsocialcaretopics/Chlamydia.

In 2002, poor nutrition accounted for 4.6% of the total disease burden in Europe,[12] whereas obesity affects up to a third of the adult population in Europe, which creates a major economic burden through loss of productivity and income, and consumes 2–8% of overall healthcare budgets.[13] In the United Kingdom, the full cost of obesity and overweight people is estimated to be in the region of £7 billion per year. The combined cost for treating conditions related to alcohol and smoking is £3.4 billion per year.[14]

These problems are escalating; for example, obesity has tripled in many countries in Europe since the 1980s, and the numbers of those affected continue to rise at an alarming rate, particularly among children. The issue is leading experts to pessimistic conclusions for the future if poor lifestyle choices are not modified.

> *... the steady increase in life expectancy that has marked the 20th Century may reverse itself in the 21st, and far too many of the next generation could end up dying before their parents.*[15]

To promote "healthy living," new strategies are called for since clearly the message is not getting through. Increasingly, developments such as the informed patient (Lorig et al., 1999b) and expert patient (Detmer et al., 2003) are helping to mobilize the biggest resource available to tackle this problem, the public themselves. These reflect the growing adoption of a partnership approach to health care between the patient and the healthcare system, and will help shift the responsibility and control from the clinical practitioner to the patient/individual.

One of the observable issues currently taking place across Europe is the struggle to develop and implement national patient databases fed by primary, secondary, and community care systems. These are ambitious and challenging programs requiring an enormous investment of resources over many years. At the same time, technological developments are being harnessed in small but innovative ways to bring quick beneficial results improving care processes and the lives of patients. These require relatively small amounts of resources and increasingly employ current consumer equipment such as mobile phones, SMS, and the existing skills and knowledge of the patient/individual in using these devices.

The key attributes of SMS applicable to health promotion are identified in Table 13.1, categorized into receiving and sending attributes, and these are then highlighted with reference to innovative uses of SMS. These applications are smoking cessation programs in the United Kingdom, sexual health advice in Australia, and weight loss and nutrition guidance in the United Kingdom.

---

[12] Proposed Second WHO European Action Plan for Food and Nutrition Policy 2007–2012.

[13] World Health Organization. www.euro.who.int/obesity/import/20060220_1.

[14] Partnerships for Better Health, Small change, big difference: healthier choices for life. UK Department of Health.

[15] Dr Philip James, Chairman of the London-based International Obesity Task Force.

Table 13.1: Key receiving and sending attributes.

| Key Receiving Attributes | Key Sending Attributes |
|---|---|
| K1 Widespread ownership of mobile phones and SMS usage | K5 Speed of transmission |
| K2 Convenience and storage ability | K6 Low cost to send message |
| K3 Personal and private | K7 Ease of administration |
| K4 Social communication | K8 Ability to integrate with applications |
| K5 Speed of transmission | K9 Ability to target population segments |
| K6 No cost to receive message | |

## 13.6 Characteristics of SMS for Use in Healthcare Promotion

### 13.6.1 K1 Widespread Ownership of Mobile Phones and SMS Usage

Although it is known that some people have more than one phone, it is clear from the figures in Table 13.2 that in the developed world the vast majority of people own and use only one. Table 13.3 shows the staggering number of mobile phones in existence.

Given the widespread ownership of mobile phones, the equally impressive figures on the use of SMS are possibly not surprising. In the United Kingdom in 2006, 41.8 billion text messages were sent, and on New Year's Day 2007, 214 million were sent. The year 2006 saw a 38% increase on the previous year and while the forecast for 2007 had been 45 billion text messages, the latest estimation is now 52 billion. In September 2007, texts were being sent at an average of 4000 per second.[16] All across Europe, the statistics for SMS growth are very similar. Table 13.4 shows the SMS figures for Finland (where in 1987 Nokia produced the first mobile phone blueprint).

SMS usage has surpassed all expectations and is now one of the most widely used methods of communication in the developed world and an essential personal item for all age groups. According to Statistics Finland, in 2004, only four 4 of 100 among 15- to 64-year-old individuals did not use a mobile phone.

Mobiles and SMS are used across all social levels but the age range is skewed toward the younger ages, which has advantages for certain types of health promotion such as sexual health. In London, the strategy is to reduce smoking prevalence, paying particular attention to achieving a reduction in those groups most at risk (unemployed people, manual laborers, pregnant women, younger people, and specific communities with high incidence). Text messaging is an excellent method of uniformly addressing a population across different socioeconomic strata because ownership of mobile phones is evenly spread across these groups.

[16] Mobile Data Association (MDA).

**Table 13.2: Number of mobile connections per 100 people in 2010.**

| UK | France | Germany | Italy | USA | Japan | China |
|---|---|---|---|---|---|---|
| 130 | 99 | 133 | 148 | 98 | 92 | 64 |

*Source: Ofcom International Communications Market Report 2011.*

**Table 13.3: Total number of mobile phones (million) in 2012.**

| UK | France | Germany | Italy | USA | Japan | China |
|---|---|---|---|---|---|---|
| 81 | 64 | 105 | 82 | 279 | 121 | 859 |

*Source: CIA World Factbook.*

**Table 13.4: Numbers of outgoing short messages from mobile phones and short messages per subscription on average from mobile phones in 2002–2006.**

| Year | Short Messages (Thousands) | Change (%) | Short Messages/ Subscription |
|---|---|---|---|
| 2002 | 1,324,668 | | 293 |
| 2003 | 1,647,218 | 24.3 | 347 |
| 2004 | 2,193,498 | 33.2 | 439 |
| 2005 | 2,728,230 | 24.4 | 507 |
| 2006 | 3,087,998 | 13.2 | 544 |

*Source: Telecommunications 2006, Statistics Finland.*

### *13.6.2 K2 Convenience and Storage*

Given the portability of mobile phones, the receiver of the message can be located anywhere at any time, and this increases the efficiency by the fact that the individual can be reached directly. The phone can be on silent in certain situations, for example at work, and the SMS can still be read. The message can be stored for reference and the content can be retrieved many times at the user's convenience. Furthermore, the messages sent to a mobile phone that is switched off are stored at an SMS Center and delivered when the handset is switched on again. Until mobile Internet devices are more advanced, and widespread health programs employing only websites will need to have a computer switched on, this will be a barrier to usage. In the Australian Sexual Health campaign, young people are encouraged to send an SMS message "sexinfo" to the number 19SEXTXT. They receive a menu of sexual health topics from which they can then select what information to receive. Alternatively, they can skip the menu by texting the choice directly.

In the United Kingdom, the One Mobile program allows weight watchers to keep food diaries and find healthy alternatives to food items by typing in part of the product's barcode. It links to a database containing nutrition information on 30,000 food items and products, and retrieves the fat and calorie content of the item. The system helps users to track their weight against a target and can be downloaded onto compatible mobile phones by reverse-billed SMS. The system has been given clinical guidance by Dr Campbell, who emphasizes the motivational aspect of SMS in that "When trying to lose weight, it's very hard to maintain that initial motivation. The mobile system makes weight loss easier because it's always there and always ready, keeping you on track, a bit like a weight loss coach in your pocket."

### 13.6.3 K3 Personal and Private

A mobile phone belongs to the individual, and increasingly is an essential item that people carry with them. From a privacy perspective, the owner can ensure that no one else reads the message. SMS messages can be sent anonymously to further protect the confidentiality of the individuals involved. In Australia, an initiative of Marie Stopes International has developed sextxt™ to provide accurate information and advice for youth on sex, STIs, contraception, and sexual health services. Sexual health information is delivered via SMS text message in a private and timely way. The sextxt™ website provides additional information on sexual health topics to ensure that young people are fully equipped with the facts they require. Sextxt™ aims to educate young people on sexual health issues and to reduce the rates of unplanned pregnancy and STIs. The Royal Free Hospital in North West London implemented an "Out-of-Hours" booking service for their Genito-Urinary Medicine (GUM) and sexual health clinics to provide a 24-hour appointment service that is totally confidential. Previously, clients were unable or unwilling to call or visit their local clinic during working hours and faced long queues if they phoned the hospital. In response to the problem, the hospital in partnership with an ICT company implemented the use of SMS for appointment. Clients are given a specific number to text and are offered appointments by text. Patients respond by selecting the most convenient time and their phone number is used as a temporary identifier. The appointment is confirmed by text with a booking reference.

### 13.6.4 K4 Social Communication

The Social Issues Research Centre in the United Kingdom carried out an interesting piece of research on how mobile phones contributed to social communication. The main conclusion was that texting helped restore our sense of connection to a community, recreating the brief, frequent, spontaneous "connections" with members of our social network that characterized the small communities of preindustrial times.

In the fast-paced modern world, we had become severely restricted in both the quantity and the quality of communication with our social network. Mobile gossip restores our sense of connection and community, and provides an antidote to the pressures and alienation of modern life. Mobiles are a "social lifeline" in a fragmented and isolated world.[17]

Texting was identified as having particular importance in maintaining contact with a wide social network, allowing us to maintain social relationships when we do not have the time for phone calls or visits. The report found that texting helps people to overcome their inhibitions and aids social skills to develop communication. More often, individuals can communicate with more people, with the help of mobile phones, than before. Text messaging is often used as a "trailer," alerting friends to the fact that you have something to say, but saving the details for a phone call or meeting. These characteristics add an interesting dimension to SMS that can be exploited by health promotion. Some people often distrust messages from "official sources" and are more likely to listen and respond to the same message when received from a more culturally familiar source.[18] In addition, health promotion using this vehicle for nonsensitive general material may have the additional advantage of being shared with friends, increasing the number of people receiving the message.

### 13.6.5 K5 Speed

SMS messages usually reach the recipient within seconds and a delivery receipt can be added to the message to confirm that it has been delivered. The recipient can also respond quickly, vastly speeding up feedback loops. In North London, a GP practice manager said "Many of our own surgery staff were amazed at the rapid response, as the majority of patients responded within minutes. If we had run the same campaign using letters we would generally have to wait a few days to get any response."[19] Using text messaging, the response rate was reported to be six times more effective than traditional methods.[20]

### 13.6.6 K6 Cost

Texting is relatively cheap compared with land mail and increasingly in the United Kingdom service providers are offering large numbers of free texts as part of the contract. It is even more cost-effective when the administration and material costs associated with sending a letter are taken into account. Text messages also have the comforting feature of being fixed in price, whereas phone calls are usually charged by time units consumed.

[17] Evolution, alienation and gossip: The role of mobile telecommunications in the 21st century by Kate Fox.

[18] Minister of State for Public Health, in Partnerships for Better Health, Small change, big difference: healthier choices for life. UK Department of Health.

[19] Connecting for Health, August 14, 2007.

[20] GPs use text to help patients stub out the ciggies. www.itpro.co.uk/news/106407.

### 13.6.7 K7 Application Integration

SMS messages can be integrated with other computerized applications and messages can be sent from computer to person and vice versa. The restrictions of SMS, which are a function of its size, are almost an advantage in simplifying the programming task. A survey conducted by SMS is almost forced to simplify requests to "Yes/No" answers and simple numeric data entry, and this has the benefit of minimizing the burden on the respondent. In the London area covered by the City and Hackney Primary Care Trust, the SMS messaging system is integrated with the GP electronic record system, enabling them to target patients who smoke, to obtain updates about patient's smoking habits, invite them to Stop Smoking Clinics, and provide ongoing support. In Manchester,[21] health authorities have used SMS text messaging campaigns to increase accessibility to the Stop Smoking Service. People can now contact the service outside normal working hours, and receive information immediately. They can send an SMS message to the Stop Smoking Service, and then receive a call advising them of various support options, such as the nearest drop-in session and pharmacies able to offer advice. There are different media used to promote the service, including the Galaxy radio station, and using beer mats in pubs. By asking people to text a key word such as "Galaxy," the health authorities are able to analyze the success of each method of promotion.

### 13.6.8 K8 Ease of Administration

The "one-to-many" feature of SMS systems allows messages to be sent to many recipients simultaneously and potentially could be in different languages for areas with multiethnic populations. These messages can also be from a prewritten list, reducing administrative time and effort, but bespoke messages can also be sent. The London practice manager quoted in K5 also said: "Using letters and even calling up patients to invite them to the surgery is a huge drain on surgery resources and staff time… we can contact hundreds of patients instantly without having to stuff a single envelope…."[22]

### 13.6.9 K9 Targeting

The ability to integrate with software applications enables effective targeting of certain health promotion campaigns. The primary care system supplier, In Practice Systems, has software that allows patient databases to be searched for particular demographics, such as new mothers. A subsearch in the results can then be done for those with mobiles. The patient list

[21] Manchester NHS uses SMS for Stop Smoking Campaign. http://publictechnology.net/modules.php?op=modload&name=News&file=article&sid=9948.

[22] http://www.text.it/case_study.cfm.

can be reviewed manually and a text message can then be sent to each mother, alerting them to services such as postnatal clinics.[23]

A recent study concluded that SMS is an acceptable medium for targeting young people to receive sexual health information. The majority of participants considered it a very good or good communication method (Wilkins & Mak, 2007).

## 13.7 Conclusions of SMS for Use in Healthcare Promotion

An observation has been made that "Human biology is ill-equipped to cope with twenty-first-century lifestyles."[16] It is widely recognized that action is required to deal with this, "Objectives to improve health outcomes and tackle key risk factors, such as smoking and obesity, need to be given equal weight."[24]

One method that can be employed is to use the strengths of twenty-first-century lifestyles to overcome the health issues arising from twenty-first-century lifestyles. Many individuals carry their mobile phones 24/7, and it has been stated that in the United States, a mobile phone is considered to be just a tool, whereas in Europe, a mobile phone is a lifestyle and that many people look to the mobile as a central source of innovation.[25]

Speaking about the unprecedented increase in obesity, the World Health Organization said that, "The epidemic now emerging in children will markedly accentuate the burden of ill health unless urgent steps with novel approaches are taken ..."[26] Although there are many innovative uses of SMS in health promotion, there are still major benefits to the health of the population to be achieved but to realize the potential SMS communication needs to be embedded within a far greater range of applications. As recommended by Atun and Sittampalam, "...the policy assessment should include consideration of how to introduce promising SMS applications at scale and in a systematic way, in order to ensure that their fullest potential is realized."[27]

Health authorities alone are unlikely to have the necessary skills and resources for this, "Member States' failure to achieve nutrition and food safety goals is due to a lack of resources, expertise, political commitment or intersectoral coordination preventing proper implementation of action plans."[28] To tackle the problem stakeholders outside the health

[23] InPS SMS upgrade allows for care service alerts—Ehealth Insider, January 11, 2005.

[24] Securing Good Health for the Whole Population, 2004 Informing Healthier Choices.

[25] Buongiorno and M:Metrics present a comparative analysis of mobile entertainment usage and attitude in Europe and the United States. http://x.jmailer.com/uk/attachments/b_mmetrics_us.pdf.

[26] The challenge of obesity in Europe and the strategies for response. http://www.euro.who.int/nutrition/20040812_1.

[27] A Review of the Characteristics and Benefits of SMS in Delivering Healthcare.

[28] Proposed Second WHO European Action Plan for Food and Nutrition Policy 2007–2012.

arena need to be actively encouraged to develop innovative systems as stressed by the Health Select Committee in the United Kingdom, "…future health care will be underpinned through working in partnerships—between individuals, communities, business, voluntary organisations, public services and government."[15]

## References

Detmer, D. E., Singleton, P. D., MacLeod, A., Wait, S., Taylor, M., & Ridgwell, J. (2003). The informed patient: An independent study supported by Johnson & Johnson and the Nuffield Trust. Cambridge University Health, UK.

Eckert, S., Gleichmann, U., Zagorski, O., & Klapp, A. (1997). Validation of the Omron R3 blood pressure self-measuring device through simultaneous comparative invasive measurements according to protocol 58130 of the German Institute for Validation. *Blood Pressure Monitoring, 2*, 189–192.

European Commission ICT (2007). Information and Communication Technologies. *European Commission C*, 2460.

Lorig, K. R., Sobel, D. S., Stewart, A. L., Brown, B. W., Jr, Bandura, A., Ritter, P., … Holman, H. R. (1999a). Evidence suggesting that a chronic disease self-management program can improve health status while reducing hospitalization: a randomized trial. *Medical Care, 37*(1), 5–14.

Wilkins, A. & Mak, D. B. (2007). Sending out an SMS: An impact and outcome evaluation of the Western Australian Department of Health's 2005 chlamydia campaign. *Health Promotion Journal of Australia, 18*(2), 113–120.

CHAPTER 14

# Construction of Artificial Grid Systems Based on ACP Approach

**Xisong Dong, Gang Xiong, and Jiachen Hou**
*State Key Laboratory of Management and Control for Complex Systems, Institute of Automation, Chinese Academy of Sciences, Beijing, China*

## 14.1 Introduction

Electric power is the backbone of a national economy and the basis for the normal operation of social life, economic prosperity, national security, and social stability. Communities that lack electric power, even for short periods, have trouble meeting basic needs for food, shelter, water, law, and order. With the interconnection of regional power networks, modern power grids have already become multilevel complex giant systems consisting of physical infrastructures, human operators, and financial resources, involving natural, political, economic, social, ecological, climatic, and human factors, and man-made systems with the widest coverage and more and more complex structures (Wang, Zhao, & Lun, 2008). Modern power grids can improve their operation efficiency and promote the optimal resource distribution. On the other hand, the increasing system scope and complexity present new challenges (Zhao, 2009). In recent years, many large-scale blackouts in different countries have happened and caused enormous social and economic losses (Zhao, 2009).

Smart grids, i.e., intellectualization of power grids, integrate advanced sensors, measurement, information, computer, and control and decision support subsystems into current power grids, to form a new type of intelligent power grids. They can fully meet specific demands of power consumers, optimize power resources, ensure the security, reliability, quality, and economy of power supply, satisfy the constraints of environmental protection, and adapt to developments and changes in the power market (Wang, Scaglione, & Thomas, 2010). The key features of smart grids include self-healing, interactivity, optimization, integration, safety, compatibility, and high quality. They will not only significantly change existing power grids, but also generate in-depth influences on various aspects such as power generation, transmission, distribution, and user demand. They also allow the internal and external factors of power grids to be coupled more closely in multiple dimensions, such as time, space, object, objective, and information

Service Science, Management, and Engineering.


DOI: 10.1016/B978-0-12-397037-4.00014-4

(Metke & Ekl, 2010). Various parameter indexes, such as topology structure and network characteristics, will also be changed significantly.

With the development of power grids and smart grids, their intelligent and emergency management needs to consider engineering, social, natural, and human factors at the same time. Unfortunately, there are no systematic and accurate models in these cases, which is a new challenge for theoretical analysis and simulation research of modern power grids and smart grids:

1. The existing control theories and methods of power grids, based on reductionism, cannot provide good enough guidance for their operation and management or give scientific theoretical explanations and analytical methods for various complex phenomena, such as evolution characteristics and consequent accident development mechanisms.
2. For power grids themselves, traditional simulation methods have many limitations: their accuracy and reliability are difficult to maintain along with the increase of complexity, and the power products have specific characteristics, such as nonstorage, intangible, uncertain random demand, and price diversity, and areas affected by power load changes include weather, holidays, society, economy, politics, and so on (Wang et al., 2008).
3. The conventional power simulation systems are built with physical models or natural phenomena, without deep consideration of the influences of human behavior, the natural environment, and social factors. It is difficult to carry out the operation assessments and adequate guidance for operation and management of modern power grids. Historic statistics given in Table 14.1 shows that the main reasons for blackouts are natural disasters, social factors, and human factors, which are all beyond power grids themselves (Zhao, 2009).
4. Future smart grids are a major advance from traditional power grids, with more complex configurations (see Fig. 14.1). Their complexity will be more obvious and serious, and traditional theoretical analysis methods cannot be applied in this area of research.

**Table 14.1: The statistics of root causes of 110 blackouts.**

| Fault | Type | Number | Proportion | Total Proportion |
|---|---|---|---|---|
| | Storm | 13 | 23.03 | |
| | Wind | 33 | 55.93 | |
| Natural disaster | Lighting | 6 | 10.17 | 53.64 |
| | Others | 7 | 11.86 | |
| Human factors | Protection | 2 | 20.00 | |
| | Misoperations | 5 | 50.00 | 9.09 |
| | Others | 3 | 30.00 | |
| Equipment failure | Lines | 12 | 41.38 | |
| | Stations | 15 | 51.72 | 26.36 |
| | Others | 2 | 6.90 | |
| Energy crises | | 4 | 100 | 3.64 |
| Unknown factors | | 8 | 100 | 7.27 |

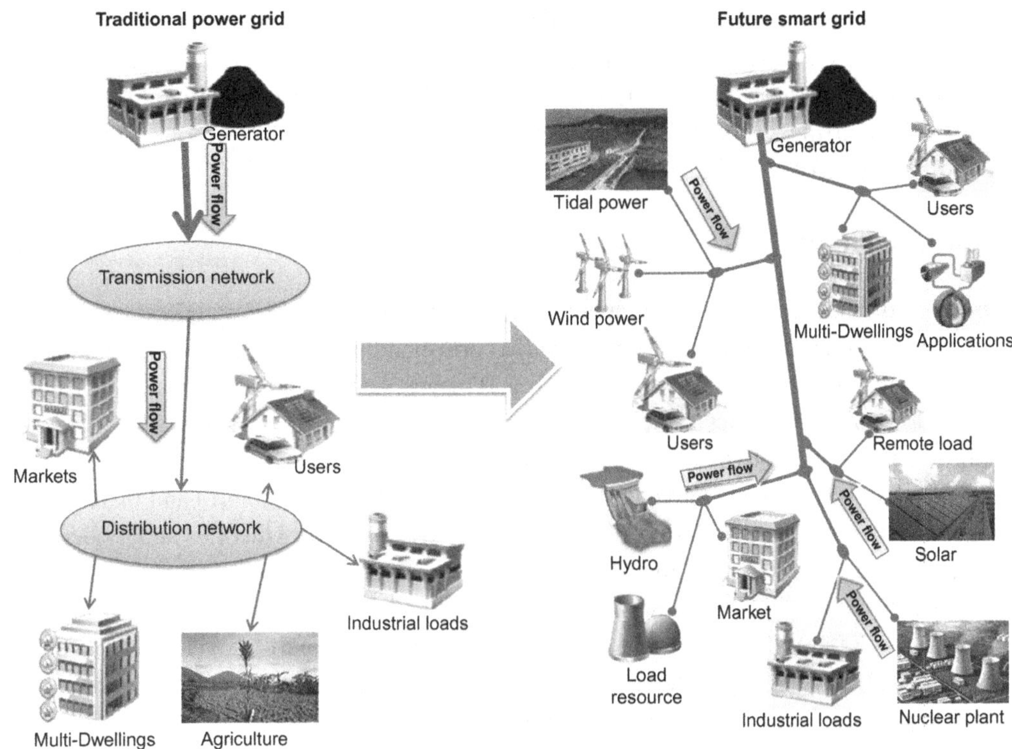


**Figure 14.1: From traditional power grid to future smart grid.**

In summary: (i) the development of power grids makes their operation and management more and more complex; (ii) the interaction between power grids and their social, natural, and engineering factors becomes closer and closer; (iii) existing research theories and methods are inadequate for a thorough study of power grids. In most cases, it is impossible to establish systematic, sufficient, and accurate models, and reductionist methods cannot effectively achieve this. Hence, new solutions need to be sought, new concepts and methods need to be introduced, and new theoretical systems need to be set up.

To overcome the limitations of traditional theories and methods, this chapter presents a novel solution based on the ACP approach and complex system theory, which is integrated with complex network theory and multiagent methods. The solution can establish artificial grid systems, which are "equivalent" to actual power grids, and realize such functions as control and management, experiment and evaluation, learning and training, fault diagnosis and optimization, and then discover an effective way to realize the optimal operation of power grids in normal conditions and emergency management in abnormal conditions. The solution can also research characteristics of smart grids, such as self-healing and strong resistance, to provide theoretical support and reference for their development.

The remainder of the text is organized as follows: Section 14.2 introduces the ACP approach; Section 14.3 gives complex network characteristics of power grids and the construction of a complex power grid network model; Section 14.4 presents the design, construction, and case study of the artificial grid systems, which are based on multiagent complex networks; Section 14.5 gives a case study; conclusions are drawn in Section 14.6.

## 14.2 ACP Approach

ACP means Artificial systems (A), Computational experiments (C), and Parallel execution (P), where "A" is the base and core and a virtual world corresponding to a real system is created based on modeling and simulation, "C" is the methods and means, and "P" is the purpose. The ACP approach allows us to construct artificial systems running parallel to the actual system, through parallel execution of these two systems and computational experiments in artificial systems, to provide a qualitative and quantitative basis for decision making, and to control and manage the actual systems better (Wang & Ranson, 2004). In fact, classical control theories and modern control theories are all versions of the ACP approach (Wang, 2007b). The ACP approach is more advantageous because it provides a new way of thinking and a different perspective for complex systems (Wang, 2004c). Based on traditional "small" closed-loop control, social and human factors are added to form "large" closed-loop control. On the basis of this, the actual systems and their "equivalent" artificial systems can construct double closed-loop control systems (see Fig. 14.2). The first characteristic of the ACP approach is to change the nondominant position of artificial systems. It can change their roles from passive to active, from static to dynamic,

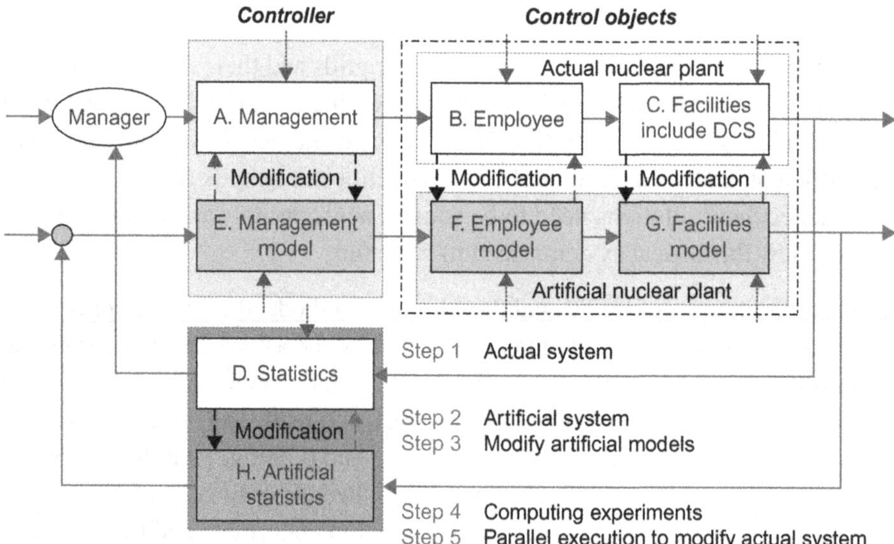


**Figure 14.2: The basic framework of the ACP approach.**

from offline to online. Finally, they would have equal status with actual systems. Thus, artificial systems can play their roles fully in the control of actual systems (Wang, 2004a).

Considering all factors such as engineering, society, human, and environment, the ACP approach combines theoretical modeling, experience modeling, and data-driven modeling to overcome the limitations of traditional methods of modeling complex actual systems. Then, artificial systems "equivalent" to actual systems can be developed. The interactions among various factors of actual systems and their evolution laws under normal and abnormal conditions can be studied by computational experiments or "trials" based on artificial systems (Wang, 2004a). The understanding from artificial systems is ensured to be "equivalent" to that from actual systems (Wang, 2004b; Wang et al., 2008). In addition, actual systems and artificial systems can be compared and analyzed through their connections, and "reference" and "estimation" of their future status can be studied. Then, their control and management methods can be adjusted accordingly. Finally, parallel execution can be implemented by the following laws: in normal circumstances, artificial systems are used to understand various evolution laws of actual systems, then their control objectives can be continuously optimized, and the possibility of abnormal situations occurring can be reduced; in abnormal circumstances, emergency control methods of actual systems can quickly recover to their normal state, so as to reduce their losses (Wang, 2004a, 2006).

### *14.2.1 Artificial Systems*

In the ACP approach, "A" denotes development of artificial systems "equivalent" to actual systems. The accuracy of approximation to an actual system is important in traditional computer simulations. However, it is no longer the only objective of an artificial systems approach. Instead, the system represented by the artificial system is considered "real," i.e., another possible reality of the target system. Based on this assumption, the real system is just one of all possible realities of artificial systems, and the behaviors of the real system and artificial system are different but are considered to be "equivalent" for their evaluation and analysis (Wang, 2006).

There are no effective and widely accepted modeling methods for complex systems, especially those involving human behaviors and social organizations. Artificial systems could be the most promising methods among the new modeling approaches.

### *14.2.2 Computational Experiments*

Traditionally, the research on complex systems has been very difficult. For example, the tests of research objects are usually impossible because of factors such as nature, economics, laws, and ethics. The research on complex systems often uses passive observations and statistical methods, because active tests, active evaluations, and repeatable experiments are difficult to carry out. Even though some experiments have been done, too many subjective, uncontrollable, and unobservable process factors have affected the validation and usage of the results and conclusions. Because analytical reasoning can only solve a limited number of problems, it is critical to find an effective way to conduct experiments for further research into complex systems (Wang, 2004b).

Modeling based on artificial systems shows promise for this purpose. Using artificial systems, the behavior of complex systems can be predicted and analyzed. Controllable experiments that are easy to manipulate and repeat can be designed and conducted. The various factors involved in computational experiments, which are a natural extension of computer simulations, can be evaluated and quantitatively analyzed. They require attention to basic design issues relevant to calibration, analysis, and verification. They also follow design principles such as replication, randomization, and blocking, similar to those experiments done in the physical world. A comparison between normal power simulation and an artificial grid system is shown in Table 14.2.

### *14.2.3 Parallel Execution*

Parallel execution of actual systems and artificial systems can be implemented by using the results of computational experiments. On the one hand, the goals of actual systems can be achieved when they are guided by various results of computational experiments. On the

**Table 14.2: Comparison between normal power simulation and computational experiments based on artificial grid systems.**

| | Normal Power Simulation | Computing Experiments Based on Artificial Grid Systems |
|---|---|---|
| Object | Physical power system | Physical power system, its society and people |
| Modeling method | Mechanism modeling | Mechanism, data and experience modeling |
| Modeling scope | Engineering complexity, such as equipment and process | Engineering complexity, social complexity |
| Simulation type | Simulation and real-time simulation | Real-time simulation and super-real-time simulation (predict) |
| Dynamics type | Transient | Transient, medium- and long-term dynamics |
| Object scale | Small scale | Large scale |
| Object scope | Specific local part (plant, transformer substation, etc.) | Whole power system |
| Object involved | Physical power system only | Physical power system, its natural, economic, social, and human factors |
| Control means | Plan response | Scenario response |
| Control manner | Prior plans | Active defense and prevention |
| Control format | Offline plan | Online response |
| Control methods | Simulate actual situation | Computational experiments based on artificial grid systems |

other hand, the implementation results of actual systems are fed back to artificial systems for amendment. Then, artificial systems and actual systems can progress and improve at the same time (Wang, 2004a).

Parallel execution means one or more artificial systems run simultaneously with actual systems, and they provide a control and management mechanism of complex systems through comparison, evaluation, and interaction with their artificial systems. Parallel execution can also apply those methods and algorithms developed in simulation-based optimization and adaptive control (Wang, 2006).

At present, the ACP approach has been successfully applied in the control and management of emergencies (Wang, 2007a), traffic systems (Wang, 2010; Wang & Tang, 2004; Xiong, Wang, Zou, Cheng, & Li, 2010), and ethylene production (Xiong, Wang, Zhu, & Cheng, 2010). Among them, PtMs (parallel traffic management system) has been successfully applied in TaiCang and the 16th Asian Games held in Guangzhou, China (Wang, 2010).

## 14.3 Complex Network Characteristics of Power Grids

Over a long period, in the research of power grids, researchers often only considered natural environmental factors, equipment factors, and manipulation factors, but rarely analyzed the characteristics of the whole power grids. Traditional analytical methods often pay attention to individual dynamic characteristics of all parts, and the dynamic behaviors of the power grids are analyzed through modeling simulation methods or multidimensional partial differential algebraic equations. In essence, these methods, which are still reductionist, do not reveal the grids' whole dynamic behavioral characteristics.

Power grids have many typical characteristics of complex systems and complex networks. The novel complex network theory provides a new perspective in the research of grids (Rosas-Casals, 2010; Watts & Strogatz, 1998; Xu, Zhou, Li, & Yang, 2009, Yu, Dwivedi, & Sokolowski, 2009).

### 14.3.1 Complex Network Theory

The most complex systems in the world can be described in the form of complex networks. At the primary stage of complex network research, the connection topology was assumed to be completely regular, but a regular network model is not sufficient to describe the networks in the real world. In 1960, Erdös and Rényi put forward the concept of random networks, which greatly stimulated network research (Erdös & Rényi, 1960). Watts and Strogatz proposed the "small-world" network model in 1998 and analyzed the complexity of power grids (Watts & Strogatz, 1998). In 1999, Barabási and Albert revealed the "scale-free" characteristic (Barabási & Albert, 1999). These advances overcame the shortage of random networks and revealed many characteristics of complex networks.

For a complex network G with $N$ nodes and $M$ edges, its main parameters are as follows:

1. *Degree, average degree, and degree distribution*: The degree $k_i$ of node $i$ is the number of edges connecting to this node. It can indicate the importance of the node to some extent. The average degree $<k>$ of the network is defined as the average value of the degree of all nodes, where $<k> = \frac{\sum_{i=1}^{N} k_i}{N}$. The degree distribution $P(k)$ is defined as the probability that a randomly chosen node has degree $k$. The topological characteristics of the whole network can be found using $P(k)$. $P(k)$ describes the degree distribution of all nodes, and indicates the topological characterization of the whole network $d$.
2. *The shortest paths*: The shortest path is defined as the path with the smallest weight (or the fewest nodes from the specific starting node to the end node among all paths).
3. *Distance and average path length*: The distance $d_{ij}$ between the nodes $i$ and $j$ is defined as the number of edges that compose the shortest path connecting the two nodes. If the nodes $i$ and $j$ are unconnected, then $d_{ij} = N$. The average path length $L$ is defined as the mean of the distance of all pairs of nodes:

$$L = \frac{1}{N(N-1)} \sum_{ij=1, i \neq j}^{N} d_{ij}$$

4. *Clustering coefficient*: Assuming the degree of node $i$ is $k_i$, these $k_i$ nodes are called the neighboring nodes of node $i$'s neighboring nodes. $E_i$ is defined as the number of actual edges among these $k_i$ nodes. The possible maximum number of links among the neighboring nodes of node $i$ is $\frac{k_i(k_i - 1)}{2}$. The clustering coefficient $C_i$ is defined as $C_i = \frac{2E_i}{k_i(k_i - 1)}$. It is an important parameter to measure the degree of agglomeration among the nodes. The clustering coefficient of the whole network $C$ is the average value of the clustering coefficients of all nodes:

$$C_i = \frac{1}{N} \sum_{i=1}^{N} C_i.$$

5. *Betweenness*: Line betweenness $B_{ij}$ is defined as the number of times the line $(i, j)$ is crossed by the shortest paths between any two nodes in G. Node betweenness $B_i$ is defined as the number of times the node $i$ is crossed by the shortest paths between any two nodes in G. The betweenness indicates the importance of a node or line in the network. Larger betweenness of the node or the line is more important because the shortest paths are passing through them.
6. *Giant component size*: Giant component size is defined as the number of nodes in the biggest connected subgraph.
7. *Network redundancy*: Network redundancy is defined as the shortest path length of the two nodes after removing the edges of any two nodes that are connected directly. If there is no such path, we let this be infinity.

With the development of power grids, especially smart grids, their nodes, edges, and their connections become more and more complex, and the network topological structure and main parameters have changed tremendously.

### 14.3.2 Complexity of Power Grids

Power grids are the biggest man-made systems, which are multidimensional, nonlinear, time-varying, and large-scale systems. The nonlinearity, variety, hierarchy, integrity, statistics, self-similarity, self-organization, and criticality of power grids all satisfy the general characteristics of complex systems. Power grids also have the general characteristics of complex networks, such as large-scale and dynamic behavior complexity of network nodes, sparse and structural complexity of connections, unpredictability and complexity of network spatial-temporal evolution, and so on (Bai et al., 2006; Chen, Sun, Cao, & Wang, 2007; Rosas-Casals, 2010; Sun, 2005; Watts & Strogatz, 1998; Xu et al., 2009; Yu et al., 2009; Zhang et al., 2008).

The complex network theory provides a novel perspective in the research of grids, by which the characteristics of the whole grid network can be grasped and the corresponding dynamic characteristics can be explored. The research of network dynamic problems, particularly dissemination and cascading, is an important reference to explain the inherent mechanisms of blackout and cascading failures and to design appropriate preventive measures. Vulnerability assessment of power grids based on complex network theory is an approach used for exploring the inherence of cascading failures.

Power grids are strong, large-scale, nonlinear, dynamic systems, and they can be considered as complex networks constructed by power plants, transformer substations, and high-voltage transmission lines with different connection modes (Bai et al., 2006; Chen et al., 2007; Rosas-Casals, 2010; Sun, 2005; Watts & Strogatz, 1998; Xu et al., 2009; Yu et al., 2009; Zhang et al., 2008). A complex network theory based on holism, including structural characteristics, transmission dynamics, and synchronous theory development, can provide new insights into research of power grids. It can scan the dynamic characteristics from a new angle, and provide new ideas for all kinds of complex characteristics and evolution laws of complex power grids (Bai et al., 2006; Chen et al., 2007; Sun, 2005; Xu et al., 2009; Yu et al., 2009; Zhang et al., 2008).

### 14.3.3 Construction of Complex Power Grid Network Model

Based on the characteristics of complex networks and power grids themselves, a complex networks topology model of power grids can be established. Concrete algorithms are as follows:

1. The nodes of power grid models include power plants, substations, and transformers, regardless of ground zero points, which are considered consistent with no differences.

2. High-voltage transmission lines (100 kV or above normally) and transformer branches are edges, which are considered undirected.
3. The transmission networks and substations of distribution networks and low-voltage transmission lines are not considered.
4. Topological properties of all lines are considered to be the same. The differences of the characteristic parameters of the transmission lines and voltage are not taken into consideration.
5. Merging the power transmission lines on the same tower, with the branch parallel capacitor excluded to remove self-loops and multiple edges, the model becomes a simplified graph.

### 14.3.4 Some Results of Research on Power Grids Based on Complex Networks

Recent research proves that power grids have the characteristics of complex networks, such as "small-world" and "scale-free." Therefore, many researchers have established their complex network models to study actual power grids based on the data of actual grids. Some research results of power grids based on complex networks are summarized as follows (Bai et al., 2006; Chen et al., 2007; Sun, 2005; Xu et al., 2009; Yu et al., 2009; Wei & Liu, 2010; Zhang et al., 2008):

1. Power grids have often been self-organized to a critical state in different evolution models.
2. Power grids generally have "small-world" properties.
3. Power grids generally have different levels of "scale-free" properties.
4. Power grids have specific higher clustering coefficients and lower average path lengths, which easily expand the breadth and depth of cascading failures and easily cause blackouts.
5. The average interruption times and scales of "small-world" grids are longer than those of "scale-free" grids.
6. The "scale-free" grids may easily cause smaller breaking chains, but the loss is smaller. This breaking of lines can release the pressure, so as to reduce the probability of massive outage. "Scale-free" power networks can release pressure through small grid outages.

## 14.4 Construction of Artificial Grid Systems

The basic idea of the artificial grid system is the application of the ACP approach. Firstly, comprehensively considering various factors in power systems, including the equipment characteristics, human factors, and social factors, the artificial grid system can be constructed. Secondly, by computational experiments based on the artificial grid system, comprehensive, accurate, and timely assessments and amendments for the planning, design, operation, and management scheme of actual power systems can be carried out without high cost and risk. Finally, a parallel system is built to connect the artificial grid system with actual power systems, and therefore their control, operation, and management can be carried out through parallel execution and feedback (Wang et al., 2008).

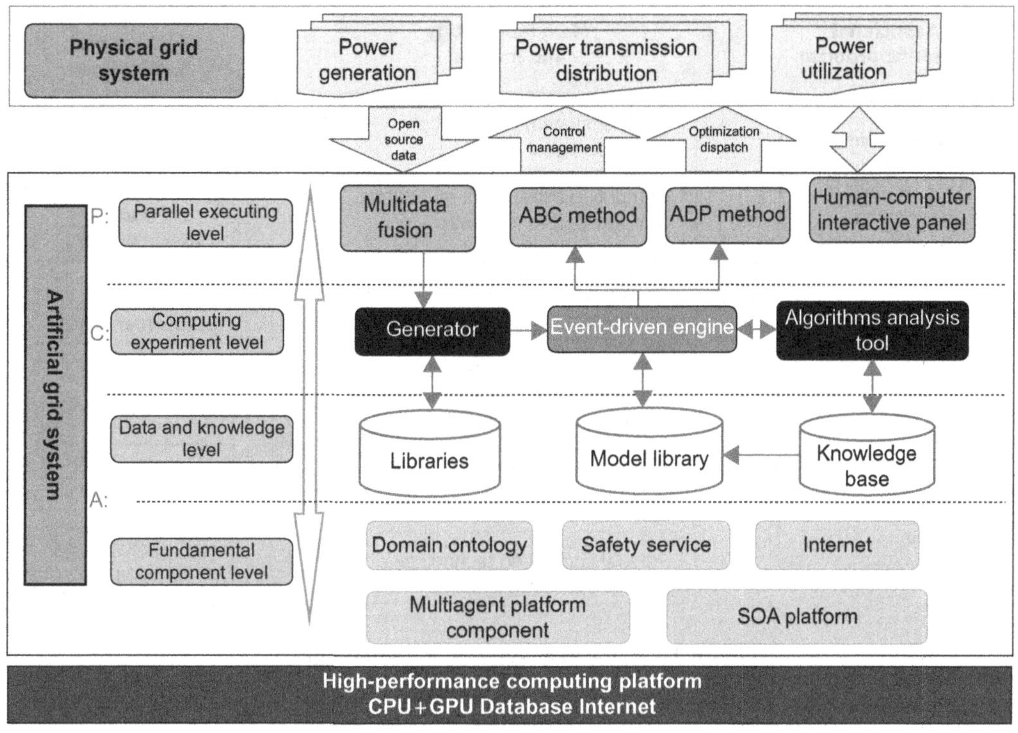


**Figure 14.3: Architecture of artificial grid systems.**

The architecture of artificial grid systems has four levels: fundamental component level, data and knowledge level, computational experiment level, and parallel execution level (see Fig. 14.3).

## 14.4.1 Artificial Grid Systems

The modeling principles of an artificial grid system are to break through the traditional concepts and it can be taken as a kind of reality, which is one of the future states into which actual power systems will evolve. In addition, the actual power systems can be taken as one of many possible future states, whose behaviors are different but "equivalent" to the behaviors of the artificial systems.

The artificial grid system is a revolutionary improvement of power simulations. Power system simulation is a reductionist method using computers and numerical modeling technologies to simulate various characteristics, states, and development of actual power systems. Fundamentally, it is a kind of up-down passive method. However, artificial grid systems are a bottom-up active integrated research method. The artificial grid system is composed of all kinds of artificial components, especially human and social behaviors. Furthermore, economic, temporal, and feasibility factors are also possible reasons to transform traditional power simulations into an artificial grid system. In most cases, the conclusion and effective

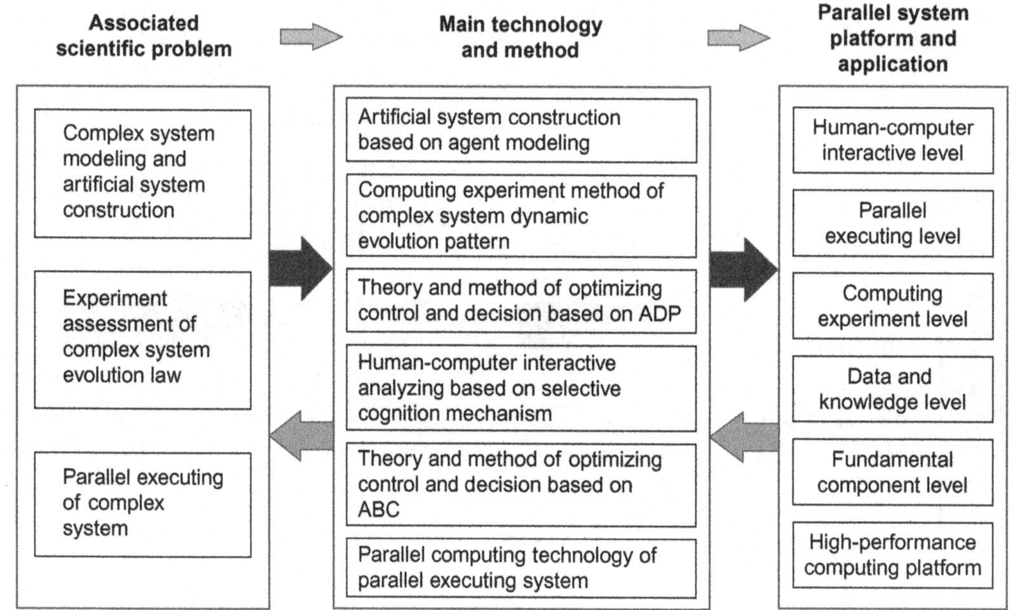


**Figure 14.4: Parallel system of a smart grid.**

range of the artificial grid system aims at specific targets and accuracy requirements. At present, many methods used in traditional system modeling can directly be applied in the construction of the artificial grid system.

In conclusion, the artificial grid system can model actual power systems over a broader range and higher logic level, and carry out computing and analysis from qualitative to quantitative complexity. It integrates mechanism modeling, data modeling, and experience modeling, and considers engineering, social, and human factors. It can describe actual power systems and smart grids well enough to simulate, assess, and improve their operational evolution and emergency management. Therefore, the artificial grid system can provide an innovative and practical way to realize the intelligent operation and management of power systems.

Based on the artificial grid system, computational experiments can be designed and analyzed. Then, parallel execution can be implemented by using the results of computational experiments integrating active power systems. Parallel power systems can also be constructed. Figure 14.4 shows the framework of a parallel system of a smart grid.

### *14.4.2 Design and Construction of Artificial Grid Systems*

The characteristics of power grids can be reflected in complex network models by introducing physical parameters. However, the existing network statuses can only reflect the performance of the grids from the perspective of network connectivity and cannot reflect their own characteristics well. The differences between nodes are not considered, which might be power stations, substations, transformers, or other high-voltage loads. Moreover, the effects of various factors, such as political, societal, economical, geographical, environmental, weather, equipment, and human, which are dynamic, complicated, variable, and may have more influence than the power grids themselves, are also not considered.

To solve the above problems, there are two aspects of work to be done. On the one hand, the description of topological structure and the definition of parameters for a complex power grid network should be improved. Some new parameters are introduced to describe the characteristics of power grids, such as net-ability, path redundancy, survivability, entropic degree, and transmission betweenness. In addition, the complex network models can be improved to become weighted directed graphs so that current transfer can be calculated better (Wei & Liu, 2010). On the other hand, agent theory is introduced to construct multiagent complex network models. Agent theory is a new development in the field of artificial intelligence. The agent, with its responsiveness, initiation, and sociality, can recognize its environment and take independent action to achieve its design targets (Karnouskos & Holanda, 2009; Pang, Gao, & Xiang, 2010; Pipattanasomporn, Feroze, & Rahman, 2009). The interaction among agents and their communication agreements can realize the control targets of the entire systems.

Complex network theory and agent-based simulation can provide more satisfactory reality, more effective modeling means, and guidance theory to describe and study complex power grids, and can greatly enhance their capacities of understanding, research, and control. Compared with traditional modeling and simulation methods, agent-based modeling and simulations can not only provide more realistic modeling methods, but also find the emerging characteristics of complex systems from micro- to macrobehaviors, and reveal their inner micromechanism macroscopic characteristics. Therefore, the artificial grid systems are established based on multiagent complex network theory, which integrates the advantages of agent theory and complex networks. They can then reflect the essential characteristics of actual power grids (Dong et al., 2011).

The agent structure of an artificial grid system is shown in Fig. 14.5:

1. Sensing modules can detect the information transmitted from an external environment.
2. Data processing and decision support modules can receive the information from sensing modules, and then make their decision-making plans by combing the information with agents' own internal state, knowledge base, and inference rules.
3. Communication modules can transfer the decision-making plans to affect the external environment, and directly receive the information from the sensing modules, simply process the execution information, and apply it in the external environment.

Based on multiagent complex network theory (processing grids' mechanism themselves), combining the Monte Carlo (processing history and real-time data) and fuzzy neural

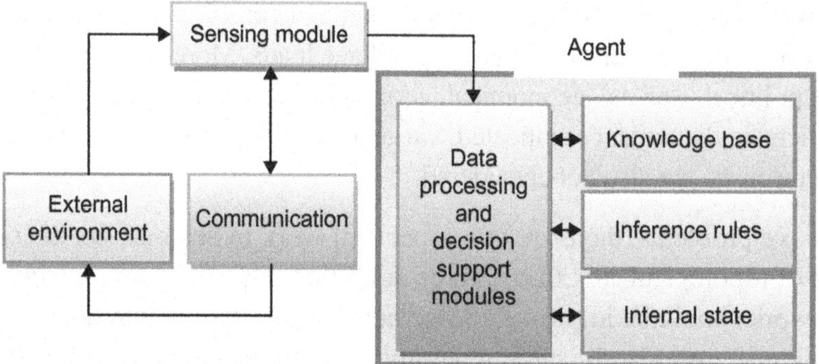


**Figure 14.5: The structure of an agent of an artificial grid system.**

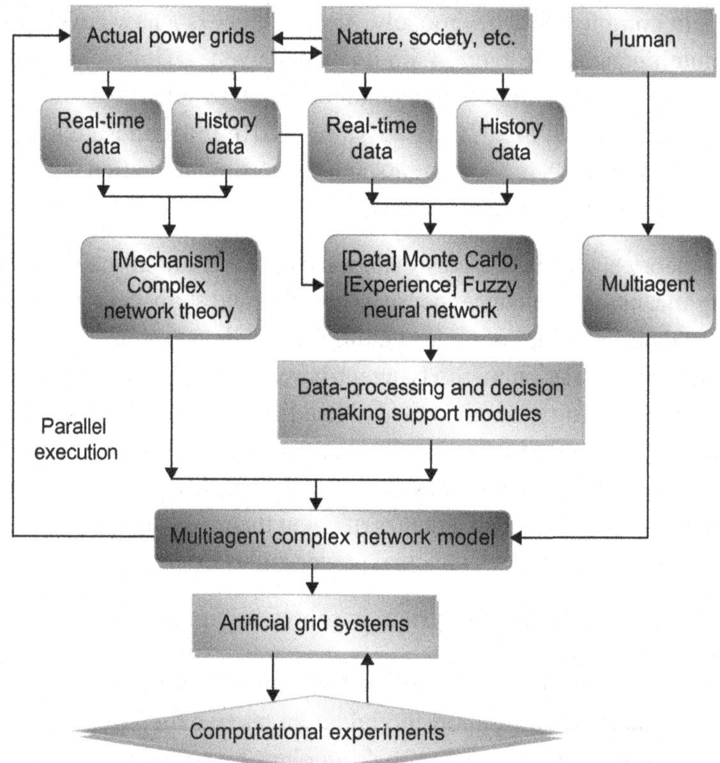


**Figure 14.6: Artificial grid systems.**

network methods (processing experience) with multiagents (processing human factors), artificial grid systems "equivalent" to actual power grids can be developed (see Fig. 14.6). Using computational experiments executed on the artificial grid systems (see Fig. 14.7), the nonlinear dynamics evolution of power grids, including prediction, generation, secondary

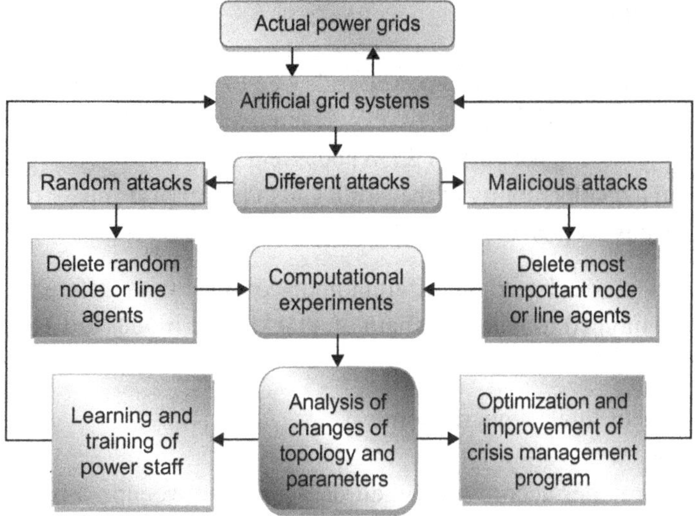


**Figure 14.7: Computational experiments based on artificial grid systems.**

changes, derivation and coupling, and variation under various unexpected incidents, can be tested. Then, network topology, parameter variation, forecast, analysis, decision making and recovery program and blackouts prediction, occurrence, development, and crisis management can be studied. Finally, emergency plan formulation, site management, and decision support can be provided for actual power grids, and the interaction between various social and human factors and power grids can be studied. The effects and anti-effects of human and social behaviors and reaction can be analyzed under various unexpected situations, and the effects of different solutions can be evaluated and optimized to provide reference, guidance, advice, and theoretical support for normal and abnormal operating conditions.

The multiagent complex power grids can be constructed as follows:

1. Construct the regular complex network (see Section 14.3.3).
2. The network is directed based on the direction of the current grids.
3. The nodes of the network are the agents.
4. Simplify the structure of the agents, because the number of grid nodes is very large.
5. The agents of the network can only communicate to the nearby nodes.
6. Limit the communication of the agents in the connection and direction of node agents.
7. Determine the directions of lines by the communication of the agents.

Using computational experiments in the artificial grid system, the control, operation, scheduling, security, optimization, and management of power grids can be researched, and their rolling optimizations in normal situations and emergency management in abnormal situations can be studied.

## 14.5 A Case Study

The artificial grid system of North China power grids (see Fig. 14.8) is built according to the ACP approach and multiagent complex networks. The model is based on the actual data, which represents the power grids as a network of 2556 nodes and 2892 edges. The vulnerability indexes are attained, and then the vulnerability of the North China power grids under different attack modes is analyzed in this section.

Three different attack modes based on different nodes or lines with different degrees or betweenness were combined to launch a chain of attacks. The modes are as follows:

1. Random attacks
   Delete some node or line agents randomly.
2. Static attacks
   Delete the lines or nodes according to their degrees or betweenness from large to small.
3. Dynamic attacks
   Delete the lines or nodes with the largest degrees or betweenness, and recalculate their degrees or betweenness after every attack; repeat this process.

A total of 10 such links are removed one after the other, and the system's efficiency is calculated after every attack, which will reflect the system's robustness to such attacks.

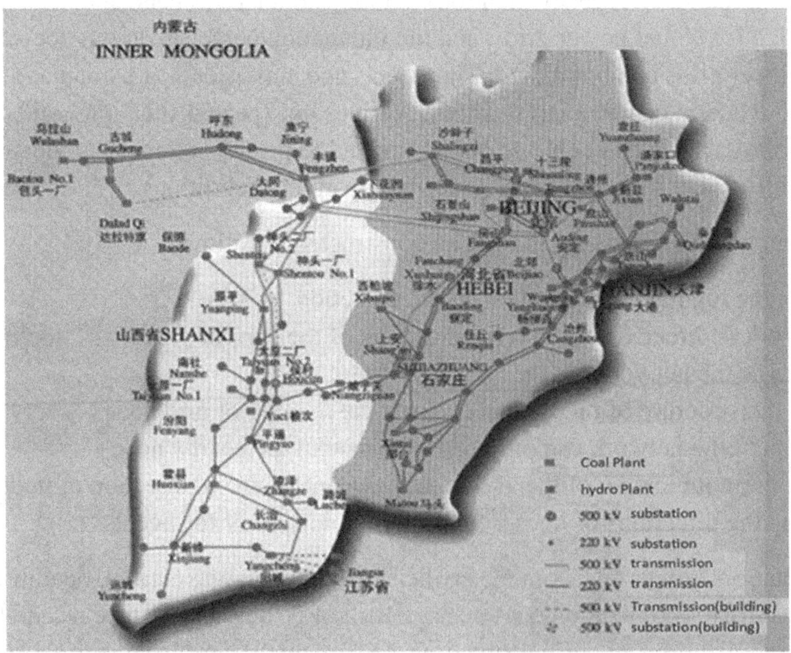


**Figure 14.8: North China power wiring diagram (220 kV and above).**

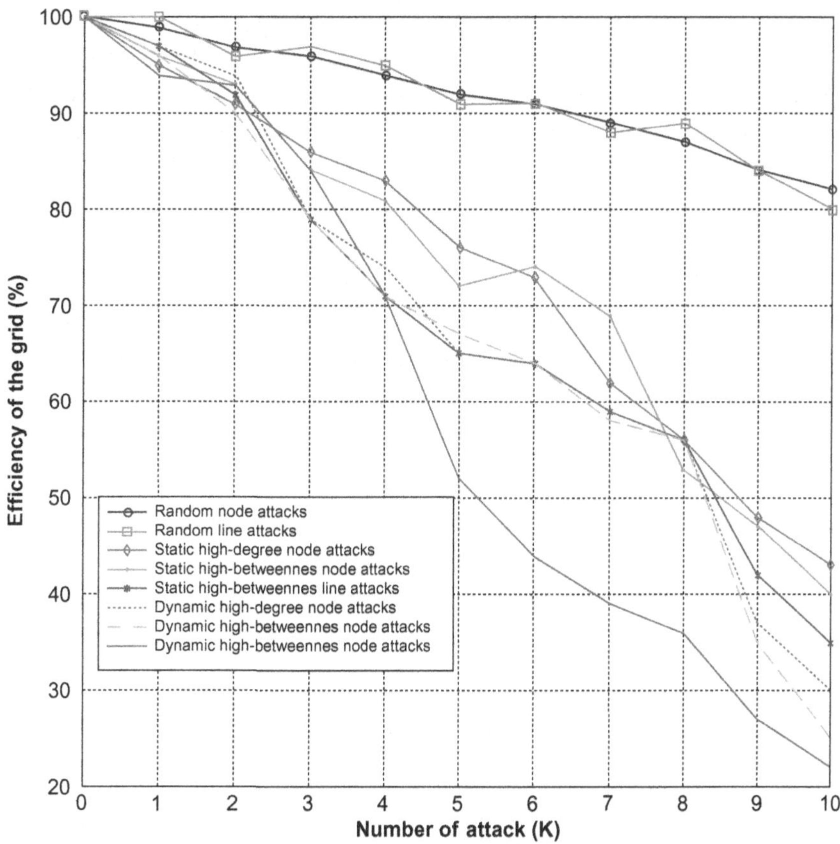


**Figure 14.9: The efficiency of the artificial systems of North China power under different attacks.**

The efficiency of the artificial systems after each attack is shown in Fig. 14.9, and the following conclusions can be drawn:

1. The nodes or lines with the highest degrees or betweenness, which are usually the transmission lines and substations at 500 kV based on actual power grids, play the roles of transmitting and distributing high power capacity, and they are much more important than others. It is also validated that the proposed vulnerability indexes agree with the operating condition of actual power grids.
2. Under random attacks, the curve changes slowly. However, under the dynamic or static attacks, when those nodes or lines with higher degrees or betweenness are removed, the network performance decreases drastically. This implies that the system is very robust against random attacks and there is no protection ability against deliberate attacks. This shows that these nodes or lines with higher degrees or betweenness guarantee the superimportance in the system. The scale of cascading failures will increase rapidly if failures occur at these nodes or lines. Sometimes this type of failure can eventually result in large-scale blackout.

3. For nodes and lines with higher betweenness, after dynamic and static attacks, the network performance after the attacks to lines decreases more drastically than for nodes. In other words, the lines with higher betweenness are more sensitive than the corresponding nodes, which implies that lines may be more important in the security of power grids.
4. For nodes with higher degrees or betweenness, after dynamic and static attacks, the network performance after the attacks to nodes with higher betweenness decreases more drastically than for those with higher degrees. In other words, the parameter betweenness may be more sensitive than the degree, which implies that betweenness may be more important in the architecture of power grids.
5. At the same time, the performance after dynamic attacks is lower than static attacks, which again shows the importance of nodes or lines with higher degrees or betweenness.

In brief, the system is very robust against random attacks. However, when the nodes or lines with higher betweenness or degrees are maliciously attacked, the system will suffer worse destruction than the damage from random attacks. To update the whole grid's reliability, the protection and safeguard of these nodes or lines must be strengthened, which is closely associated to network structure. The topological structures of power grids have an important influence on their vulnerability.

Using computational experiments in the artificial grid systems, topology and various parameter index changes and the corresponding effects on a smart grid's security can also be researched. Its characteristics, such as self-healing, security, and strong resistance, can be studied, and theoretical guidance and reference for smart grid development can be provided.

## 14.6 Conclusions

Modern power grids are typical multilevel complex giant systems. Conventional analytical theory and methods based on reductionism cannot provide sufficient guidance for their operation and management.

This chapter puts forward and studies how to construct artificial grid systems based on the ACP approach using multiagent complex networks. Based on the artificial grid systems, the control, scheduling, optimization, and management of modern power grids can be improved. In particular, their rolling optimizations in normal situations and emergency management in abnormal situations can be guided theoretically and supported technically. The proposed approach can provide scientific guidance for security and stability, quality, and economical operation of power grids. It can improve the operations and management level and economic efficiency of power companies, reduce production costs, and enhance their innovation, adaptability, and comprehensive competitiveness. The topology and various parameter index changes and their effects on smart grids can be theoretically studied to guarantee self-healing, security and strong resistance. As a case study, the artificial grid system with actual data from the North China power grid is constructed, simulated, and analyzed under different attacks.

## References


Bai, W. J., Zhou, T., Fu, Z. Q., Chen, Y. H., Wu, X., & Wang, B. H. (2006). Electric power grids and blackouts in perspective of complex networks. In *Proceedings of the 4th International Conference on Communications, Circuits and Systems* (pp. 2687–2691). New York: IEEE Press.

Barabási, A. L. & Albert, R. (1999). Emergence of scaling in random networks. *Science, 286*(5439), 509–512.

Chen, X. G., Sun, K., Cao, Y. J., & Wang, S. B. (2007). Identification of vulnerable lines in power grid based on complex network theory. In *Proceedings of the IEEE Power Engineering Society General Meeting* (pp. 1–6). Piscataway, NJ: IEEE Computer Society.

Dong, X. S., Xiong, G., Hou, J. C., Fan, D., Nyberg, R. T., & Hämäläinen, P. (2011). Analysis of vulnerability of power grid based on multi-agent complex systems. In *Proceedings of the 2011 IEEE International Conference on Service Operations and Logistics, and Informatics* (pp. 186–190). Piscataway, NJ: IEEE Computer Society.

Erdös, P. & Rényi, A. (1960). On the evolution of random graphs. *Publications of the Mathematical Institute of the Hungarian Academy of Sciences, 5*, 17–61.

Karnouskos, S. & Holanda, T. N. (2009). Simulation of a smart grid city with software agents. In *Proceedings of the 3rd UKSim European Symposium on Computer Modeling and Simulation* (pp. 424–429). Piscataway, NJ: IEEE Computer Society.

Metke, A. R. & Ekl, R. L. (2010). Security technology for smart grid networks. *IEEE Transactions on Smart Grid, 1*(1), 99–107.

Pang, Q. L., Gao, H. H., & Xiang, M. J. (2010). Multi-agent based fault location algorithm for smart distribution grid. In *Proceedings of the 10th IET International Conference on Developments in Power System Protection* (pp. 1–5). Stevenage, UK: Institution of Engineering and Technology.

Pipattanasomporn, M., Feroze, H., & Rahman, S. (2009). Multi-agent systems in a distributed smart grid: Design and implementation. In *Proceedings of the IEEE/PES Power Systems Conference and Exposition* (pp. 1–8). Piscataway, NJ: IEEE Computer Society.

Rosas-Casals, M. (2010). Power grids as complex networks: Topology and fragility. In *Complexity in Engineering* (pp. 21–26). Piscataway, NJ: IEEE Computer Society.

Sun, K. (2005). Complex networks theory: A new method of research in power grid. In *Proceedings of the IEEE/PES Transmission and Distribution Conference and Exhibition: Asia and Pacific* (pp. 1–6).

Wang, F. Y. (2004a). Artificial societies, computational experiments, and parallel systems: An investigation on computational theory of complex social-economic systems. *Complex Systems and Complexity Science, 1*(4), 25–35.

Wang, F. Y. (2004b). Computational experiments for behavior analysis and decision evaluation of complex systems. *Journal of System Simulation, 16*(5), 893–897.

Wang, F. Y. (2004c). Parallel system methods for management and control of complex systems. *Control and Decision, 19*(5), 485–489.

Wang, F. Y. (2006). On the modeling, analysis, control and management of complex systems. *Complex Systems and Complexity Science, 3*(2), 27–34.

Wang, F. Y. (2007a). Systemic framework of PeMS (Parallel emergency management system) and its applications. *Chinese Emergency Management, 12*, 22–27.

Wang, F. Y. (2007b). Toward a paradigm shift in social computing: The ACP approach. *IEEE Intelligent Systems, 22*(5), 65–67.

Wang, F. Y (2010). Parallel control and management for intelligent transportation system: Concepts, architectures and application. *IEEE Transactions on Intelligent Transportation Systems, 11*(3), 1–10.

Wang, F. Y. & Ranson, S. (2004). From artificial life to artificial societies-new methods for studies of complex social systems. *Complex Systems and Complexity Science, 1*(1), 33–41.

Wang, F. Y. & Tang, S. M. (2004). Concepts and frameworks of artificial transportation systems. *Complex Systems and Complexity Science, 1*(2), 52–59.

Wang, Z. F., Scaglione, A., & Thomas, R. J. (2010). Generating statistically correct random topologies for testing smart grid communication and control networks. *IEEE Transactions on Smart Grid, 1*(1), 28–39.

Wang, F. Y., Zhao, J., & Lun, S. X. (2008). Artificial grid systems for the operation and management of complex power grids. *Southern Power System Technology, 2*(3), 1–6.

Watts, D. J. & Strogatz, S. H. (1998). Collective dynamics of "small-world" networks. *Nature, 393*(6684), 440–442.

Wei, Z. B. & Liu, J. Y. (2010). Research on the electric power grid vulnerability under the directed-weighted topological model based on complex network theory. In *International Conference on Mechanic Automation and Control Engineering* (pp. 3927–3930). Piscataway, NJ: IEEE Computer Society.

Xiong, G., Wang, K. F., Zhu, F. H., & Cheng, C. (2010). Parallel traffic management for 2010 Asian Games. *IEEE Intelligent Systems, 24*(5), 81–85.

Xiong, G., Wang, F. Y., Zou, Y. M., Cheng, C. J., & Li, L. F. (2010). Parallel evaluation method to improve long period ethylene production management. *Management Control, 13*(3), 401–406.

Xu, S. Z., Zhou, H., Li, C. X., & Yang, X. M. (2009). Vulnerability assessment of power grid based on complex network theory. In *Proceedings of the Asia-Pacific Power and Energy Engineering Conference* (pp. 1–4). Piscataway, NJ: IEEE Computer Society.

Yu, X. H., Dwivedi, A., & Sokolowski, P. (2009). On complex network approach for fault detection in power grids. In *Proceedings of the IEEE International Conference on Control and Automation* (pp. 13–16). Piscataway, NJ: IEEE Computer Society.

Zhang, G. H., Wang, C., Zhang, J. H., Yang, J. Y., Zhang, Y., & Duan, M. Y. (2008). Vulnerability assessment of bulk power grid based on complex network theory. In *Proceedings of the 3rd International Conference on Electric Utility Deregulation and Restructuring and Power Technologies* (pp. 1554–1558). Piscataway, NJ: IEEE Computer Society.

Zhao, W. W. (2009). *Research on analyzing model and prevention & emergency system for power system blackouts* (Ph.D. thesis). North China Electric Power University, Beijing, China.

CHAPTER 15

# Influence of Electric Vehicles on After-Sales Service

**Sven Schulze, Christian Engel, and Uwe Dombrowski**

*Institut für Fabrikbetriebslehre und Unternehmensforschung, Technische Universität Braunschweig, Langer Kamp 19, 38106 Braunschweig, Germany*

## 15.1 Introduction

Over several years, electric vehicles have become one of the main concerns in economics and research (Chan, 2002; Chan, Wong, Bouscayrol, & Chen, 2009). Many politicians, scientists, and entrepreneurs foresee a booming market for electric vehicles. One reason for this is the limited amount of fossil fuels, which may lead to an end to combustion technology. Moreover, it is estimated that electric mobility has a positive effect on carbon dioxide emissions, in cases where the energy can be gained from renewable resources (Frischknecht & Flury, 2011; Galus & Andersson, 2008). An electric vehicle can be defined as a ground vehicle that uses an electric propulsion engine. Although there are different electric vehicles (e.g., battery, hybrid, or fuel-cell), which use different technologies (Chan, 2002), this chapter will not distinguish among them since their impact on after-sales service is comparable. A definite prediction of which technology will succeed cannot be made at this time. Even though electric mobility has currently not yet taken hold in the market, the estimated growth rates are significant. McKinsey already expects market shares of new registrations of electric vehicles between 5% and 16% in large cities in 2015 (McKinsey, 2010). According to McKinsey's estimations, two-thirds of the approximately 114 million cars sold will have an electric engine in 2030; however, 75% will still have an additional internal combustion engine (McKinsey & Company, 2011). Governmental decisions will have a high impact on the future development of electric mobility, for example subsidies and research promotions (Valentine-Urbschat & Bernhart, 2009).

At the moment there are several obstacles preventing common use of electric mobility. The main deficit of electric mobility is the lack of high-power energy batteries that allow an adequate operation distance with little charging time. Furthermore, an intelligent energy infrastructure is necessary for market penetration of electric cars (Larminie & Lowry, 2003). Another actual disadvantage of electric vehicles is the high initial costs (Chan et al., 2009).

Service Science, Management, and Engineering.


DOI: 10.1016/B978-0-12-397037-4.00015-6

Due to the lack of standardization of technology, parts, and interfaces, enterprises follow different development trends (Larminie & Lowry, 2003).

With the changes in market shares of electric and combustion engine-powered vehicles, significant changes will occur in the automotive industry (Chan et al., 2009). The transformation of the automotive industry will change the whole value chain to a high degree (McKinsey & Company, 2011). The value share of electronics and electrics in the car, which today in many cars already exceeds 30%, will further increase. Original equipment manufacturers (OEMs) and original equipment suppliers (OESs), which have core competencies, for example in engines, clutches, and gears, have to realign their strategy and identify new business opportunities (Dombrowski, Schulze, & Engel, 2011; Valentine-Urbschat & Bernhart, 2009). Furthermore, new business models will arise, and new market entrants need orientation and a profound strategy to use these opportunities (McKinsey & Company, 2011).

However, the trend toward electric vehicles not only has a tremendous impact on manufacturers and suppliers in the automotive industry, but the after-sales service is also significantly affected (Diez, 2010; Dombrowski, Engel, & Schulze, 2011a). The after-sales service has a marked influence on customer satisfaction and is the basis for a long-term buyer–seller relationship. In particular, companies can use after-sales service to differentiate themselves from competitors (Rigopoulou, Chaniotakis, Lymperopoulos, & Siomkos, 2008). Stakeholders in the automobile market (e.g., OEM, regulated and independent aftermarket, OES) have to provide a customized and value-added portfolio of services to ensure a long-lasting relationship (Cavalieri, Gaiardelli, & Ierace, 2007). Furthermore, the after-sales service is highly profitable. In spite of a small share of less than one-quarter of total sales in the German automotive industry today, over 50% of the profits are made in this segment (Little, 2008). The after-sales service ensures continuous revenue throughout the product life cycle, while the sale of new cars is highly cyclical. Maintenance and services often are contracted on a long-term basis, and spare parts have to be available even in an economic crisis. The sale of a new product offers only a single or very few opportunities to convince the customer of the new product. In contrast, the customer contact in after-sales service is more intensive and on a long-term basis. Thus, with a convincing performance in after-sales service, the customer may buy the next product from the same manufacturer with a higher probability. Due to the intensive contact and the possibility to communicate with the customer in the after-sales service, the customer requirements can be determined more easily. These described reasons result in the strategy of many companies to increase their market shares in the after-sales service (Dombrowski, Schulze, & Wrehde, 2007; Rigopoulou et al., 2008). In addition to OEM and OES, a variety of other stakeholders are present in this business: regulated and independent workshops or manufacturers of gray market parts for example.

The new spectrum of parts, along with the electric mobility, will affect the after-sales service market significantly (Dombrowski, Engel, & Schulze, 2011b). Today, many of the necessary maintenance and repair works are related to the internal combustion engine and power transmission. These and many other activities will be dispensed in the future. New tasks will be created, but enterprise strategies have to adapt to the new conditions. For example, the requirements concerning employee qualifications will change dramatically. In future, the electric mobility may also have a marked impact on regional power generation (Galus & Andersson, 2008; Hadley & Tsvetkova, 2008). However, the specific development of the after-sales business is difficult to forecast at present (Dombrowski, Engel, & Schulze, 2011a) because the crucial influencing factor of the after-sales service market and its activities is the automobile population. Its size and structure directly affect the number of potential service orders, the size of the spare parts market, and the accessories business. Since the use of electric automobiles is still rare, basic changes have not yet started. This chapter deals with the effects of increasing share of electric vehicles on the after-sales service market, focusing, in particular, on the opportunities in the future after-sales service market.

## 15.2 After-Sales Service in the Automotive Industry

In order to analyze the changes initiated by the increasing market share of electric mobility, three segments of after-sales market are defined. Furthermore, the current stakeholders and the main service operations in the automotive after-sales market will be presented.

The after-sales service comprises all activities of an enterprise that are provided to the customer after the purchase and delivery of a product (e.g., maintenance and repair activities or a helpdesk for technical or functional questions). Figure 15.1 shows the typical organizational structure of an automotive after-sales service. It can be divided into three segments: spare parts service, customer service, and accessories business.

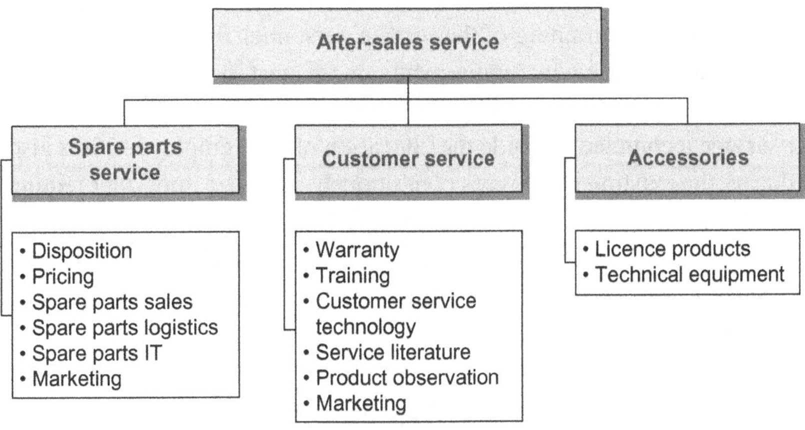


**Figure 15.1: Automotive after-sales service.**

The field of spare parts service comprises tasks like disposition, pricing, sales, marketing, and logistics. It has to ensure the availability of spare parts throughout the product life cycle to guarantee high customer satisfaction and loyalty (Dombrowski, Engel, & Schulze, 2011a). Besides the availability, the costs of spare parts are of great importance and depend on the right supply strategy chosen (Dombrowski et al., 2007).

A distinctive feature of the automotive industry is the fact that parts cannot be used in several end products without customization. The majority of parts are specifically designed for a single model. This obligates suppliers to the OEM and limits economies of scales (Sturgeon, Memedovic, van Biesebroeck, & Gereffi, 2009). One of the current developments is the significant change in the spectrum of spare parts. Electric and electronic components are increasingly used in automobiles. Due to the replacement of mechanical by electronic and electric components, the value share of electronic components grows continuously. While initially only simple electronic components were used, for example for the electronic injection or ignition, nowadays electronic and electric components are used in almost every area. Not only mechanical components are being replaced, but a lot of new functions are being included, for example in the area of safety or infotainment. Since 1995, the value share of electric and electronic devices has increased by 6.5% on average per year. Today, the average value share of electronics in an automobile is over 30%. Due to this trend, new problems arise in spare parts services. The demand for spare parts is difficult to predict in electronic devices. Furthermore, the possibility of remanufacturing is often not clear. Despite these problems, the parts should be available continuously until the end of the life cycle of the product. The main reason for the equivocal possibility of remanufacturing electronic components is the short time of series production of semiconductor components, which is, for example, about 2 years for microprocessors, in contrast to most primary products (Dombrowski & Schulze, 2008).

The second segment of after-sales service, the customer service, focuses on service operations and includes qualification and training of the service personnel. The employees have to be qualified in accordance with the new requirements. In the past, new occupational profiles with a focus on the handling of electronic components were developed (e.g., mechatronics technician or automotive service technician). Beside the education of new employees, it is also essential to train and educate the existing employees continuously. Another important requirement, which results from the increasing share of electronics, is the use of new diagnostic instruments. These instruments combine specific hardware and software and offer the opportunity to gather information for customer relationship management. For example, it is possible to determine the next date for necessary inspections of a customer and to create special offers for further services. An additional task of the customer service is the writing of customer service literature, for example manuals. Moreover, the customer service has the responsibility to observe product acceptance and other marketing tasks.

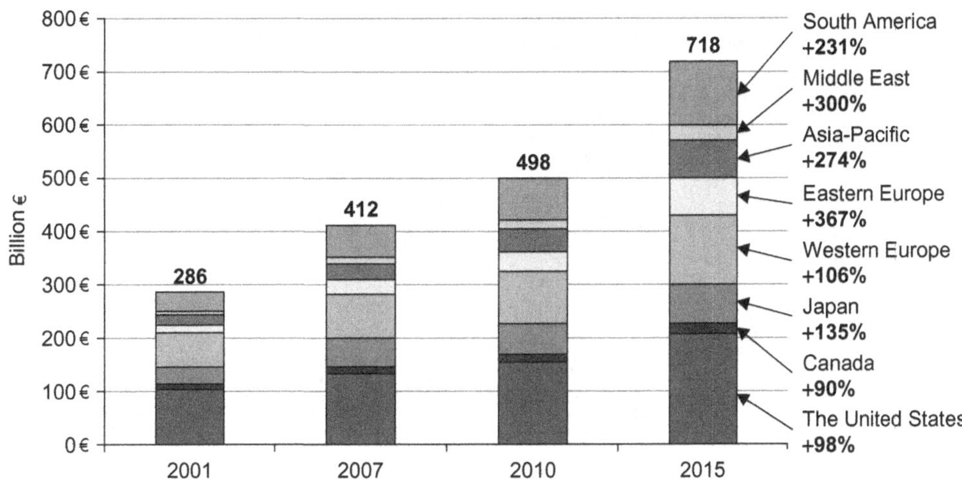


**Figure 15.2: Growth of automotive after-sales service.**

The third segment of after-sales service, the accessories business, includes licensed products and also technical equipment. In this segment, the highest profit margins of the after-sales service can be skimmed. This segment will not change dramatically with the shift to electric mobility.

The remarkable growth rates in the after-sales service in the recent past are also expected in the future. Figure 15.2 shows the growth of the worldwide automobile aftermarket. In the year 2001, the aftermarket comprised 286 billion; up to 2010 this market had grown by over 200 billion. In the next 5 years, the market is expected to grow again by 200 billion. Furthermore, it can be seen that in the period from 2001 to 2015, this market will grow in the industrialized countries (e.g., in the United States by 98%) as well as in emerging countries, where the growth in the same period has been even higher (e.g., Asia 274%). Therefore, in a long-term strategy, stakeholders must consider an internationalization of their services.

### *15.2.1 Stakeholders of the Automotive Aftermarket*

Each of the three segments of after-sales service has a variety of stakeholders with different motivations and goals. Driven by their interests, the automotive aftermarket is a complex network of dependencies (Dombrowski et al., 2007). Figure 15.3 shows the general structure of the automotive aftermarket. There are several possible distribution channels, how parts can reach the final customer. In particular, if the customer repairs his car on his own, he can purchase the spare parts directly from the OEM, the OES, or the gray market via the Internet or by a specialized reseller. If the customer goes to a service station, necessary

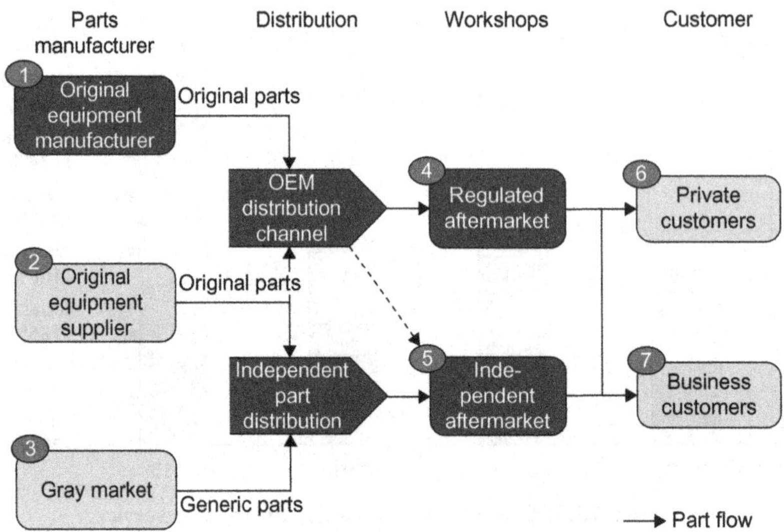


**Figure 15.3: Structure of the automotive aftermarket.**

spare parts for maintenance and repair will be bought by the service stations of the regulated or independent aftermarket. The regulated aftermarket will focus on original parts from the OES or OEM while the independent aftermarket will mostly purchase the spare parts from the gray market. Beside spare parts provision, the stations will also fulfill service operations like maintenance or repair tasks. The requirements and interests of the various stakeholders in the automotive aftermarket are described in the following parts of the chapter.

#### *15.2.1.1 Original Equipment Manufacturers*

The first group of stakeholders in the automotive aftermarket is the OEMs. The OEMs produce cars in their manufacturing plants and sell them either directly through their own dealerships or through a distributor to the customers. In the past, the automotive sales continuously increased due to the greater importance of individual mobility. The world's car production grew from 36.5 million in 1996 to more than 60.3 million in 2010 (International Organization of Motor Vehicle Manufacturers, 2010). Due to the concentration on core competences, the depth of added value has decreased to 22% in the German automobile industry over the last decades (Verein deutscher Automobilindustrie, 2009). While the dependency of the suppliers increases, the OEM depends on the OES for spare parts production and long-term supply. However, the OEMs are usually not the direct provider of after-sales service operations. This is done by contracted service providers in the regulated aftermarket. The OEMs have a high interest in selling their own original spare parts that are normally more expensive than spare parts sold by vendors in the independent aftermarket (Wokl, 2010).

In the past, a concentration of OEMs could be seen. Eleven leading firms from the United States, Germany, and Japan dominate the production in the main markets today. High entry barriers like high initial costs for the design of a new car hinder new competitors to enter the market (Sturgeon et al., 2009). Compared to the other actors in the after-sales service market, the OEMs have a high number of strategic options in the regulated aftermarket due to their system leadership, their size, and financial power (Diez, 2010).

#### *15.2.1.2 Original Equipment Suppliers*

The OESs are manufacturer of components and spare parts and deliver the parts directly to the OEMs during serial production (Dombrowski & Schulze, 2008). In the last decade, the largest supplier could shift the balance of power away from the OEM due to an increased outsourcing of production and development (Sturgeon et al., 2009). In most cases, a supply chain in the automotive sector consists of several tiers of suppliers that deliver parts to the OES. The OES is also interested in selling original spare parts to service stations or to the final customers, like the OEM. Often the OESs are not bound to sell the spare parts just through the OEM distribution channel but also via the independent aftermarket. By selling the spare parts directly to the final customers, the profit margin is much higher as no other stakeholders share the profit. However, for the automotive suppliers, the biggest problem is the lack of contact with the final customer because direct sale of the spare parts is not the normal case (Diez, 2010).

Legislation changes in the past allowed the OESs to sell spare parts as original parts in the European Union, if they are identical to those used in the series product. This declaration is a big advantage to gain customers and should strengthen competition (European Union, 2010). In addition, the OESs are interested in a high forecast accuracy of spare parts demand and a standardization of the spare parts that leads to higher production lots (Wokl, 2010). The OESs are contractually bound to deliver spare parts throughout the product life cycle of the automobile, even though the demand decreases and the production of some spare parts is no longer economical.

#### *15.2.1.3 The Gray Market*

Increasing competition is a current trend in after-sales service. Due to the high margin, new competitors try to position themselves in this market. These new competitors are independent producers or service suppliers, the so-called gray market. The potential savings for customers can be very high. Depending on the type of car and the spare part, it can be between 5% and >30%. The manufacturers are selling generic parts that are not approved and authorized by the OEMs (Dombrowski & Schulze, 2008). Suppliers of the products sold on the gray market are often called generic part manufacturers; they have low development costs because they mainly copy or pirate the original parts. The gray market can therefore provide spare parts at a significantly lower price than the original parts (Antia, Bergen, & Dutta, 2004;

Klostermann & Günnel, 2010). In addition, the vendors in the gray market concentrate on economically viable spare parts while the OEMs are obliged to provide every part of the automobile throughout the product life cycle. Many OEMs and OESs try to impede the gray market through the use of patents and property rights on their parts.

#### 15.2.1.4 The Regulated Aftermarket

Workshops of the regulated aftermarket have binding contracts with at least one OEM and provide a closely defined bundle of services. Thereby the offered services vary and depend on the contract. Often also new and used cars are sold by the service stations of the regulated market. In addition, these workshops represent the brand of the OEM and have to display an appropriate brand image (Hecker & Seeba, 2010). The strategic latitude of the different authorized dealers and authorized repair shops is restricted due to their contractual relationship with one or more OEMs. The main target of the regulated aftermarket is to reinforce their position as "brand champion" in order to strengthen their respective regional or local competitive environment (Diez, 2010).

#### 15.2.1.5 The Independent Aftermarket

The free workshops do not have a contractual relationship with any OEM (Klostermann & Günnel, 2010). Repair shop chains and specialist stores are considered as classical discounters in the automotive aftermarket (Diez, 2010). Participants in this market segment differ widely with the range of services offered. On the one hand, there are universal workshops, which have a wide range of services starting from a simple tire change to difficult repair works (Wokl, 2010). On the other hand, there are specialty shops, which specialize in certain services like window repair or paint work (Hecker & Seeba, 2010). Through the high volume of services and the specialization, they often have lower prices than the workshops in the regulated aftermarket. The OEMs try to push the regulated aftermarket and to reduce competition in the market for automotive repair and maintenance. The increased use of information technology in automobiles can be seen as a new opportunity. In particular, diagnostic tools and repair codes have become necessary for maintenance work but require a large amount of technical information from the OEM. Thus, the competition can be diminished by holding back this information from the independent aftermarket (Hawker, 2008).

The independent aftermarket has started to use the Internet as a marketing channel. This is an adequate possibility for improving the distribution of spare parts or services. The major advantage of the Internet as a marketing channel is the worldwide accessibility for customers at moderate costs. In addition, it is possible to inform the customer continuously about the availability and the delivery time of the wanted spare parts.

The market entry barriers for new competitors decrease, thus the competitors can enter the spare parts market more easily and faster. By using the Internet as a marketing channel, customers are able to compare prices easily, and thus they can obtain the spare parts from the cheapest supplier with an adequate lead time. Indeed, the acquisition of a repair service is more complex than the acquisition of spare parts or a new or used car, but especially for standard repair services or maintenance tasks prices can easily be compared by Internet users. The personal relationship between a driver and his repair shop will be weakened if new and customer-oriented business models in the e-mobility part are implemented (Dombrowski, Engel, & Schulze, 2011a).

#### 15.2.1.6 Customers

Customers are the most important group of stakeholders in the automotive aftermarket (Dombrowski et al., 2007). Nonetheless, they are also the most heterogeneous group, consisting of private, fleet, and business customers with different requirements concerning quality, price, and velocity of service. Customers, in general, are looking for cheap spare parts of good quality and for a quick and good service. In many countries, a structural change from private to commercial customers has been observed in the business of new cars. In the last few years, a significant increase in the number of registrations from vehicle fleet management providers, leasing companies, and renting companies can be noticed. This structural change also influences the after-sales service. In general, leasing enterprises, car rental companies, or fleet managers have large car pools, which are built up, organized, and controlled with professional structures. By these means, they have the power to regulate who gets how many customer orders and under which conditions. So, what was a B2C business becomes a B2B business. The "intermediates" not only put pressure on the repair shops but also on the manufacturer and supplier in the aftermarket. In the case of the manufacturer especially the spare parts business is involved (Diez, 2010).

During the product life cycle, the requirements of private customers toward after-sales service are changing. In the first 2 years after the purchase of an automobile, the customers take it for granted that all occurring defects and spare parts are covered by the guarantee. Even in the following years, the customers focus on good services and new original spare parts. Later in the vehicle's life, the requirements of the customer change. With regard to the low residual value of the automobile, low prices of spare parts are steadily gaining more importance for customers. At this stage, the customer is willing to buy cheap parts provided by vendors in the independent aftermarket or from the gray market (Dombrowski et al., 2007). In accordance with a study by Plus, the most important criterion in the choice of a repair shop for German private vehicle owners is the price level, as shown in Fig. 15.4. The second criterion is the quality of work and parts. The quality of consulting, date availability, or opening hours is less important.

In addition to the stakeholders described above, there are even more stakeholders in the automotive aftermarket. Insurance companies, which have to regulate damages in accidents, are becoming more and more aware of their market power and try to push orders to the

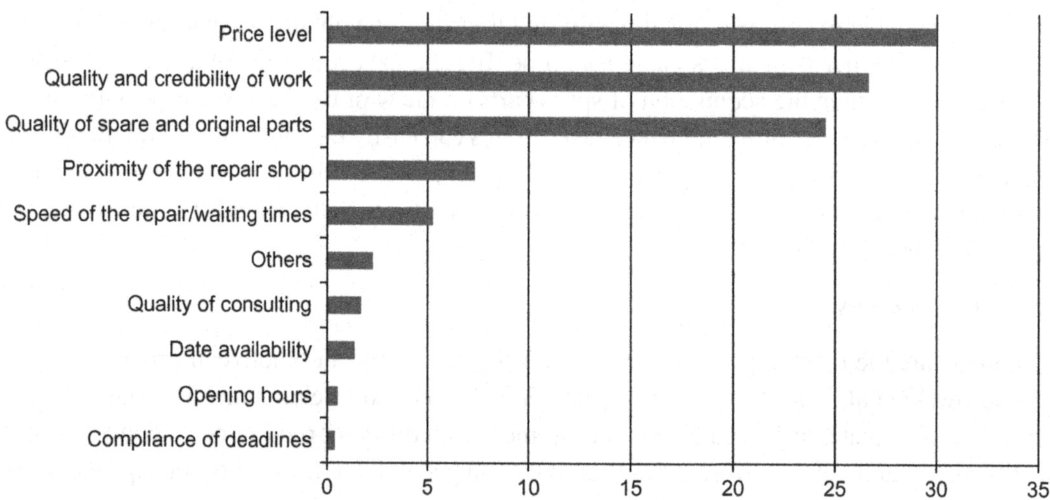


**Figure 15.4: Decision criteria in the choice of a repair shop (Plus Marktforschung GmbH/ Deutsche Post AG 2010).**

cheaper independent aftermarket (Diez, 2010). Scrap dealers can also sell used parts, especially to customers with old cars. Because of their lower impact on the after-sales service, they are not considered in the following text.

The consequences of legislative measures in the after-sales market should not be neglected; therefore, the government is also an important stakeholder (Dombrowski, Engel, & Schulze, 2011b). Besides the attempt to increase the competition further, influential factors exist in order to establish lower prices for the customer. For example, economy stimulating measures like cash for clunkers cause significant changes in the automobile population with regard to class and age. The spare parts market can also be affected by substance bans (e.g., lead-free soldering). An additional example of changing legal framework conditions is the liberalization of the block exemption regulation of cars in Europe. This regulation aims at lower prices of spare parts for the customer by intensifying the competition. Hence, manufacturers are no longer allowed to base their warranty on the condition that maintenance work is done in authorized repair shops of the regulated aftermarket. Independent repair shops have to be given access to more technical information by the OEM as well.

### *15.2.2 Typical Services in the Automotive Aftermarket*

As mentioned above, the after-sales service is divided into three segments: the spare parts service, the customer service, and the accessories business. Each of these three segments contains processes that are influenced in different ways by the rising share of electric vehicles.

The services in the segment of customer service can be divided into two main categories: maintenance and repairs. While the maintenance work occurs periodically based on a schedule, repairs occur sporadically due to an accident or defect and are not predictable. In both maintenance and repair activities, an overall decreasing trend can be identified. In particular, the higher quality of the cars results in lower maintenance and longer service intervals. The average number of maintenance activities per car in Germany has declined over the last 20 years from 1.3 to 0.9 yearly. In the same period, the number of repair activities has decreased from 1.0 to 0.6 per car and year (DAT, 2010).

The reasons for maintenance and repair work and their frequency of occurrence are shown in Fig. 15.5. In 14% of the cases, an exchange of the brake pads is the reason for maintenance. This simple task can be done at nearly every service station without any specialized knowledge. Due to the high amount of maintenance incidents and resulting spare parts demand combined with the relatively low technical requirements of manufacturing brake pads, original parts and generic parts compete on a large scale. Other common causes for attendance at a service station are failures of electric components and the engine. Due to the high complexity of the electrical network in an automobile and the necessary technical equipment, like diagnostic equipment with specialized software, the number of service stations that offer this service is significantly lower. Furthermore, the variety of electronic control units is relatively high and know-how of manufacturing is necessary. Therefore, the competition decreases and the margins in this area can be higher. A third common reason for maintenance and repair is the engine. Repair tasks concerning the engine often are expensive due to the high amount of manual labor and expensive spare parts, although in many cases just a change of oil is necessary. Overall, the 10 reasons presented cause 60% of all repair and maintenance measures.

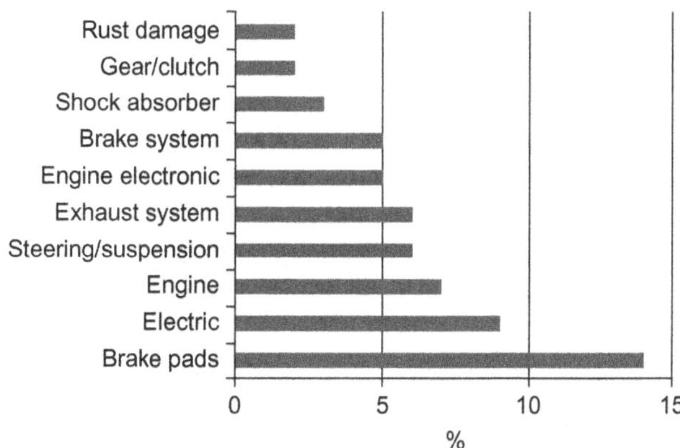


**Figure 15.5: Reasons for repair and maintenance.**

## 15.3 Changes Due to the Increasing Share of Electric Mobility

In the previous section, some typical service operations of the current automotive aftermarket have been presented. The increasing share of electric mobility will lead to a tremendous change. There will be companies from different branches, especially the electronic sector, who will enter the automotive aftermarket to benefit from the high profit margins in this sector. In the coming years, the automotive value chain will be restructured and next to the old actors new participants will also appear (Diez, 2010). Figure 15.6 shows the five main drivers for change in the automotive after-sales service.

One of the main differences between cars with internal combustion engines and electric vehicles is the decreasing share of mechanical and moving parts. Experts estimate that electronic vehicles have about 90% less moving parts than vehicles with a combustion engine (Diez, 2010). The significantly lower number of these parts in electric vehicles will have an enormous impact on the after-sales service. As there are no moving cylinders in an electric engine, an oil cycle for cooling and lubricating becomes obsolete. Therefore, oil change services and regular screenings of mechanical wear parts (e.g., timing belts and timing chains) are not necessary. Since a large number of the services performed currently are connected to components of the internal combustion engine and aligned parts of the power train, these services will become obsolete. Furthermore, the decreasing share of mechanical and moving parts leads to a decreasing sale of wear parts. In particular, spare parts like spark plugs, cylinder head gaskets, and lubricants will no longer be necessary.

Moreover, the operating temperature of an electric engine is significantly lower compared to an internal combustion engine. Cooling with a water cooling system is essential for a combustion engine but not for an electric engine. However, the electric vehicles will need

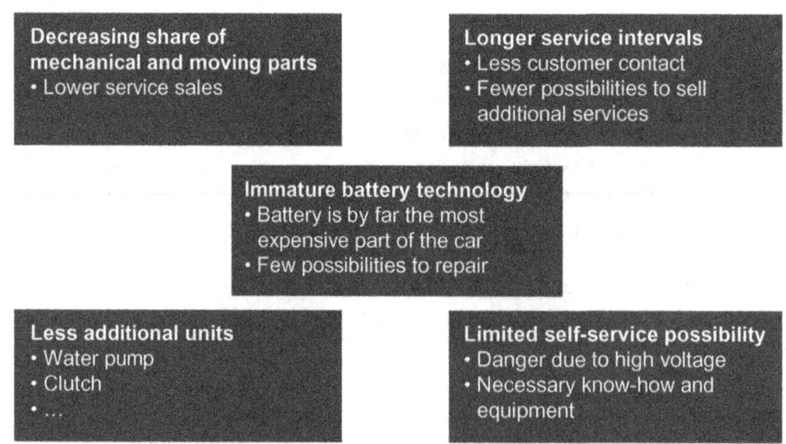


**Figure 15.6: Change drivers in the after-sales service due to electric mobility.**

a cooling system for the battery packs. The first generation of batteries becomes very hot when in use. Therefore, an air-conditioning system needs to be installed. In this air-conditioning system, potential new spare parts are included. Altogether an increasing share of electric vehicles will, however, cause a decrease in the maintenance and repair volume (Diez, 2010).

Another main difference between an electric vehicle and a combustion engine vehicle is the lower number of parts, as shown in Table 15.1. Some examples are (1) mechanical components such as gear boxes, clutches, automatic transmissions, and turbocharger; (2) spare parts like exhaust systems with mufflers and center mufflers, which are obsolete; (3) electronic components such as starters, alternators, and fuel pumps, including its sensors, which are not needed in electric vehicles.

The outcome for after-sales service will be a massive slump in the sale of spare parts and fewer necessary service operations, as shown exemplarily in Table 15.2.

Another characteristic of electric vehicles compared to cars with internal combustion engines is the longer service intervals. This is a result of the already mentioned smaller number of wear parts in the electric motor. It is also a consequence of the limited possibility of maintaining and repairing an electric motor. Thus, only a few mechanical components of the electric motor (e.g., carbon brushes) offer the possibility to change wear parts.

**Table 15.1: Dispensed components in the drive train (Wallentowitz, Freialdenhoven, & Loschewski, 2010).**

| System | Component |
|---|---|
| Combustion engine | Crankcase, crankshaft, cock, connecting rod, cylinder barrel, cylinder head, valves, camshaft, cooling circuit, turbo compressor, and engine control |
| Fuel supply | Tank vessel, fuel pump, injection system, and pipeline |
| Exhaust system | Exhaust manifold/tubes and catalytic converter |
| Clutch | Disk clutch and high dynamic converter |
| Transmission | Body, gear wheel, switchgear, and roll bearing |

**Table 15.2: Dispensed maintenance tasks (Wallentowitz et al., 2010).**

| Part | Activity |
|---|---|
| Vehicle from bottom | Visual check/repair of the transmission, change of clutch oil/filter, control of the conventional brake system, control of the steering system, and visual check/repair of the exhaust system |
| Engine compartment | Change of the engine oil/oil filter, adjustment of valve play, adjustment of the v-belt, replacement of timing belt, correction of coolant level, replacement of the air/fuel filter, change of the spark plugs, and control of all pipelines and hoses |

Probably the biggest technical problem of electric mobility at the moment is the energy supply and storage. One of the key components of alternative vehicles is the energy storage device acting as energy buffer (Conte, 2006). The battery technology used in the first generation of electric vehicles is still immature. Lead–acid batteries and electric double layer capacitors have a relatively low energy density. Lithium-ion batteries with a high energy density meet the requirements of electric cars, but they are far more expensive (Conte, 2006). So far, batteries have been developed specifically for use in laptops or mobile devices. These products, however, have a much shorter product life cycle compared to an automobile that normally has a life cycle of more than 12 years (DAT, 2010). The installed batteries in electric vehicles of the first generation, therefore, have a shorter life cycle than the rest of the car. Experts estimate the lifetime performance of a battery between 75,000 and 150,000 km with a weight of up to 400 kg (Frischknecht & Flury, 2011). Furthermore, the batteries have a limited amount of charge cycles. In some cases there are only 700–1000 charge cycles possible before the batteries have to be replaced (Chan & Wong, 2004). To establish the ability of the customer to use the electric vehicle continuously, it is necessary to replace the battery during product life. Since batteries are the most expensive part in an electric vehicle, the exchange may not be economical for the customer. Therefore, customers might hesitate to replace the battery pack of an electric vehicle and buy a completely new car instead. This would lead to a decrease in after-sales revenue but to an increase in sales of new cars.

The last major difference between electric vehicles and cars with internal combustion engines in terms of after-sales service is the fact that customers cannot fix problems in the car on their own at home, because high-voltage technology-specific knowledge about the product is needed to repair the engine of the electric vehicle. Do-it-yourself repair without this specific knowledge may be very dangerous because of the high voltage. Furthermore, specific technical equipment is needed to repair and maintain electric vehicles. The commercial stakeholder in the aftermarket will benefit from this fact.

## 15.4 The Impact on Stakeholders

The impact of the identified changes to the automobile aftermarket on the previously named stakeholders will be derived in the following section. Furthermore, possible strategies are presented, which can be used to meet the identified challenges. Figure 15.7 provides an overview of the impacts of the main changes on the various stakeholder groups.

### 15.4.1 Decreasing Share of Mechanical and Moving Parts

The decreasing share of mechanical and moving parts in electric vehicles initially has a direct impact on the OESs and the upstream supplier levels (Diez, 2010). The decrease in revenues of selling wear parts will be noticeable. For the OEM, this decline is also associated with a reduction in the profit margins of the spare parts business. The gray market in particular

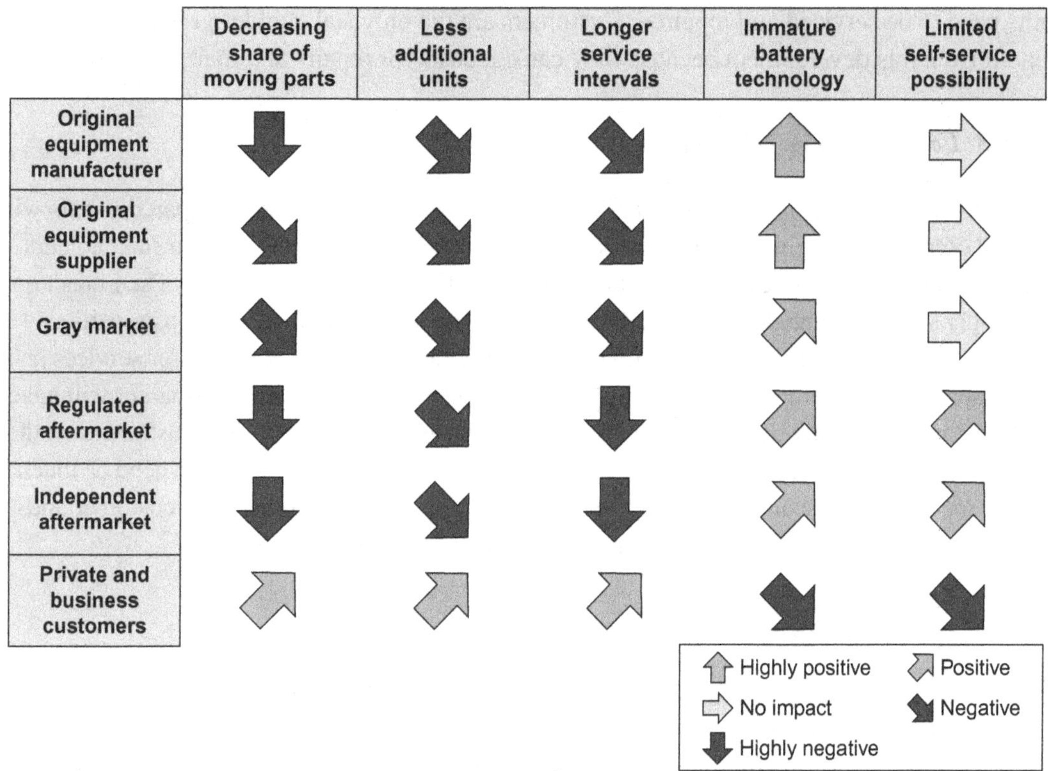


**Figure 15.7: Impacts of the change drivers on stakeholders.**

will suffer a clear decline in the number of spare parts because the mechanical and moving parts are provided by the gray market at the moment because they are comparatively easy to produce (Hesselbach et al., 2004). The service providers of the regulated and independent aftermarket will also be affected because mechanical and moving parts are the main reasons for cyclical inspections and therefore for visits to a service station. The decreasing number of repair and maintenance operations generated by these parts has a negative effect on the service stations in the regulated and independent aftermarket. In contrast to the other stakeholders, the customers benefit from the decreasing share of mechanical and moving parts as they have to spend less money on maintenance and repairs of their vehicles.

### *15.4.2 Fewer Additional Units*

For both the OEM and OES, sales decline will be the result of the smaller amount of additional units in electric vehicles. Also, the gray market will be affected by this development since customers no longer need generic spare parts for the missing units. The regulated and independent aftermarket will be affected negatively by fewer additional units because fewer units need to be serviced and repaired. Customers are the only stakeholder group that will benefit from this development because they can cut costs for repairs and maintenance.

### 15.4.3 Longer Service Intervals

The third change driver is the longer service intervals for electric vehicles compared to cars with internal combustion engines. The longer service intervals have similar effects on stakeholders in the automotive after-sales service as the previously described change drivers. The impact on the supply side is also negative. OEM, OES, and the gray market will sell less spare parts or additional units because customers will visit service stations more infrequently for services and to buy spare parts. Therefore, the actors of the independent and regulated aftermarket will have problems in maintaining their business volume. In contrast to the supply side, customers on the demand side will profit from the longer service intervals. They will not need to spend as much money for maintenance and repair services as before. The relationship between customer and service stations becomes loose due to the infrequent and seldom contacts.

### 15.4.4 Immature Battery Technology

In most cases, the changes caused by the increasing share of electric mobility lead to a deteriorating market situation on the supply side. However, the OEM, OES, independent aftermarket, and regulated providers in the aftermarket can also benefit from the increase in electric mobility. If having an adequate strategy, these stakeholders might benefit from immature battery technology. The batteries are the most important and most expensive spare part of an electric vehicle. If they are no longer operating, the customer has no other choice than replacing the battery pack or to scrap the car. When the electric car is in an early stage of its life, it is likely that the customer will replace the batteries, which means that the OEM, the OES, and the gray market benefit. In addition, the demand is foreseeable due to the simple calculation of the amount of replacement batteries based on the average status of sold electric vehicles of the OEM. As the batteries are a highly profitable spare part, new manufacturers of the gray market will try to provide cheaper generic batteries for electric vehicles. The service stations from the regulated and independent aftermarket will profit to a lesser degree. Usually, a battery change has to be done in a workshop by a specialist since there is a need for know-how and special tools. In particular, when the battery weighs up to 400 kg and has to be replaced, this task cannot be performed by the customer on its own. Besides the service operations, the stations can also sell the battery to the final customer and will receive a high margin. The customers are influenced negatively because expensive services are needed in order to remain mobile with the electric car. In a later stage of the life cycle of the vehicle, a critical point will be reached. At this point, the customer will no longer be willing to replace the batteries of the existing car. The customer would rather buy a new electric vehicle. Subsequently, the OEM and OES will profit not in the after-sales service but in the new car business.

The batteries have a key position in the automotive aftermarket since they are by far the most expensive spare part (Larminie & Lowry, 2003). The OEMs have to develop a strategy to force the customers to use original batteries throughout the entire life cycle of the electric vehicle. A possible strategy would be the separate sale of the automobile and a service contract for the batteries. The OEM would be in charge of the exchange of batteries, and the customer does not need to calculate whether he should buy a new battery or an already used battery or sell the car. The advantages for the OEM are continuous payments for the batteries and possible positive pooling effects. In addition, the customer would benefit because instead of high investments, which are not completely predictable, there will be a low continuous payment and the certainty of always being able to operate with functionable batteries. As a side effect of this strategy, the OEMs are able to control the risks by new battery-vendors from the gray market.

### 15.4.5 Limited Self-Service Possibility

The change driver limited self-service possibility is the last factor to be considered. Many customers, not only those with older cars, do the necessary maintenance (e.g., oil change) and smaller repairs (e.g., exchange of engine units) on their own (Little, 2008). In this regard, electric vehicles are different. On the one hand, they need to be maintained less than cars with internal combustion engines, which is an advantage for customers. On the other hand, the self-service possibility is limited for customers due to the high-voltage technology and the required special equipment and know-how. The OEM, the OES, and the gray market are neither impacted positively nor negatively by this change driver. For these stakeholder groups, it is more or less irrelevant if they sell their spare parts directly to the customer or to the service stations.

In contrast, for the service stations, there is a positive effect due to the limited self-service possibility. They will profit because the customers, who would normally self-service their cars, now have no other opportunity than coming to the service stations. The impact on the stakeholder group of the customers has to be differentiated. There is no change for customers who go to service stations anyway. The group of customers who did the service on their own, however, have to pay much more money to the service stations.

## 15.5 Future Perspectives

Electric mobility will have an increasing market share in the automotive business. It is not yet apparent, which of the currently competing technological concepts will become the new standard (Chan, 2002). But regardless of the choice of standard to be established, the rising share of electric vehicles will have a profound impact on the automotive after-sales service. This chapter focuses on the changes in the automotive aftermarket caused by the rising share of electric mobility. The impacts for different stakeholders of the after-sales market are shown.

A possible strategy to cope with these challenges is the bundling of products and services. By applying this strategy, the OEMs can increase their sales and bind the customers. An example of such bundling is the combined sale of the product with services like financing, long-term guarantee, maintenance, and insurance. By these means, the seller benefits from a reduction of costs and a simultaneous increase of sales and profit. Furthermore, it is possible to decrease the transparency concerning the effective costs for individual products and services, thus reducing direct comparison. From a customer's point of view, bundling is positive because the costs of the combination of product and service are easier to assess and the customer will be more comfortable with this. This leads to higher customer satisfaction and thus to higher loyalty to the producer.

A further strategy is based on the assumption that a customer is more interested in the use of a product than the possession of the product. Thus, more and more leasing contracts for automobiles are sold. Customers buy a defined usage, for example a vehicle with an availability of 95%. By these means, they have to pay a flat rate and they do not need to take care of it. The producer endeavors to render the assured usage. Therefore, the producer will maintain and repair the car at regular periods. The biggest advantage for a customer is the significant improvement of downtimes. An additional advantage can be the lower costs of spare parts, of planning, of organization, of rejects, of material, and of energy.

## References

Antia, K., Bergen, M., & Dutta, S. (2004). Competing with gray markets. *MIT Sloan Management Review, 46*(1), 62–68.

Cavalieri, S., Gaiardelli, P., & Ierace, S. (2007). Aligning strategic profiles with operational metrics in after-sales service. *International Journal of Productivity and Performance Management, 56*(5&6), 436–455.

Chan, C. C. (2002). The state of the art of electric and hybrid vehicles. *Proceedings of the IEEE, 90*(2), 247–275.

Chan, C. C. & Wong, Y. S. (2004). Electric vehicles charge forward. *IEEE Power and Energy Magazine, 2*(6), 24–33.

Chan, C. C., Wong, Y. S., Bouscayrol, A., & Chen, K. (2009). Powering sustainable mobility: Roadmaps of electric, hybrid, and fuel cell vehicles. *Proceedings of the IEEE, 97*(4), 603–607.

Conte, F. (2006). Battery and battery management for hybrid electric vehicles: A review. *Elektrotechnik & Informationstechnik, 123*(10), 424–431.

DAT (2010). Report 2010, Autohaus, 9/2010.

Diez, W. (2010). *Zeitenwende im automobilservice*. http://automechanika.messefrankfurt.com/frankfurt/de/besucher/events/AM-Academy.html

Dombrowski, U., Engel, C., & Schulze, S. (2011a). *Changes and challenges in the after sales service due to the electric mobility*. IEEE International Conference on Service Operations and Logistics, and Informatics 2011, Peking, China, 10–12 June 2011.

Dombrowski, U., Engel, C., & Schulze, S. (2011b). *Scenario management for sustainable strategy development in the automotive aftermarket*. Third CIRP International Conference on Industrial Product Service Systems, Braunschweig, 5–6 May 2011, pp. 285–290.

Dombrowski, U. & Schulze, S. (2008). *Repair of automotive electronics as a module of life cycle orientated spare parts management*. 15th CIRP International Conference on Life Cycle Engineering, Sydney, Australia, 17–19 March 2008, pp. 464–469.

Dombrowski, U., Schulze, S., & Engel, C. (2011). Zukunftsgerechte gestaltung des after sales service. *ZWF Zeitschrift für wirtschaftlichen Fabrikbetrieb*, *106*(5), 366–371.

Dombrowski, U., Schulze, S., & Wrehde, J. (2007). *Efficient spare part management to satisfy customers need.* International Conference on Service Operations and Logistics, and Informatics, Philadelphia, pp. 304–309.

European Union (2010). Supplementary guidelines on vertical restraints in agreements for the sale and repair of motor vehicles and for the distribution of spare parts for motor vehicles. 2010/C 138/05.

Frischknecht, R. & Flury, K. (2011). Life cycle assessment of electric mobility: Answers and challenges. *The International of Life Cycle Assess*, *16*, 691–695.

Galus, M. D. & Andersson, G. (2008). Demand management of grid connected plug-in hybrid electric vehicles (PHEV). *IEEE Energy 2030*, Atlanta, GA, 17–18 November 2008.

Hadley, S. W. & Tsvetkova, A. A. (2008). Potential impacts of plug-in hybrid electric vehicles on regional power generation. *The Electricity Journal*, *22*(10), 56–68.

Hawker, N. (2008). Under threat: Competition in the automotive service aftermarket. AAI Working Paper 08-05.

Hecker, F. & Seeba, H. (2010). *Aftersales in der Automobilwirtschaft: Konzepte für Ihren Erfolg*. Munich: Autohaus Buch und Formular.

Hesselbach, J., Dombrowski, U., Bothe, T., Graf, R., Wrehde, J., & Mansour, M. (2004). Planning process for the spare part management of automotive electronics. *WGP Production Engineering*, *XI/1*, 113–118.

International Organization of Motor Vehicle Manufacturers (2010). http://oica.net/wp-content/uploads/ranking-2010.pdf.

Klostermann, L. & Günnel, S. (2010). Competitive terms of trade for European automotive suppliers – Overcoming traditional strategies in the independent aftermarket. *Journal of Revenue and Pricing Management*, *11*, 127–136. Published online 17 September 2010.

Larminie, J. & Lowry, J. (2003). *Electric vehicle technology explained*. Chichester, UK: Wiley.

Little, A. D. (2008). *Automotive INSIGHT – Automotive after sales 2015*. Online: http://www.adl.com/uploads/tx_extthoughtleadership/AMG_Automotive_after_sales_2015_01.pdf.

McKinsey (2010). *Elektromobilität in Megastädten: Schon 2015 Marktanteile von bis zu 16 Prozent*. Study by McKinsey Consultants, Düsseldorf, Germany.

McKinsey & Company (2011). *Boost! Transforming the powertrain value chain, a portfolio change*. Study by McKinsey Consultants, Düsseldorf, Germany.

Rigopoulou, I., Chaniotakis, I., Lymperopoulos, C., & Siomkos, G. (2008). After-sales service quality as an antecedent of customer satisfaction: The case of electronic appliances. *Managing Service Quality*, *18*(5), 512–527.

Sturgeon, T., Memedovic, O., van Biesebroeck, J., & Gereffi, G. (2009). Globalisation of the automotive industry: Main features and trends. *International Journal of Technological Learning, Innovation and Development*, 2(1/2), 7–24.

Valentine-Urbschat, M. & Bernhart, W. (2009). *Powertrain 2020—The future drives electric*. Study by Roland Berger Strategy Consultants.

Verein deutscher Automobilindustrie (2009). Jahresbericht 2008. Berlin: VDA.

Wallentowitz, H., Freialdenhoven, A., & Loschewski, I. (2010). Strategien zur Elektrifizierung des Antriebstrangs. Technologien, Märkte und Implikationen. Vieweg.

Wokl, H. (2010). Servicetrends – heute und morgen. *Der freie Kfz-Servicemarkt*, 10–13.

CHAPTER 16

# Service Modeling Optimization and Service Composition QoS Analysis

**Sheng Liu, Gang Xiong, and Dong Fan**
*State Key Laboratory of Management and Control for Complex Systems, Institute of Automation, Chinese Academy of Sciences, Beijing, China*

## 16.1 Introduction

Web service, a software system designed to support interoperable machine-to-machine interaction over a network (Fan & Li, 2003; Wang & Tang, 2004), is of great importance in service-oriented architecture (SOA). Since the evolution of cloud computing, which commonly utilizes SOA, Web service has become a hot topic and has attracted much attention from both industry and academia, and the number of public Web services is increasing. Hence, Web service discovery has been extensively studied, mainly as regards its functional properties (Kyriazis, Tserpes, Menychtas, Litke, & Varavarigou, 2008), i.e., locating services that can meet a specified functional description. However, because of the large amount of services with identical or similar functionalities, users will be overwhelmed by the choice. Since Web service discovery alone cannot tackle this problem, effective approaches to Web service selection and recommendation have become necessary, which is a key issue in the field of service computing (Huang & Fan, 2003; Rudolph, Kuntze, & Velikova, 2009; Zhang, Wang, Zhu, Zhao, & Tang, 2008).

BPEL4WS is a de facto standard for specifying and executing Web services. It comprises different types of Web services defined by the Web Services Description Language (WSDL) (Miao et al., 2011; Nurcan, 1998). This is a language with rich expressivity in comparison to other languages for business process modeling, in particular those which are supported by workflow management systems (Zhao & Fan, 2003; Zhu, Li, Li, Chen, & Wen, 2011). It contains a number of primitive activities as well as structured activities. On the other hand, this leads to complexity of BPEL4WS and low-performance modeling and running (Cao, Chan, & Chan, 2005; Lin & Fan, 2003; Wang, 2005). To improve the ease of use and running performance, we need a method and tool that can simplify business process modeling and optimize the BPEL4WS model for running.

In recent years, SOA turns to the software development standard and enterprise application systems programming trends. Web service has become the most promising technology of

Service Science, Management, and Engineering.


DOI: 10.1016/B978-0-12-397037-4.00016-8

distributed software integration in the Internet environment (Curbera et al., 2002). Service providers provide Web services, register the Web service's WSDL description to a Universal Description, Discovery, and Integration (UDDI) registry on the Internet, then service users access the UDDI search service, and invoke it to perform specific tasks.

Web service assessment is usually divided into two aspects, namely functional assessment and nonfunctional assessment (Curbera, Khalaf, Mukhi, Tai, & Weerawarana, 2003). Functional assessment studies the function to provide conditions to meet the needs of users, and so on (Andrews et al., 2003). Nonfunctional assessment includes the cost of service calls, service availability, service execution time, and other indicators of assessment, collectively called the quality of service (QoS). Users can select and configure the service to provide a basis (Zhang, Su, & Chen, 2006).

Business processes are usually faced with the costs of execution, execution time, and other constraints (Ardagna & Pernici, 2005; Barros, Dumas, & Oaks, 2006; Johansson, 1993; Xiong, Fan, & Zhou, 2008, 2009). In an SOA environment, the use of BPEL2WS to describe the enterprise's business processes, cost, and execution time constraints are converted into Web service (described with BPEL2WS) cost of service composition and execution time constraints. The cost of Web service composition and execution time analysis of Web service composition are the scope of the quality of analysis (Menascé, Casalicchio, & Dubey, 2010). The quality of Web service composition depends on the quality of each Web service and each structure in the BPEL2WS model (Liu & Fan, 2005). At present, many literature reviews on Web composition quality problems are concerned with the average service execution time, execution cost, and performance reliability properties (Ardagna & Pernici, 2005; Hwang, Wang, Tang, & Srivastava, 2007; Menascé, Ruan, & Gomma, 2007; Menascé et al., 2010; Xiong et al., 2008, 2009), which in practice is often not sufficient. Because Web service execution is influenced by the machine load, network status, and other factors, its execution time is within a certain range of random variation (Liu & Fan, 2005). When the execution time randomly changes in value, about 50% of the actual value is more than the average execution time in the presence of demanding applications, the average execution time based on the quality of analysis often overestimates the performance of the system, so it is difficult to meet the actual needs (Li, Li, & Qian, 2006; Pozewaunig, Eder, & Liebhart, 1997). Not only does the average execution time of Web service composition needs focusing, but the distribution of execution time for single Web service and the execution time for Web service composition also need to be analyzed. The distribution of the QoS can accurately reflect the real situation as regards the execution time. But the Web service composition execution time distribution is an open research question. A Critical Path Method/Program/Project Evaluation and Review Technique (CPM/PERT) graph based method (Pozewaunig et al., 1997) can only obtain the execution time interval. Since BPEL2WS is widely used to describe the Web service composition modeling language, this research will be based on BPEL2WS. The Web service composition execution time distribution is an important factor for users to select and configure the Web service provider. In the following, the BPEL2WS model is considered to be equivalent to Web service composition.

## 16.2 BPEL4WS-Based Modeling Optimization

In this section, an optimized modeling method based on BPEL4WS is presented. It provides a graphical modeling interface that is designed for people with no knowledge of BPEL4WS. It also provides a tool to convert the graphical model to BPEL4WS. The graphical model for business processes is much simpler than the corresponding BPEL4WS, which greatly improves the efficiency of the modeling. The running speed of the BPEL4WS generated by using our methods is higher than that generated using other methods.

### 16.2.1 Comparison of Modeling of SMEE and Mainstream Methods Based on BPEL4WS

The modeling and output of BPEL4WS is termed SMEE (Simplified Modeling and Efficient Execution). Nowadays, mainstream modeling methods include Active BPEL Designer (ABD), Oracle BPEL Designer (OBD), and so on, which can explain BPEL4WS and generate a graphical model interface that can produce a BPEL4WS model based on text. The design and approval process can be described by SMEE, ABD, and OBD, as shown in Fig. 16.1. The process shown in Fig. 16.1 is a simple document approval process. In this process, after the manager starts the *beginning* activity, the designer will design requirements, technical standards, and design-related information according to *design* activity. When the designer completes the *design* activity, the related documents will be submitted to the approver. The approver examines the submitted documents according to *approval* activity and outputs a variable whose value is either "passed" or "not passed." The computer will check the variable value from the *approval* activity according to *pass* activity. If the variable value is "passed," the process will go to *end* activity else the process should turn to *design* activity. In the *end* activity, the process-related documents will be collected and archived.

Figures 16.2 and 16.3 show the design approval process described by SMEE and ABD/OBD. By comparison, it is obvious that the method of SMEE, which uses business language during model definition, is more concise than the mainstream method. However, OBD/ABD uses a

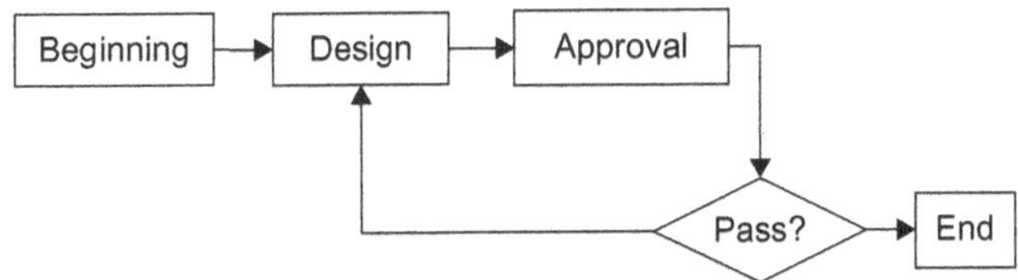


**Figure 16.1: Document design approval process.**

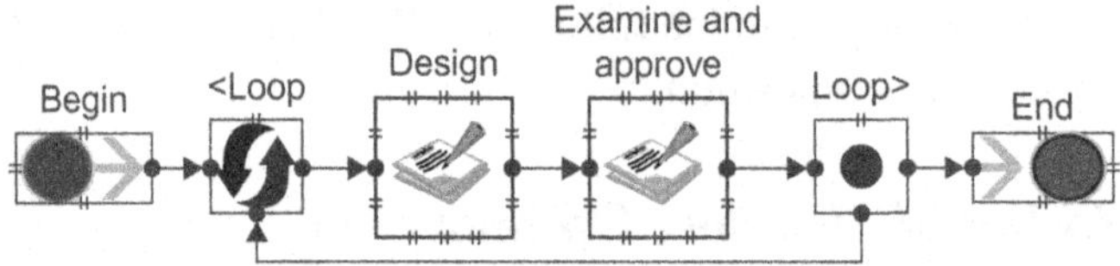


**Figure 16.2: Document design and approval process using the SMEE method.**

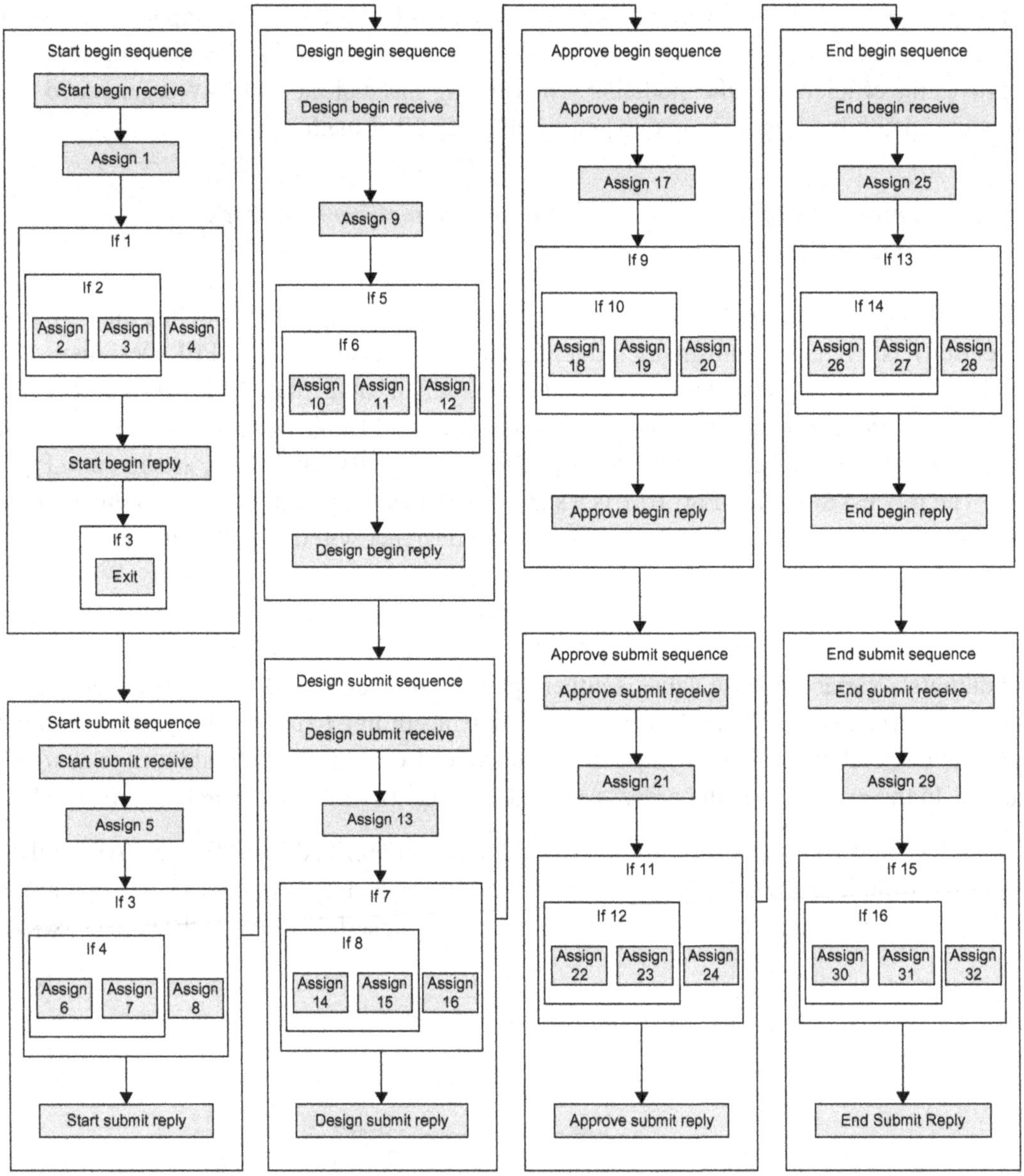

Start begin sequence
Start begin receive
Assign 1
If 1
If 2
Assign 2
Assign 3
Assign 4
Start begin reply
If 3
Exit

Start submit sequence
Start submit receive
Assign 5
If 3
If 4
Assign 6
Assign 7
Assign 8
Start submit reply

Design begin sequence
Design begin receive
Assign 9
If 5
If 6
Assign 10
Assign 11
Assign 12
Design begin reply

Design submit sequence
Design submit receive
Assign 13
If 7
If 8
Assign 14
Assign 15
Assign 16
Design submit reply

Approve begin sequence
Approve begin receive
Assign 17
If 9
If 10
Assign 18
Assign 19
Assign 20
Approve begin reply

Approve submit sequence
Approve submit receive
Assign 21
If 11
If 12
Assign 22
Assign 23
Assign 24
Approve submit reply

End begin sequence
End begin receive
Assign 25
If 13
If 14
Assign 26
Assign 27
Assign 28
End begin reply

End submit sequence
End submit receive
Assign 29
If 15
If 16
Assign 30
Assign 31
Assign 32
End Submit Reply

**Figure 16.3: Document design and approval process using the OBD/ABD method.**

**Table 16.1: Comparison of modeling methods.**

| | SMEE | OBD/ABD | Comparison |
|---|---|---|---|
| Number of model elements | 6 | 49 | 88% |
| Modeling time (s) | About 50 | About 320 | About 84% |

number of computer terms. They are very complicated and require high computer proficiency. Users are likely to commit errors when modeling processes using OBD/ABD.

By comparing Figs 16.2 and 16.3, we get the results shown in Table 16.1. By contrast, it is not difficult to find that SMEE has great advantages in model simplicity and modeling speed.

### *16.2.2 Definition of SMEE*

The SMEE has five nodes, namely starting activity, ending activities, general activities, changing nodes, and a logical connection. An SMEE is denoted by SSME = $(S, E, A, T, L)$, which represents a collection of the above five groups. A correct SMEE model should satisfy the following conditions:

1. $S = (s)$, an SMEE model can have only one beginning activity.
2. $s = (out\_files\ out\_values)$, denotes beginning activity.
3. $E = (e)$, an SMEE model can have only one ending activity.
4. $e = (in\_files)$, denotes ending activity.
5. $A = (a_1\ a_2 \ldots a_n)$, an SMEE model should have at least one general activity.
6. $a_i = (in\_files_i\ out\_files_i\ out\_values_i)$, denotes every activity.
7. $T = (ta_1\ ta_2 \ldots ta_m\ ta_1\ tb_2 \ldots tb_m)$, changing needs a paired node, where $ta_i$ and $tb_i$ are changing pairs.
8. $L = (l_1\ l_2 \ldots l_s)$, an SMEE model should have some logical connections.
9. $l_i = (from_i\ to_i\ activate\_value_i)$, denotes every logical connection to meet
   $from_i \in S \cup A \cup T,\ to_i \in E \cup A \cup T,\ from_i \neq to_i,$
   $\forall i, j \in (1,2,\ldots,s),\ from_i \cap from_j \cap S = \varnothing \quad (i \neq j),$
   $\forall i, j \in (1,2,\ldots,s),\ from_i \cap from_j \cap A = \varnothing \quad (i \neq j),$
   $\forall i, j \in (1,2,\ldots,s),\ to_i \cap to_j \cap E \neq \varnothing \quad (i \neq j).$
10. $activate\_value_i \in (\text{'any'}) \cup out\_values_{from\_i}$, where $out\_values_{from\_i}$ denotes $out\_values$ represented by starting, ending, and general activities of $from_i$.

### *16.2.3 Research Method for Converting SMEE Model into BPEL4WS*

In an SMEE model, each activity (beginning activity, ending activity, and common activity) is divided into two steps (launching the task and submitting the task) during the execution period. Therefore, two <*receive*> activities should be provided in the BPEL4WS in agreement with every activity in the SMEE model. Similar to the transition node in the

SMEE model, it will be converted into an *<if>* activity if the condition is defined for it, else it will be converted into an *<flow>* activity. The BPEL4WS model is generated in six steps:

1. The total SMEE model is converted into a *<sequence>* activity named *Total_SEQ*. The beginning activity is converted into the first activity in *Total_SEQ*. The ending activity is converted into the last activity in *Total_SEQ*.
2. Traverse backwards along the logic links from the beginning activity. If the linked object is a common activity, convert it into BPEL4WS and insert the BPEL4WS into *Total_SEQ*. If the linked object is a transition node, seek its coupled back transition node, then convert it into BPEL4WS according to whether it defines the condition, then insert the BPEL4WS into *Total_SEQ*.
3. Convert the activities in every coupled transition node into BPEL4WS according to the methods mentioned above.
4. Define *<CorrelationSet>* so as to synchronize all the *<receive>* activities in a BPEL4WS model.
5. Define task executor, task output description, and in/out documents in the *.wsdl* file. Refer them in the *.bpel* file.
6. Package the BPEL4WS model into BPR, which can run on an *ActiveBpel®* Server.

Using the above six steps, the definition and output of the SMEE model are achieved. The total process is shown in Fig. 16.4.

Similar to common activity, the detailed generating process is expressed in Fig. 16.5.

### 16.2.4 Running Performance Evaluation for Generated BPEL4WS Model

In order to validate the SMEE method, we create a product design process as shown in Fig. 16.1 using the SMEE modeling tool. Then, we generate a BPEL4WS model from the SMEE model. The generated BPEL4WS model can run properly on the Web server (see Fig. 16.6).

Let the BPEL4WS model generated using the SMEE method run on the computer (E7200 CPU, 4g memory) together with the models generated by other methods. The running performance of BPEL4WS activity generated using the SMEE is compared with that generated by other methods, as shown in Fig. 16.7. It is obvious from the figure that the model generated by our method runs faster than the one generated by other methods.

## 16.3 Web Service Composition Quality Analysis with Stochastic Service Times

Our goal is to solve the Web service composition execution time distribution function when each single Web service execution time is a random distribution. According to the distribution function, we can obtain the probability with which Web service composition can be finished within any given time. So we can accurately measure the quality of Web service composition.

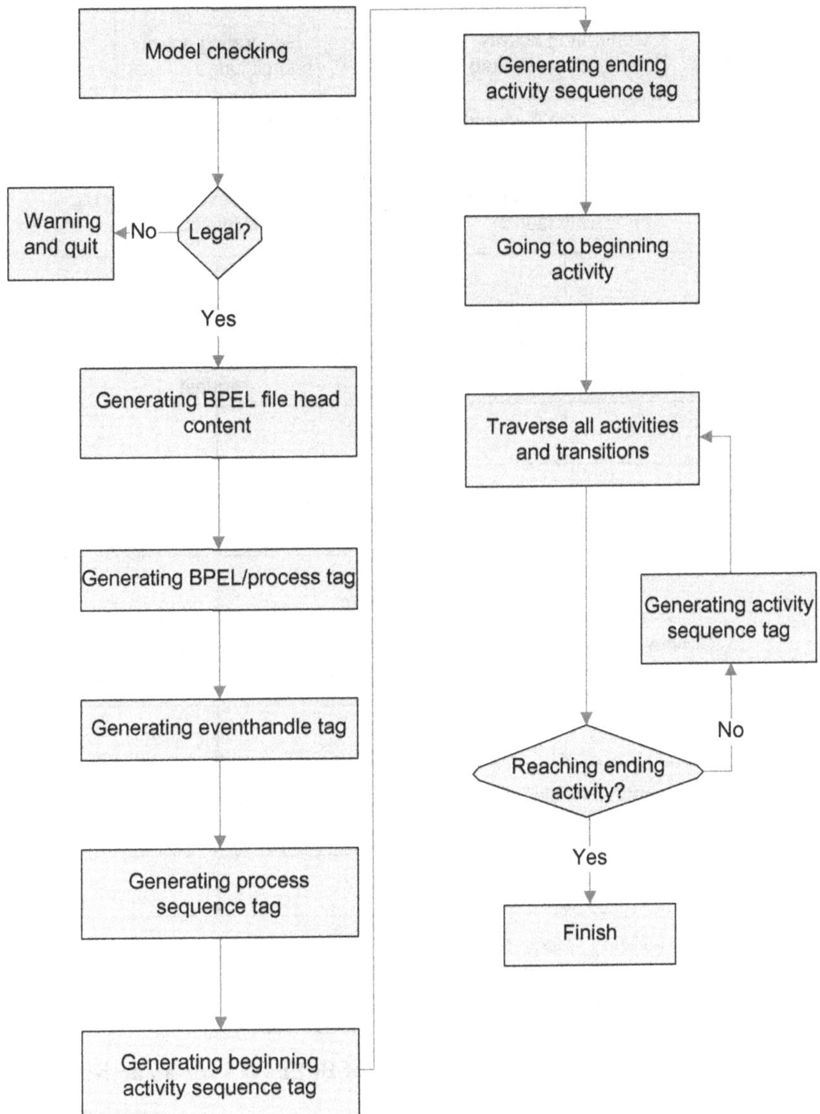


**Figure 16.4: Total process to output BPEL4WS.**

### 16.3.1 Problem Description

Suppose a Web service composition described by BPEL2WS $B$ is composed of $n$ Web services ($WS_1$, $WS_2$, … , $WS_n$). Let $d_{BPEL}$ denote the Web service composition execution time. Because a Web service in a BPEL2WS model is invoked through the *<invoke>* activities, the activity *<invoke$_i$>* corresponds to Web service $WS_i$. Let $d_{WS_i}$ denote the execution time of $WS_i$. Because the execution time of *<receive>*, *<reply>*, *<assign>*, and other types of activities

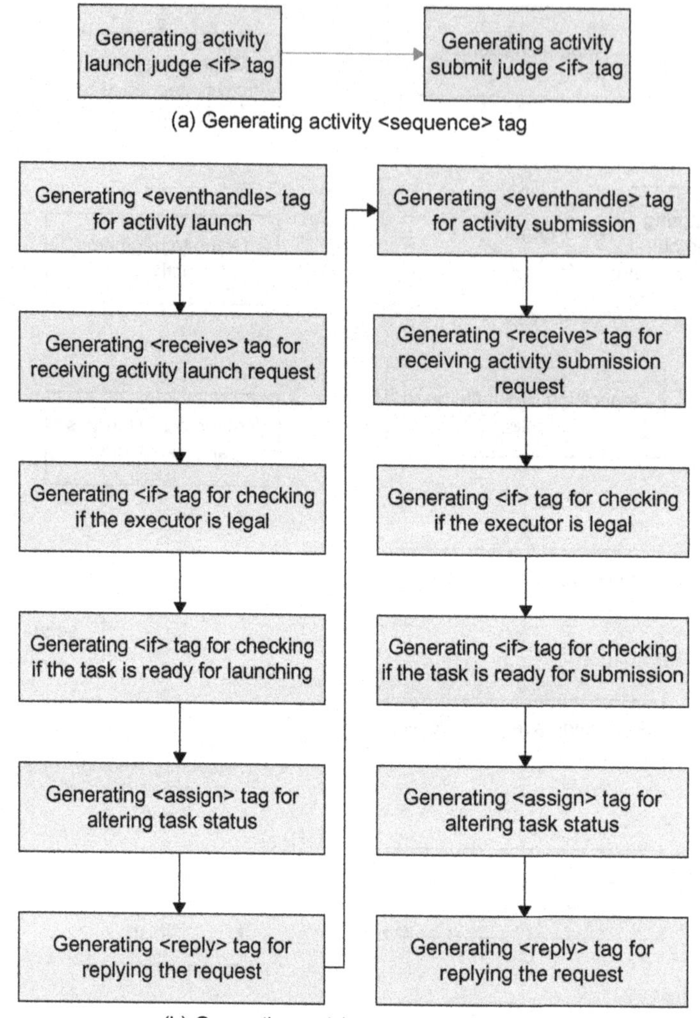


**Figure 16.5: The output process of BPEL4WS for an activity.**

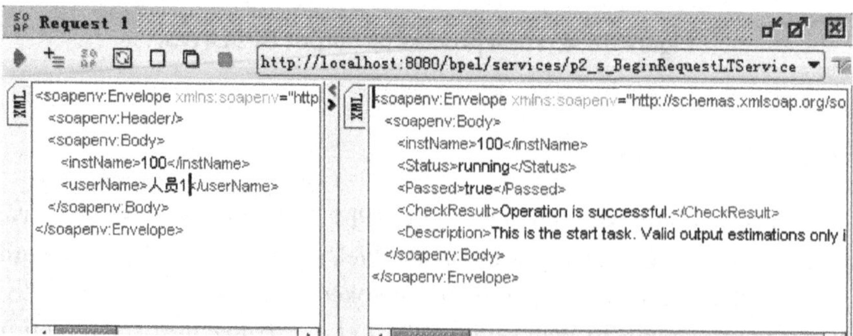


**Figure 16.6: Running interface of BPEL4WS model generated using the SMEE model.**

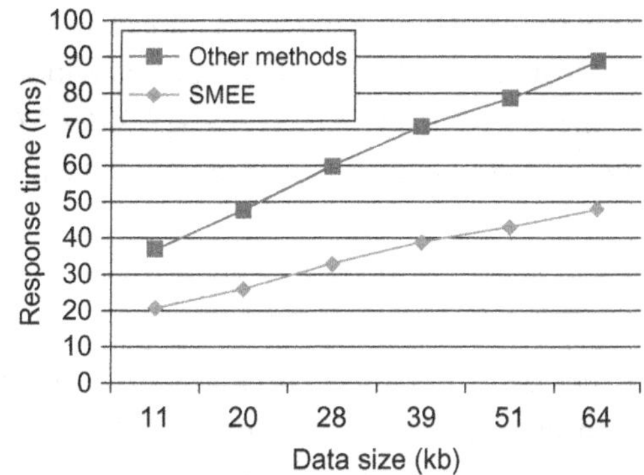


**Figure 16.7: Running performance comparison between the models generated using the SMEE method and other methods.**

can be ignored according to the <*invoke*> activity execution time, their execution time is not taken into account when computing the execution time of the model. For the <*wait*> activities, <*error-handling*> activities, and <*pick*>, <*scope*> activities, which are mainly used for abstract computer processing, there is no business meaning; therefore, this chapter does not discuss their execution time. Let $d_{FLOW_i}$, $d_{SEQ_i}$, $d_{IF_i}$, $d_{WH_i}$, $d_{REP_i}$, and $d_{FOR_i}$ denote the execution time of Web services in the containers $FLOW_i\,(i = 1,2,\ldots,n\_flow)$, $SEQUENCE_i\,(i = 1,2,\ldots,n\_seq)$, $IF_i\,(i = 1,2,\ldots,n\_if)$, $WHILE_i\,(i = 1,2,\ldots,n\_wh)$, $REPEATUNTILE_i\,(i = 1,2,\ldots,n\_rep)$, $FOREACH_i\,(i = 1,2,\ldots,n\_for)$. $U_i$ is used to represent common containers.

Let $F(d)$ represent execution time probability distribution function, then $F(d)$ denotes the probability when the execution time is greater than 0 and less than $d$. Let $f(d)$ denote the execution time distribution probability density function (PDF). Then according to probability theory, we obtain

$$F(d) = \int_0^d f(t)\,\mathrm{d}t \tag{16.1}$$

$$f(d) = F'(d) \tag{16.2}$$

Let $f_{WS_i}(d)$ represent the execution time distribution PDF and $f_{BPEL}(d)$ denote the Web service composition execution time distribution PDF, then $F_{FLOW_i}(d)$, $F_{REP_i}(d)$, $F_{IF_i}(d)$, $F_{WH_i}(d)$, $F_{REP_i}(d)$, and $F_{FOR_i}(d)$ denote cumulative probability distribution functions of execution time in Web service containers $FLOW_i$, $SEQUENCE_i$, $IF_i$, $WHILE_i$, $REPEATUNTILE_i$, and $FOREACH_i$, respectively.

### 16.3.2 Analysis of the QoS of Each Kind of BPEL2WS Container

The BPEL2WS model is composed of <*RECEIVE*>, <*REPLY*>, <*ASSIGN*>, <*INVOKE*>, and other basic activities through <*FLOW*>, <*SEQUENCE*>, <*IF*>, <*WHILE*>, <*REPEATUNTIL*>, and <*FOREACH*> containers (Rudolph et al., 2009). If $f_{WS_i}(d)$ of each $WS_i$ is known, by solving the relationship between $f(d)$ of the container and $f(d)$ of subactivities within the container (or subcontainer), then from $f_{WS_i}(d)$ onwards, we can obtain the execution time PDF expressions of the BPEL2WS model. We first solve $f(d)$ of *<FLOW>*, *<SEQUENCE>*, *<IF>*, *<WHILE>*, *<REPEATUNTIL>*, and *<FOREACH>* containers and the related expression of the internal subactivity (or subcontainers).

#### *16.3.2.1 Flow Container*

For a BPEL2WS model that performs in parallel relationship with *<FLOW>*, *<FLOW>* inside the container for all subactivities (or subcontainers) while starting, you must have finished executing *<FLOW>* container before the operation can end. Therefore, the longest subexecution activities (or subcontainers) *<FLOW>* execution time is the execution time of the container. Assumptions are contained within the container sub-*<FLOW>* activity (or subcontainers) $U_i\,(i = 1,2,\ldots,m)$, so we obtain

$$d_{FLOW} = \mathrm{Max}(d_{U_1}, d_{U_2}, \ldots, d_{U_m}) \tag{16.3}$$

According to probability theory, we can easily get

$$F_{FLOW}(d) = \prod_{i=1}^{m} F_{U_i}(d) \tag{16.4}$$

According to Eq. (16.3),

$$f_{FLOW}(d) = \sum_{i=1}^{m}\left(\frac{F'_{U_i}(d)}{F_{U_i}(d)}\prod_{k=1}^{m} F_{U_k}(d)\right) \tag{16.5}$$

Using Eqs (16.1) and (16.5), the relationship between the PDF of the *<FLOW>* container and its internal subactivity (or subcontainers) can be obtained as

$$f_{FLOW}(d) = \sum_{i=1}^{m}\left(\frac{f_{U_i}(d)}{\int_0^d f(t)\,\mathrm{d}t}\prod_{k=1}^{m}\int_0^d f_{U_k}(t)\,\mathrm{d}t\right) \tag{16.6}$$

#### *16.3.2.2 Sequence Container*

The BPEL2WS SEQUENCE container models the logical structure of that sequence, the subcontainer activities (or subcontainers) according to chronological order once the order that, before a subactivity (or subcontainers) is executed after a start of full implementation to conditions, the last container is finished is necessary and sufficient conditions. Assuming that subactivities are contained within the container SEQUENCE (or subcontainers) $U_i\,(i = 1,2,\ldots,m)$, we have,

$$d_{SEQ} = \sum_{i=1}^{m} d_{U_i} \tag{16.7}$$

According to probability theory, we can easily get

$$F_{SEQ}(d) = \int_0^d f_{seq}(t)\mathrm{d}t = \underset{\sum_{i=1}^{m} t_{U_i} \le d}{\iint \cdots \int} \prod_{j=1}^{m} (f_{U_j}(t_{U_j})\mathrm{d}t_{U_j}) \tag{16.8}$$

Expanding Eq. (16.8) we get

$$F_{SEQ}(d) = \int_0^d \int_0^{d-t_{U_1}} \cdots \int_0^{x-\sum_{i=1}^{m-1} t_{U_i}} \left(\prod_{i=1}^{m} f_{U_i}(t_{U_i})\mathrm{d}t_{U_i}\right) \tag{16.9}$$

Divide Eqs (16.2) and (16.9) by $d$, we get

$$F_{SEQ}(d) = \left(\int_0^d \int_0^{d-t_{U_1}} \cdots \int_0^{x-\sum_{i=1}^{m-1} t_{U_i}} \left(\prod_{i=1}^{m} f_{U_i}(t_{U_i})\mathrm{d}t_{U_i}\right)\right)' \tag{16.10}$$

#### 16.3.2.3 If Container

The BPEL2WS model uses the *<IF>* container to denote logic conditional selection structure. *<IF>* containers select one branch to execute at each time. The probability of subactivities (or subcontainers) $U_i\,(i = 1, 2, \ldots, m)$ is usually obtained by statistical methods. The probability $(i = 1, 2, \ldots, m)$ of each branch is assumed to be known.

When $\sum_{i=1}^{m} p_i = 1$, namely there is one subactivity (or subcontainers) to perform in the *<IF>* container each time, it is easy to get

$$f_{IF}(d) = \sum_{i=1}^{m} (p_i * f_{U_i}(t)) \tag{16.11}$$

When $\sum_{i=1}^{m} p_i < 1$, namely the *<IF>* container does not contain the default implementation of the subcontainer (*<ELSE>* subcontainers), it is easy to find

$$F_{IF}(d) = \begin{cases} 1 - \sum_{i=1}^{m} p_i & (d = 0) \\ 1 - \sum_{i=1}^{m} p_i + \int_0^d f_{IF}(t)\mathrm{d}t & (d > 0) \end{cases} \tag{16.12}$$

$$f_{IF}(d) = \begin{cases} +\infty & (d = 0) \\ \sum_{i=1}^{m} (p_i * f_{U_i}(t)) & (d > 0) \end{cases} \tag{16.13}$$

#### 16.3.2.4 While Container

BPEL2WS uses the *<WHILE>* container to model the logical structure of a loop. *<WHILE>* containers always execute the loop under a certain condition until the condition becomes false. The cycling count of the subactivity (or subcontainers) inside the *<WHILE> container*, in general, is not too large; otherwise, it is easy to drop into a dead loop. Let $m$ denote the maximum execution time of loops and $p_i\,(i = 0, 1, \ldots, m)$ denote the probability with which subactivity (or subcontainers) inside the <*WHILE*> container is performed *i* times. Using the loop condition of the <*WHILE*> container, the value $p_i$ can be calculated. It is assumed that $p_i\,(i = 0, 1, \cdots, m)$ are known; according to probability theory, we can see that

$$\sum_{i=0}^{m} p_i = 1 \tag{16.14}$$

Let $F_{WH\_p_i}(d)\,(i = 1, 2, \cdots, \infty)$ represent the probability with which a subactivity (or subcontainers) inside the <*WHILE*> container is performed *i* times. Let $f_{WH\_p_i}(d)$ $(i = 1, 2, \cdots, \infty)$ denote the execution time PDF of subactivity (or subcontainers) inside the <*WHILE*> container. It is not difficult to see the subactivity (or subcontainers) inside the <*WHILE*> container with *i* time execution is equivalent to the <*SEQUENCE*> container with *i* subactivity (or subcontainers). Equations (16.9) and (16.10) yield

$$F_{WH\_p_i}(d) = \int_0^d \int_0^{d-t_{U_1}} \cdots \int_0^{x-\sum_{k=1}^{i-1} t_{U_k}} \left( \prod_{k=1}^{i} f_{U_1}(t_{U_k}) \mathrm{d}t_{U_k} \right) \tag{16.15}$$

$$f_{WH\_p_i}(d) = \left( \int_0^d \int_0^{d-t_{U_1}} \cdots \int_0^{x-\sum_{k=1}^{i-1} t_{U_k}} \left( \prod_{k=1}^{i} f_{U_1}(t_{U_k}) \mathrm{d}t_{U_k} \right) \right)' \tag{16.16}$$

A <*WHILE*> container can be regarded as an <*IF*> container that is composed of *m* <*SEQUENCE*> containers (listed as $f_{WH\_p_1}(d), f_{WH\_p_2}(d), \cdots, f_{WH\_p_m}(d)$).

When $p_0 = 0$, which means the elements inside the <*WHILE*> container will be performed at least once, Eq. (16.12) yields

$$f_{WH}(d) = \sum_{i=1}^{m} (p_i * f_{WH\_p_i}(t)) \tag{16.17}$$

From Eqs (16.16) and (16.17), we obtain

$$f_{WH}(d) = \sum_{i=1}^{m} \left[ p_i * \left( \int_0^d \int_0^{d-t_{U_1}} \cdots \int_0^{x-\sum_{k=1}^{i-1} t_{U_k}} \left( \prod_{k=1}^{i} f_{U_1}(t_{U_k}) \mathrm{d}t_{U_k} \right) \right)' \right] \tag{16.18}$$

When $p_0 \neq 0$, which shows that the <*WHILE*> container has a probability with which the inside subactivity (or subcontainers) will not be performed, Eqs (16.12), (16.13), and (16.16) yield

$$F_{WH}(d) = \begin{cases} p_0 & (d = 0) \\ 1 - \sum_{i=1}^{m} p_i + \int_0^d f_{WH}(t)\,\mathrm{d}t & (d > 0) \end{cases} \tag{16.19}$$

$$f_{WH}(d) = \begin{cases} +\infty & (d = 0) \\ \sum_{i=1}^{m} \left( p_i * \left( \int_0^d \int_0^{d-t_{U_1}} \cdots \int_0^{x-\sum_{k=1}^{i-1} t_{U_k}} \left( \prod_{k=1}^{i} f_{U_1}(t_{U_k}) \mathrm{d}t_{U_k} \right) \right)' \right) & (d > 0) \end{cases} \tag{16.20}$$

#### 16.3.2.5 Repeatuntil Container

The <*REPEATUNTIL*> container is a container that expresses a circular logic like the <*WHILE*> container, the difference is that the elements in a <*REPEATUNTIL*> container will be executed at least once, otherwise it is the same as a <*WHILE*> container. The <*REPEATUNTIL*> container includes an internal subactivity (or subcontainers) $U_1$. We suppose that the maximum loop number is $m$. Let $p_i$ $(i = 1, 2, \ldots, m)$ denote the probability with which the subactivity within the container is performed $m$ times. According to <*REPEATUNTIL*> container cycling conditions, we can calculate $p_i$ $(i = 1, 2, \ldots, m)$. Here, $p_i$ $(i = 1, 2, \ldots, m)$ is assumed known.

Let $F_{REP\_p_i}(d)$ $(i = 1, 2, \ldots, \infty)$ represent the probability distribution functions with which the subactivity (or subcontainers) inside the <*REPEATUNTIL*> container is performed $i$ times. Let $f_{REP\_pi}(d)$ $(i = 1, 2, \ldots, \infty)$ denote the probability distribution density functions with which the subactivity (or subcontainers) inside the <*REPEATUNTIL*> container is performed $i$ times. A <*REPEATUNTIL*> container can be regarded as an <*IF*> container that is composed of $m$ <*SEQUENCE*> containers (listed as $f_{REP\_p_1}(d), f_{REP\_p_2}(d), \ldots, f_{REP\_p_m}(d)$). Using Eqs (16.9) and (16.10), we get

$$F_{REP\_p_i}(d) = \int_0^d \int_0^{d-t_{U_1}} \cdots \int_0^{x-\sum_{k=1}^{i-1} t_{U_k}} \left( \prod_{k=1}^{i} f_{U_1}(t_{U_k}) \mathrm{d}t_{U_k} \right) \tag{16.21}$$

$$F_{REP\_p_i}(d) = \left( \int_0^d \int_0^{d-t_{U_1}} \cdots \int_0^{x-\sum_{k=1}^{i-1} t_{U_k}} \left( \prod_{k=1}^{i} f_{U_1}(t_{U_k}) \mathrm{d}t_{U_k} \right) \right)' \tag{16.22}$$

A <*REPEATUNTIL*> container can be regarded as an <*IF*> container that is composed of $m$ <*SEQUENCE*> containers (listed as $f_{REP\_p_1}(d), f_{REP\_p_2}(d), \ldots, f_{REP\_p_m}(d)$). According to Eq. (16.11),

$$f_{REP}(d) = \sum_{i=1}^{m} (p_i * f_{REP\_p_i}(t)) \tag{16.23}$$

Using Eqs (16.22) and (16.23),

$$f_{REP}(d) = \sum_{i=1}^{m} \left( p_i * \left( \int_0^d \int_0^{d-t_{U_1}} \cdots \int_0^{x-\sum_{k=1}^{i-1} t_{U_k}} \left( \prod_{k=1}^{i} f_{U_1}(t_{U_k}) \mathrm{d}t_{U_k} \right) \right)' \right) \tag{16.24}$$

#### 16.3.2.6 Foreach Container

The <*FOREACH*> container is a loop container that contains a <*SCOPE*> container named $U_1$. The <*SCOPE*> container can be regarded as a separate subprocess. The <*SCOPE*> container can be changed into a <*SEQUENCE*> container or a <*FLOW*> container by setting the execution type of the <*FOREACH*> container. The number of cycles of the <*FOREACH*> container and the completion condition can be preset. We assume that the <*FOREACH*> container completion condition is that all the cycles are completed. It is also the default condition to finish the <*FOREACH*> container.

We define a signal *ES* to indicate the type of performance. $ES = 0$ means serial implementation, $ES = 0$ means parallel execution. Let *M* denote the number of cycles; according to Eqs (16.6) and (16.10)

$$f_{FOR}(d) = \begin{cases} \left(\int_0^d \int_0^{d-t_{U_1}} \cdots \int_0^{x-\sum_{i=1}^{m-1} t_{U_i}} \left(\prod_{i=1}^{m} f_{U_1}(t_{U_i}) \mathrm{d}t_{U_i}\right)\right)' & (ES = 0) \\ m * f_{U_1}(d) * \left(\int_0^d f_{U_1}(t)\mathrm{d}t\right)^{m-1} & (ES = 1) \end{cases} \tag{16.25}$$

The BPEL2WS model is basically a combination of activities through the nesting of the containers (Rudolph et al., 2009). By obtaining the relation between the execution time PDF of each container and its internal activities with the PDF of each <*INVOKE*> activity, we can obtain the probability distribution function of the BPEL2WS model.

### *16.3.3 Analysis of the QoS of an Actual BPEL2WS Model*

The BPEL2WS model can be expressed as an execution tree and an execution chart (Menascé et al., 2010), as shown in Fig. 16.8. The map on the left-hand side corresponds to BPEL2WS model text and that on the right is the map of the executive, in which the service call is represented using oval boxes and IF container using a small round box with the arrow line representing the logical order of execution. Dotted lines denote the <*FLOW*> container. The lower right tree shows the implementation, in which the container is represented using a long box and service call using an oval frame with the arrow line representing affiliation. If the model structure is known along with the execution time of the <*INVOKE*> event PDF of the known cases, the model can calculate the execution time BPEL2WS PDF. A BPEL2WS model example is shown in Fig. 16.1. Known conditions are as follows: $p_1 = 0.4$, $p_2 = 0.6$. The execution times of the <*FOREACH1*> container is 3. The execution style of the <*FOREACH1*> container is concurrent.

$$f_{WS_1}(d) = 8 * e^{-8d}, \quad f_{WS_2}(d) = 5 * e^{-5d},$$

$$f_{WS_3}(d) = 6 * e^{-6d}, \quad f_{WS_4}(d) = 10 * e^{-10d},$$

$$f_{WS_5}(d) = 6 * e^{-6d}, \quad f_{WS_6}(d) = 2 * e^{-2d}.$$

Based on Eqs (16.1)–(16.25), the probability distribution density function of BPEL2WS, shown in Fig. 16.1, is given by the following:

$$\begin{aligned} f_{BPEL}(d) = & -\frac{168}{325}e^{-28d} + \frac{24}{11}e^{-24d} - \frac{1472}{525}e^{-23d} + \frac{396}{175}e^{-22d} \\ & - 9e^{-18d} + \frac{1088}{75}e^{-17d} - \frac{144}{35}e^{-16d} + \frac{72}{5}e^{-12d} - \frac{704}{15}e^{-11d} + 6e^{-10d} \\ & + \frac{26728}{525}e^{-8d} - 12e^{-6d} - \frac{64}{3}e^{-5d} + \frac{483473}{75075}e^{-2d} \end{aligned} \tag{16.26}$$

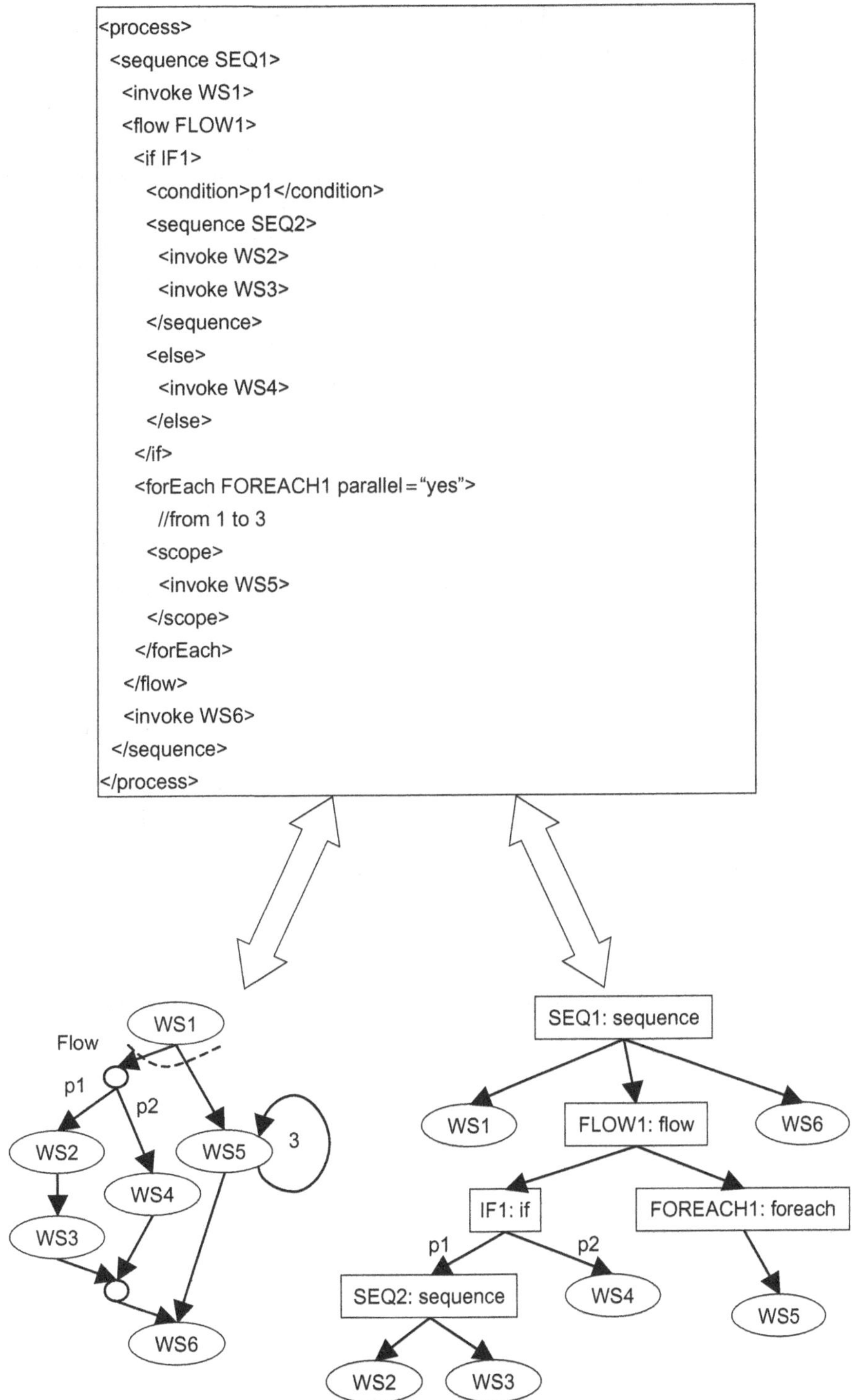

```
<process>
 <sequence SEQ1>
  <invoke WS1>
  <flow FLOW1>
   <if IF1>
    <condition>p1</condition>
    <sequence SEQ2>
     <invoke WS2>
     <invoke WS3>
    </sequence>
    <else>
     <invoke WS4>
    </else>
   </if>
   <forEach FOREACH1 parallel="yes">
     //from 1 to 3
    <scope>
     <invoke WS5>
    </scope>
   </forEach>
  </flow>
  <invoke WS6>
 </sequence>
</process>
```



**Figure 16.8: BPEL2WS model (left) with its directed execution graph and execution tree.**

Let $P(a, b)$ denote the probability with which the execution time $t$ is between $a$ and $b$. According to probability theory, it can be found that

$$P(a,b) = \int_a^b f(t)\mathrm{d}t \tag{16.27}$$

Using Eqs (16.27) and (16.28), the execution time distribution status of Web service composition is shown in Fig. 16.1 (see Table 16.1). According to the data in Table 16.2, we can use the Newton interpolation method to solve the probability with which the Web service composition execution time is in the given range. We suppose that $a$ is as in Table 16.2 and $k$ is greater than $a$ and less than $a + 0.1$, then the Newton interpolation formula is as follows:

$$P(0,k) \approx P(0,a) + (P(0,a+0.1) - P(0,a)) * (k-a)/0.1 \tag{16.28}$$

We suppose $a$ and $b$ are as in Table 16.2 and $a < b$, $k > a$, $k < a + 0.1$, $m > b$, and $m < b + 0.1$, then the Newton interpolation formula is as follows:

$$\begin{aligned} P(k,m) \approx P(m) - P(k) = (P(0,b) + (P(0,b+0.1) - P(0,b)) * (k-b)/0.1) \\ -(P(0,a) + (P(0,a+0.1) - P(0,a)) * (k-b)/0.1) \end{aligned} \tag{16.29}$$

When $a + 0.1$ or $b + 0.1$ is not as in Table 16.2, it shows that at this moment $P(0, a + 0.1)$ and $P(b + 0.1)$ are large enough to be recognized as 1, and if we replace them with 1 in Eqs (16.28) and (16.29) and calculate, we will get the result we are looking for.

We suppose that we want to know the probability with which the execution time of Web service composition is less than 2.56. From Table 16.2 and according to Eq. (16.29), we get

$$\begin{aligned} P(0,2.56) &\approx P(0,2.5) + (P(0,2.6) - P(0,2.5)) * (2.56 - 2.5)/0.1 \\ &= 0.980 \end{aligned} \tag{16.30}$$

**Table 16.2: Execution time distribution of composite web service.**

| $a$ | P(0,$a$) | $a$ | P(0,$a$) | $a$ | P(0,$a$) | $a$ | P(0,$a$) |
|---|---|---|---|---|---|---|---|
| 0.2 | 0.006 | 1.3 | 0.768 | 2.4 | 0.974 | 3.5 | 0.997 |
| 0.3 | 0.032 | 1.4 | 0.808 | 2.5 | 0.978 | 3.6 | 0.998 |
| 0.4 | 0.084 | 1.5 | 0.842 | 2.6 | 0.982 | 3.7 | 0.998 |
| 0.5 | 0.159 | 1.6 | 0.870 | 2.7 | 0.985 | 3.8 | 0.998 |
| 0.6 | 0.248 | 1.7 | 0.893 | 2.8 | 0.988 | 3.9 | 0.999 |
| 0.7 | 0.342 | 1.8 | 0.913 | 2.9 | 0.990 | 4.0 | 0.999 |
| 0.8 | 0.434 | 1.9 | 0.928 | 3.0 | 0.992 | 4.1 | 0.999 |
| 0.9 | 0.520 | 2.0 | 0.941 | 3.1 | 0.993 | 4.2 | 0.999 |
| 1.0 | 0.596 | 2.1 | 0.952 | 3.2 | 0.995 | | |
| 1.1 | 0.662 | 2.2 | 0.961 | 3.3 | 0.996 | | |
| 1.2 | 0.720 | 2.3 | 0.968 | 3.4 | 0.996 | | |

The probability with which the execution time of Web service composition is less than 2.56 is 98.0%. In other words, there are 98.0% Web service compositions whose execution time is less than 2.5.

It is supposed that we want to know the probability with which the execution time of Web service composition is bigger than 1.52 and less than 2.05. From Table 16.2 and according to Eq. (16.30), we get

$$\begin{aligned} P(1.52, 2.05) \approx P(2.05) - P(1.52) \\ &= (P(0,2.0) + (P(0,2.1) - P(0,2.0)) * 0.05/0.1) \\ &\quad - (P(0,1.5) + (P(0,1.6) - P(0,1.5)) * 0.02/0.1) \\ &= 0.947 - 0.848 = 0.099 \end{aligned}$$

The probability with which the execution time of Web service composition is greater than 1.52 and less than 2.05 is 9.9%. In other words, there are 9.9% Web service compositions whose execution time is greater than 1.52 and less than 2.05.

## 16.4 Conclusions

By using the SMEE method suggested here, the time taken for process modeling and model running is clearly reduced. This enables the enterprise to create and run its business processes more quickly than using other methods. The enterprise will realize business process automation rapidly with the SMEE method, which increases business efficiency. Accordingly, the enterprise's core competitiveness is improved.

SOA enables Web service providers to provide loosely coupled and interoperable services at different QoS levels. Business processes composed of activities that are supported by services can implement complex business functions. Business process execution languages for Web service (BPEL2WS) that are composed of service activities and logic structures such as sequence, flow, if, and while may express the structures of business processes. This chapter analyzed the total execution time of Web service composition that is described by BPEL2WS and discussed how to solve the PDF of Web service composition using the PDF of each service and the structure of the BPEL2WS model when the execution time of each service is stochastically distributed. Therefore, one can get the probability with which the business process can be finished before any given deadline using the PDF. Finally, an example is given to show that the proposed method can be effectively utilized in practice.

## References

Andrews, T., Curbera, F., Dholakia, H., Goland, Y., Klein, J., Leymann, … Weerawarana, S. (2003). Business Process Execution Language for Web Services, version 1.1. Retrieved from http://www-106.ibm.com/developerworks/webservices/library/ws-bpel

Ardagna, D. & Pernici, B. (2005). Global and local QoS guarantee in web service selection. In *Proceedings of Business Process Management Workshops, Lecture Notes in Computer Science*, 2006, *3812*, 32–46.

Barros, A., Dumas, M., & Oaks, P. (2006). Standards for Web service choreography and orchestration: Status and perspectives. In *Proceedings of 2006 Business Process Management Workshops, Lecture Notes in Computer Science, 3812*, 61–74.

Cao, J. N., Chan, C., & Chan, K. (2005). Workflow analysis for web publishing using a stage-activity process model. *The Journal of Systems and Software, 76*, 221–235.

Curbera, F., Duftler, M., Khalaf, R., Nagy, W., Mukhi, N., & Weerawarana, S. (2002). Unraveling the Web services web: An introduction to SOAP, WSDL, and UDDI. *IEEE Internet Computing, March/April, 6*(2), 86–93.

Curbera, F., Khalaf, R., Mukhi, N., Tai, S., & Weerawarana, S. (2003). The next step in Web services. *Communications of the ACM, 46*(10), 29–34.

Fan, Y. S. & Li, H. F. (2003). Current state and development trends of enterprise integration. *Manufacturing Information Engineering of China, 32*(1), 59–61.

Huang, C. & Fan, Y. S. (2003). Intelligent workflow management: Architecture and technologies. In *The 3rd International Conference on Electronic Commerce*, Hangzhou, China, October, pp. 995–999.

Hwang, S. Y., Wang, H., Tang, J., & Srivastava, J. (2007). A probabilistic approach to modeling and estimating the QoS of web-services-based workflows. *Information Sciences, 177*(23), 5484–5503.

Johansson, H. J. E. (1993). *Business Process Reengineering: BreakPoint Strategies for Market Dominance*. Chichester, UK: John Wiley & Sons.

Kyriazis, D., Tserpes, K., Menychtas, A., Litke, A., & Varvarigou, T. (2008). An innovative workflow mapping mechanism for Grids in the frame of Quality of Service. *Future Generation Computer Systems, 24*, 498–511.

Li, X. Y., Li, X. X., & Qian, Y. (2006). A web service based online optimization and monitoring system for chemical processing systems. *Computer Aided Chemical Engineering, 21*(2), 1425–1430.

Lin, H. P. & Fan, Y. S. (2003). Scheduling method based on workflow management technology. *Journal of Tsinghua University (Science and Technology), 43*(3), 402–405.

Liu, S. & Fan, Y. (2005). Method for analyzing staying-time of instances in workflow models with resources constraints. *Acta Electronica Sinica, 33*(10), 1867–1871 (in Chinese).

Menascé, D. A., Casalicchio, E., & Dubey, V. (2010). On optimal service selection in Service Oriented Architectures. *Performance Evaluation, 67*(8), 659–675.

Menascé, D. A., Ruan, H., & Gomma, H. (2007). QoS management in service oriented architectures. *Performance Evaluation Journal, 64*(7–8), 646–663.

Miao, Q., Zhu, F., Lv, Y., Cheng, C., Chen, C., & Qiu, X. (2011). A game-engine-based platform for modeling and computing of artificial transportation systems. *IEEE Transactions on Intelligent Transportation Systems, 12*(2), 343–353.

Nurcan, S. (1998). Analysis and design of co-operative work processes: a framework. *Information and Software Technology, 40*, 143–156.

Pozewaunig, H., Eder, J., & Liebhart, W. (1997). ePERT: extending PERT for workflow management systems. In *Proceedings of First European Symposium in Advances in Databases and Information Systems (ADBIS)*, St. Petersburg, Russia, September, pp. 217–224.

Rudolph, C., Kuntze, N., & Velikova, Z. (2009). Secure web service workflow execution. *Electronic Notes in Theoretical Computer Science, 236*(2), 33–46.

Wang, F. Y. (2005). Agent-based control for networked traffic management systems. *IEEE Intelligent Systems, 20*(5), 92–96.

Wang, F. Y. & Tang, S. (2004). Artificial Societies for Integrated and Sustainable Development of Metropolitan Systems. *IEEE Intelligent Systems, 19*(4), 82–87.

Xiong, P., Fan, Y., & Zhou, M. (2008). QoS-aware web service configuration. *IEEE Transactions on System, Man and Cybernetics, Part A, 38*(4), 888–895.

Xiong, P., Fan, Y., & Zhou, M. (2009). Web service configuration under multiple quality-of-service attribute. *IEEE Transactions on Automation Science and Engineering, 6*(2), 311–321.

Zhang, C. W., Su, S., & Chen, J. L. (2006). Genetic algorithm on Web services selection supporting QoS. *Chinese Journal of Computers, 29*(7), 1029–1037 (in Chinese).

Zhang, N., Wang, F. Y., Zhu, F., Zhao, D., & Tang, S. (2008). DynaCAS: Computational experiments and decision support for ITS. *IEEE Intelligent Systems, 23*(6), 19–23.

Zhao, Y. & Fan, Y. S. (2003). Enterprise information engineering utilizing enterprise modeling with quality function deployment. In *The 3rd International Conference on Electronic Commerce*, Hangzhou, China, October, pp. 977–980.

Zhu, F., Li, G., Li, Z., Chen, C., & Wen, D. (2011). A case study of evaluating traffic signal control systems using computational experiments. *IEEE Transactions on Intelligent Transportation Systems, 12*(4), 1220–1226.

CHAPTER 17

# Urban Traffic Management System Based on Ontology and Multiagent System

**Dong Shen and Songhang Chen**

*State Key Laboratory of Management and Control for Complex Systems, Institute of Automation, Chinese Academy of Sciences, Beijing, China*

## 17.1 Introduction

With the development of society and the high-speed growth of economies, as well as the upgrading of the level of urbanization, resident activities in economy, culture, etc. are increasingly frequent. This further causes a demand for higher growth of urban traffic. Many traffic problems occur because of this, such as jams, delays, accidents, etc. This motivates us to keep finding better approaches to improve and optimize urban traffic control and management. However, an urban traffic system (UTS) is a typical large complex system with the following features, as is well known:

- Large or huge systems. There are thousands of intersections, road elements, traffic facilities, and vehicles, which are dynamically in and out in the UTS. Besides, the relations between intersections, road elements, and traffic flow are very complex.
- High degree of complexity. The UTS has large dimensions, thus it has similar common characteristics as other complex systems, such as distribution, strong coupling, nonlinear, time varying, stochastic, hierarchical, and/or uncertain.
- Abundant data processing. There are a number of data to be processed, determined, and stored in the UTS. The data information includes not only static type information such as intersections and road elements but also dynamic information such as real-time traffic information, traffic accidents, traffic policies, etc.

Because of these factors, it is impossible to achieve the desired performance using traditional centralized control methods. To be specific, the following targets are difficult to achieve simultaneously:

- Simultaneous optimization of the local intersection and the whole UTS. In a UTS, each intersection has its control target or task, which further determines its control strategy.

Service Science, Management, and Engineering.


DOI: 10.1016/B978-0-12-397037-4.00017-X

However, there is also an exchange of traffic information among intersections, aiming to develop coordination control of multi-intersections or regional traffic systems, which can improve the traffic efficiency. Therefore, it is required that the traffic strategy should meet both requirements from the locality and the whole system. It is possible that the local task may not coincide with the whole optimization objective. If so, a compromise will have to be made between them.

- Combined applications of multiple control strategies. The internal structure of a UTS is very complex. For example, each intersection is different from others, thus it is difficult to control all intersections with a single strategy. Even for just one intersection, the control strategy often is different at different moments. Therefore, it is hard to improve the traffic by one control strategy when combined applications of multiple control strategies are required by the UTS.
- Adaptability to varying traffic flow and emergencies. The UTS works in combination with a number of interactional units, which makes the traffic flow time varying. Thus it is necessary that the control methods should be able to adapt to complex traffic flow. Besides, there are some emergencies that cause large disturbances to the traffic flow, such as traffic accidents and temporary traffic control. It follows that the UTS should be able to identify emergencies and adjust its strategies.

Based on these analyses, we have to develop an approach to deal with multiobjective, hierarchical, and distributed systems. The multiagent approach among others is one effective approach, which has been proved to achieve the targets stated above (B. Chen & Cheng, 2010; Liu & Wang, 2002; Wang, 2005). Multiagent technology, since its introduction in the distributed artificial intelligence domain in the 1980s, has shown a significant advantage over traditional centralized control in distributed complex system control problems (Adler & Blue, 2002). As pointed out by Adler and Blue (2002), multiagent technology can be efficiently applied to the analysis and control of complex systems when the following three conditions are satisfied simultaneously: (1) the problem domain is geographically distributed; (2) the subsystems exist in a dynamic environment; and (3) the subsystems need to interact more flexibly.

As we can see, all these three conditions are met by the UTS. This has motivated much research on how to apply multiagent technology to various traffic problems (see Adler & Blue, 2002; B. Chen & Cheng, 2010; B. Chen, Cheng, & Palen, 2009; Choy, Srinivasan, & Cheu, 2003; Gokulan & Srinivasan, 2010; He, Miao, Li, Wang, & Tang, 2006; Liu & Wang, 2002; Roozemond, 1999; Tomás & García, 2005; Wang, 2005, and references therein). However, multiagent technology has not been widely applied to UTS (Wang, 2005). The paper by Wang (2005) has pointed out that most works focus on the construction of a hierarchical structure, modeling, and real-time optimization algorithm. Notice that the control and management system for UTS should have more flexibility, scalability, robustness, fault tolerance, and adaptability, so that the system can deal with various emergencies, grow as the UTS grows, and achieve coordination control targets for multiple intersections.

However, if such a large traffic management system has been developed, it is also a big challenge to guarantee its working efficiency. This is because the exchange information among different agents will be very great. On the other hand, the UTS itself continues to generate a mass of data to be analyzed, which acts as guidelines for the choice of different control strategies. Moreover, because of the complexity of the traffic system, the exchange information is rich and heterogeneous. In order to make the traffic management system accurately understand the semantics of exchange information and to grasp the real meaning of communication, we need to apply ontological technology to multiagent systems. The ontology, shared by all agents and databases, could help us to realize more effective agreement and understanding. In other words, the ontology could provide a shared virtual world for all agents and databases, so that all agents could be clear of their targets and actions. This is also a reason why ontological technology is suitable for multiagent systems.

This chapter will propose a concept framework for urban traffic management systems based on ontological and multiagent system technology. The proposed framework can help to build an urban traffic management system with advantages such as robustness, flexibility, and scalability. Besides, this framework can (a) realize the communications among different agents; (b) perform information retrieval, judgment, and selection on the database; (c) allow users to dynamically define new agents according to their actual needs; and (d) develop the system itself with a variety of information retrieval, collection, addition, and removal features.

The advantages of the framework are as follows.

1. It has good scalability. Based on the multiagent system technology and modular design pattern, the system can easily be extended according to actual needs. Moreover, the ontological technology provides a unified specification of a virtual world, so that the system can grow with the integration of various new control algorithms, special instructions, and management policies. For example, when a new and more effective control strategy is found, then the system would analyze its basic elements and compare them with the existing ontology. If all the elements have already been included in the ontology, then the new control strategy can be added to the ontology and form a clear concept relationship. If there are some elements not included in the ontology, then they can be analyzed and classified into a suitable domain, and then the new control strategy can be added. This potential for expansion enables the system to update and adapt to various complicated traffic demands.
2. It provides an environment for easier access to and analysis of information. With the help of ontology and multiagent technologies, all kinds of information and data can be dynamically discovered, integrated, and invoked. Using ontological technology, all concepts and information can be expressed in a semantic form, so the system is user friendly. They can easily find target information and dynamically adjust the system. The proposed framework in this chapter develops standard information description in modular form that makes the transmission and sharing of information among agents more convenient.

Moreover, since our framework is based on the modularization mode, users may regulate the system according to specific needs and different objectives, for example different cities, different regions, and different time intervals.

3. It enhances the robustness of the system. When agents face novel or unforeseen events, they can try to understand the new events through relationships and commitments. Moreover, the agents further update the novel events into the ontology by learning, which will support other agents in the future.

There are few works about traffic systems based on ontology and multiagents (Merdan, Koppensteiner, Hegny, & Favre-Bulle, 2008; Merdan, Vittori, Koppensteiner, Vrba, & Favre-Bulle, 2008). In the works by Merdan, Koppensteiner, et al. (2008) and Merdan, Vittori, et al. (2008), the ontological technology is used to integrate and share traffic information, where the research is not rigorous enough. These works focus on traffic control strategy simulations based on their own simulation platform, while our framework will give an in-depth study on the ontology of urban traffic management systems. The objective is to establish an extended and improved ontological framework and further guide multiagent system design based on this ontological framework. Notice that there have been many works of combined ontology and multiagent technology and their applications (Beimejo-Alonso & Sanz, 2006; R. S. Chen & Chen, 2008; Huhns & Singh, 1997; Laclavík, 2005; Su, Matskin, & Rao, 2003; Tamma, Cranefield, Finin, & Willmott, 2005; Warden et al., 2010) although there are fewer applications for UTS. Most of these works focus on semantic Web and knowledge management. All this research has shown that the ontological technology plays an important role in unifying the related fields and has great application potential. This also motivates us to consider the application of ontological technology to UTS.

## 17.2 Literature Review

In this section, we will give a brief literature review of ontology and multiagent technology. For ontological technology, we will review the definitions, implications, construction methods, and applications of ontologies for agents. For multiagent technology, we will review the definitions and applications for traffic systems.

### 17.2.1 Ontological Technology

Ontology was originally linked to philosophy where it means explanation or description of an objective existence, or in other words, the philosophy of being or the types of existence. Thus ontology in philosophy focuses on the abstract essence of objective reality. Neches et al. (1991) proposed the first definition for ontology in the artificial intelligence (AI) field, i.e., "an ontology defines the basic terms and relations comprising the vocabulary of a topic area as well as the rules for combining terms and relations to define extensions to the vocabulary." However, a more popular definition of ontology was given by Gruber in 1993: "an ontology is an explicit specification of a conceptualization" (Gruber, 1993). On the basis of this definition, Borst defined ontology in his doctoral dissertation (Borst, 1997) as "an ontology is a formal specification of a shared conceptualization." This definition is further revised in the work by Studer, Benjamins, and Fensel (1998) as "an ontology is a formal, explicit specification of a shared conceptualization." Then ontology was gradually formed and accepted by most scholars and engineers. More discussions on ontology can be found in the works by Chandrasekaran, Josephson, and Benjamins (1999), Uschold and Gruninger (1996), and Pérez and Benjamins (1991).

Studer et al. (1998) point out that the definition of ontology has four main points, i.e., conceptualization, explicit, formal, and share. "Conceptualization" refers to an abstract model of some phenomenon in the world by having identified the relevant concepts of that phenomenon. "Explicit" means that the type of concepts used, and the constraints on their use, are explicitly defined. For example, in medical domains, the concepts are diseases and symptoms, the relations between them are causal and a constraint is that a disease cannot cause itself. "Formal" refers to the fact that the ontology should be machine readable, which excludes natural language. "Shared" reflects the notion that an ontology captures consensual knowledge, i.e., it is not private to some individual, but accepted by a group.

Many ontologies have been established according to various fields and specific applications, where the construction methods are different (Davy et al., 2007; Dolan & Blake, 2009; Fernández-López, Gómez-Pérez, Pazos-Sierra, & Pazos-Sierra, 1999; Freitas, Stuckenschmidt, Malucelli, & Pinto, 2007; Loia & Lee, 2010). Although there is no standard ontology construction method, however, many researchers have given some reference points on how to construct an ontology. For example, Gruber (1995) proposes five design rules which have great influence, namely clarity, coherence, extendibility, minimal encoding bias, and minimal ontological commitment. In addition, experts in related domains are indispensable in ontology construction, as they know more details than common people of the specific application.

Ontology has now been widely applied to various fields, such as software engineering (Calero, Ruiz, & Piattini, 2006; Happel & Seedorf, 2006), information systems (Eschenbach & Grüninger, 2008; Fonseca & Egenhofer, 1999), knowledge discovery (Loia & Lee, 2010), traffic engineering services (Davy et al., 2007), medicine (Dolan & Blake, 2009), etc. In the work by Uschold and Jasper (1999), the authors proposed a framework for understanding and classifying ontology applications, where the applications are divided into three classes, namely neutral authoring, common access of information, and indexing for search. Then they analyze the three classes of applications from multiple perspectives in order that research from different areas could be more closed coordinated and make faster progress.

Ontology can be regarded as an agreement of a shared conceptualization model in the sense of a combination with multiagent technology. It includes frameworks for modeling domain knowledge and agreements about the representation of particular domain theories (Huhns & Singh, 1997). The exchanges between agents require not only linguistic competence but also common cognition and understanding of related domain knowledge. This is what ontologies aim to realize. An ontology provides a shared virtual world among different agents, in which each agent can ground its beliefs and actions. The understanding of concepts of agents is dependent on ontology. This is also why ontological technology can be combined with multiagent technology. On one hand, ontology is the representational vocabulary on a specific domain or subject matter; on the other hand, ontology can be used as a body of knowledge describing some domain (Chandrasekaran et al., 1999). As pointed out by Bermejo-Alonso and Sanz (2006), ontology can play the following roles in a multiagent system: specification, common access to information, ontology-based search, reuse of knowledge, integration of heterogeneous information sources, and ontologies for semantic interoperability. Papers by Tamma et al. (2005), R. S. Chen and Chen (2008), Warden et al. (2010), Su et al. (2003), and Laclavík (2005) also discuss how to apply ontology to multiagent systems. A methodology for ontology-based multiagent system development is proposed in the works by Tran and Low (2008), which can be used as a guideline.

For more survey papers on ontology, please refer to the works by Chandrasekaran et al. (1999), Gómez-Pérez, Fernández-López, and Corcho (2004), and Pinto and Martins (2004).

### 17.2.2 Multiagent Technology

The concept of agent technology originates from the distributed artificial intelligence (DAI) field. There is no strict and accurate definition of agents. Most definitions actually are given by descriptions of its characteristics. Wooldridge and Jennings (1995) proposed that the agent is used to denote a hardware- or software-based computer system that enjoys some properties. The properties listed there include autonomy, social ability, reactivity, and proactivity. There are also some other attributes of agents, such as mobility, veracity, benevolence, and rationality. These characteristics help the agents to be applied in process control, manufacturing, air traffic control, information management, electronic commerce, patient monitoring, and many other fields (Jennings & Wooldridge, 1998; Wooldridge, 2009).

Compared with single agents, multiagent systems (MAS) can better describe distributed complex systems. MAS could be used to solve the target, which requires teamwork by the interaction among agents. Common interactions include coordination, negotiation, and cooperation. In order to solve complex problems, individual agents are first defined with some given actions and parameters. Then the coordination strategies, as well as the interaction rules between the agents and environments, are further defined. Then, large-scale complex problems can be efficiently solved by individual behaviors of agents and their interactions.

Because of their particular advantages, agents and MAS have been widely applied to traffic systems (Adler & Blue, 2002; B. Chen & Cheng, 2010; B. Chen et al., 2009; Choy et al., 2003; Gokulan & Srinivasan, 2010; He et al., 2006; Liu & Wang, 2002; Roozemond, 1999; Tomás & García, 2005; Wang, 2005). For example, B. Chen et al. (2009) integrate mobile agent technology and MAS for distributed traffic detection and management systems, so that the systems can adapt to dynamic and uncertain environments, while Roozemond (1999) uses autonomous intelligent agents for urban traffic control systems. The problem of traffic management and route guidance is dealt with in the works by Adler and Blue (2002), where the system has high scalability and adaptability according to various road networks and number of users on the basis of principled negotiation of cooperating multiagents. Choy et al. (2003) and Li (2007) discuss the application of MAS in urban traffic signal control. It is worth pointing out that the paper by B. Chen and Cheng (2010) reviews the applications of MAS to traffic systems from several aspects, which include architecture and platforms of agent-based traffic control and management systems, agent-based systems for roadway transportation, agent-based systems for air traffic control and management, and agent-based systems for railway transportation and multiagent traffic modeling and simulation.

However, research on multiagent technology for traffic control and management is still at the theoretical stage. In other words, we are still some way short of applying multiagent technology to actual traffic control and management. According to some hot topics in related research, as well as features of control systems under networked conditions, Wang (2005) proposes to apply a networked agent-based control strategy to traffic control and management, where the author analyzes the move from control algorithms to control agents, develops a system architecture for agent-based traffic management, and proposes the realization of software platforms and operation protocols. Based on the famous principle "increasing intelligence with decreasing precision in a hierarchical control structure" (Saridis, 1979), Wang (2005) divides the agent-based control system's structure into three levels: organization, coordination, and execution, which correspond to global traffic operation center, regional traffic operation center, and networked traffic devices, respectively, in the realization of hardware and software. This system can enable the control mode of "local simple and remote complex," which can further improve the efficiency and intelligence of traffic control systems. Practical applications (Wang, 2010; Zhu, Li, Li, Chen, & Wen, in press) have shown the effectiveness of the architecture.

## 17.3 Ontology of Urban Traffic Management System

The ontology of the urban traffic management system (UTMS) is an explicit specification of knowledge of the urban traffic domain, which mainly contains the following seven classes: road network, environment, traffic information, traffic facility, vehicle, control algorithms, and time, as shown in Fig. 17.1. We will expand all concepts and knowledge of each class as follows.

### 17.3.1 Road Network

The road network is the basic part of traffic management system ontology. In this section, we will give a detailed presentation of the road network. Before that we need to clarify some statements. The current ontology is a basic version that includes necessary elements,

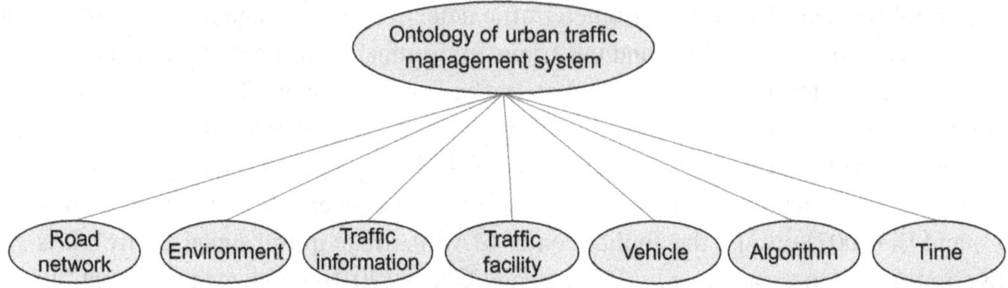


**Figure 17.1: Ontology of urban traffic management system.**

and it can be extended as we have discussed in the introduction. The extension of the ontology depends on the actual needs for a specific city. Moreover, there may be more than one name for an attribute or feature; however, here we adopt only one of them. For the specific, it is simply required that all agents should agree with a definite name for one attribute. Furthermore, sometimes a name is a complex attribute and consists of a prefix and a name body. This is unavoidable since there are so many roads and junctions in a city, even in a small city. Thus we need to place the prefix to indicate the type of attribute.

As is shown in Fig. 17.2, the road network can be further divided into two subclasses: road and road facility. In a road network, roads are the major components, which thus should be detailed in an ontology. A road is made up of some road segments and junctions. As we often see from any map, road segments can be regarded as lines and junctions can be regarded as nodes. The region fenced by some road segments is usually called a block.

#### *17.3.1.1 Road*

Roads basically construct a city's road network. Here by *road* we mean a set of road elements, such as road segments, junctions, and roundabouts, which share the same road name. For example, Zhongguancun East Road of Beijing has four road segments.

For a road, the basic attributes include *road name* and *road type*. *Road name* uniquely defines a road. However, there may be more than one name for a road. For example, a road can have both a common name and a numerical name such as "45# road." The official name and other alternative names should all be recorded, especially when the system is used by all users, which include ordinary citizens for querying information. The *road type* can be classified for specific situations. Commonly used types include *motorway, slip road, walkway, service road, single carriage way*, etc.

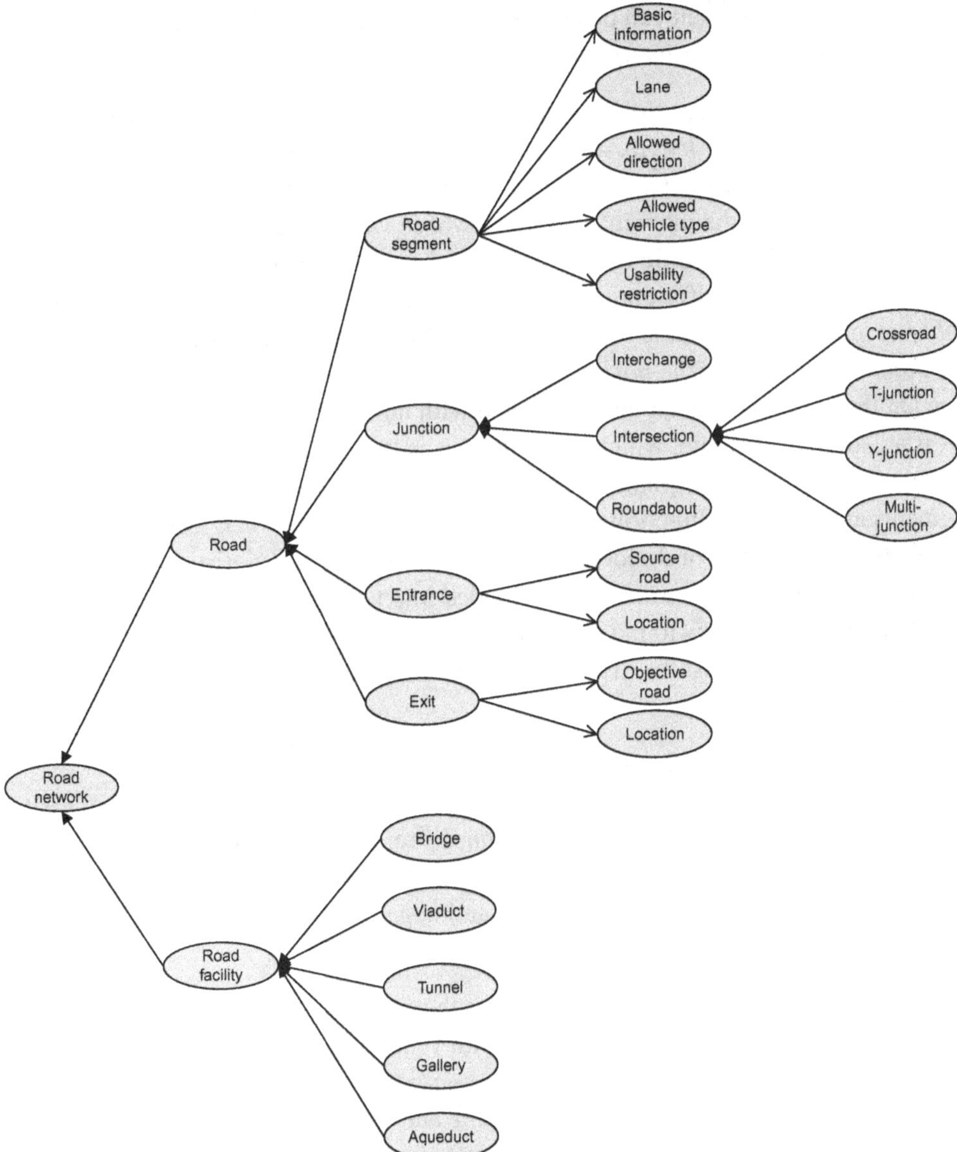


**Figure 17.2: Classification of the road network.**

#### *17.3.1.2 Road Segment*

A *road segment* is a line segment of a road. This means the part of a road between two junctions, which is a through unit of vehicles. It is obvious that a road segment begins and ends at a junction. Thus two adjacent junctions usually define a unique road segment. On the other hand, the intersection of different road segments clearly is a junction. This is the basic relationship between road segment and junction.

The attributes of a road segment can be classified as *basic information, allowed direction, allowed vehicle type,* and *usability restriction.*

The *basic information* describes the basic message of a road segment, which includes *road belonged to, identifier, width, length, number of lanes, road gradient*, and *divider*. Here, *road belonged to* and *identifier* define a road segment by its independent attributes, while *width* and *length* describe the inherent attributes, whose values are positive real numbers. *Number of lanes* reflects the traffic capacity of a road segment. Since a road segment often has more than one lane and different road segments may have different numbers of lanes, the *number of lanes* also hints at some traffic rules. The *number of lanes* can further be divided to *number of lanes of the positive direction* and *number of lanes of the negative direction*. On the other hand, each lane is labeled with number and driving direction. *Road gradient* is an additive attribute, which in most cases is "flat." If the road segment is not flat, then the degree of its maximal gradient is valued to the road segment. As is well known, road segments are usually bidirectional, thus there are dividers between lanes for different driving directions. The types of dividers include *traffic line divider, fence divider, green plants divider*, etc., which in essence can be divided into two groups, crossable or not. One could add temporal restrictions to the division, for example if an emergency happens, the management system can adjust the driving direction of some lanes more easily when the road segment is crossable.

*Allowed direction* is valued with *bidirectional* and *one-way traffic*. When it is bidirectional, then the management system needs to define which direction is the *positive direction* and which one is the *negative direction*. Here, the reason why we define the direction as positive and negative instead of east (or west, south, north) is that the road orientation may not be due east. Thus positive and negative directions would be suitable for computers.

*Allowed vehicle type* specifies the class of vehicles that is allowed to travel on a road segment. This ontology class can be connected with vehicle subontology. The value of this attribute may be *all vehicle, passenger car, lorry, non-passenger car, pedestrian only, bicycle*, and *pedestrian*. Some special examples are *bus lane* and *bus rapid transit* (BRT) lane. Furthermore, the allowed vehicle type can be detailed as a time-dependent type. For example, one road segment may allow passengers only from 8:00 to 20:00, while all vehicles are allowed for the rest of the day.

*Usability restriction* includes *height limitation, speed limitation, width limitation*, and *weight limitation*. The former three are ubiquitous in road networks. *Height limitation* has a positive real number value, which describes the maximum height. For example, if it is valued at 4 m, it means that all vehicles which do not exceed 4 m can pass the road segment. The height limitation is a necessary factor when approaching an aqueduct, bridge, tunnel, or viaduct. Considering UTS, most roads would require limit speeds, when *speed limitation* is also very common. *Width limitation* is usually connected with road width and special structures. *Weight limitation* is relatively uncommon compared with the others and it is usually required when entering a new road network or some part of it.

#### 17.3.1.3 Junction

*Junction* is the meeting of one road segment and other road segments; thus it is the turning point where vehicles enter a road segment from another one. The traffic data of junctions can be used to study and solve many traffic problems. Moreover, the traffic rules of junctions are also more complex than those of other road network elements. Junctions can be divided into *interchange, intersection*, and *roundabout*. The differences lie in the crossing of vehicles from different road segments. *Interchange* allows vehicles to keep moving since the crossing of different roads overlaps on different planes. *Roundabout* means that all vehicles entering the junctions should follow one appointed direction, which usually is anticlockwise or clockwise. *Intersection* is the most common junction in road networks, whose traffic rules are also the most complex. The intersection is further divided into *crossroad, T-junction, Y-junction*, and *multi-junction*.

*Crossroad* is the most common type of intersection. For a crossroad, we need to define the geographic coordinates, the corresponding road segments, the number of entry lanes, the number of exit lanes, the driving direction of each entry lane, having traffic lights or not, having crosswalk or not, and other attributes. For example, the driving direction of an entry lane may be going straight, turning left, turning right, going straight and turning left, going straight and turning right, and U-turn. Also the order of each lane should be given.

#### 17.3.1.4 Entrance and Exit

Both *entrance* and *exit* are needed when a side road merges with a main road. *Entrance* is attributed with *source road* and *location*, where the former defines which side road is merged in and the latter defines the position of the entrance. Similarly, *exit* is attributed with *objective road* and *location*.

#### 17.3.1.5 Road Facility

*Road facility* defines other additional structures that are physical parts of some roads. Common facilities include *bridge, viaduct, tunnel, gallery*, and *aqueduct*. These facilities have general attributes and special attributes. One example of special attributes is vertical position, valued by "over" or "under," and it describes the vertical position of the structures according to other roads.

### 17.3.2 Environment

Here by environment we mean the outside conditions that may affect the traffic except for roads which have been described in detail in the previous subsection. The environment considered in our system mainly covers three classes, i.e., meteorology, land coverage and use, and service. These three kinds of environment have major influence on drivers as regards driving strategy. For example, if it is raining, then the driver must slow his/her speed to keep safe. The drivers' actions are restricted by the traffic environment, thus environment is an important subontology of the traffic management system ontology.

#### 17.3.2.1 Meteorology

*Meteorology* describes the weather information that directly influences drivers' decisions. This subontology is divided into four classes: *temperature, wind, visibility*, and *weather type*.

*Temperature* is attributed with *degree*, which is measured in degrees Celsius. The degree is often valued by integers. Actually, according to specific city, the temperature range can be given on the basis of historical temperature data.

*Wind* has two common attributes. One is *wind direction*, which describes the direction that the wind blows from. It is valued by geographic directions, such as east, north, northeastward, and north by east 20°. The other attribute is *wind force*, which is valued by wind scale. In China, the wind scale is usually divided into 12 grades, denoted by integers 1 to 12. If it is stated as "north wind with level 4–5," the wind scale here means average wind force, while if it is stated as "wind gust with level 7," it means the maximum of the wind force is 7.

*Visibility* describes the maximum distance that a driver with normal vision can see. Visibility is affected by meteorological conditions, for example fog, snow, and/or wind could make the visibility deteriorate. The visibility is attributed with degree, where the latter is valued by grades or distances. Moreover, the visibility can be further divided into daylight visibility and night visibility.

*Weather type* is an important factor of the traffic system. As we have said earlier, bad weather can produce serious effects on traffic and cause more traffic accidents. Weather can be further divided into the following subclasses – *clearness, cloud, snow, rain, hail, sleet*, and *fog*. All these have attribute *degree*, which measures the grades of any specific weather. For example, *rain* can be attributed with different degrees according to light rain, moderate rain, heavy rain, shower, and/or thunder shower.

#### 17.3.2.2 Land Coverage and Use

*Land coverage and use* provides information of areas. The enclosed area is bounded by some road segments. As is well known, most addresses are often marked with a road name and their location number. However, some large regions may have their own name.

*Land coverage and use* can be classified as follows: *residential quarter, urban green space, construction site, buildings, city park, airport*, and *railway station*. Here by *buildings*, we mainly mean office buildings. Land coverage and use describes a kind of property of the enclosed area, which further has great influence on the status of nearby traffic. These traffic statuses are important for the traffic management system to make control decisions, especially in unexpected situations. For example, there are numerous people around railway stations, including arrivals and departures, thus the flow of people is of high density and great quantity. If heavy rain is falling, there should be more buses and taxies dispatched to the railway station to avoid traffic jams and lead the traffic. This is an obvious idea arising from consideration of land coverage and use.

Each class of *land coverage and use* has attributes. For example, *urban green space* is attributed with location, size, shape, individual place or not, people gathering degree, and others.

#### 17.3.2.3 Service

The types of buildings in the city are classified by different services. This motivates us to add subontology of *service*. Note that a building may have multiple services. The subontology *service* can help us to have a clearer picture of the traffic environment.

Typical services are classed as *public service, emergency, education, shopping, food and housing*, and *entertainment* (see Fig. 17.3).

*Public service* means the kind of service that is basic for all people, which includes *post office, bank, court, city hall*, etc. *City hall* here means government services, which may be separated into many individual departments, like education bureau, tax bureau, industrial and commercial bureau, justice bureau, labor and social security bureau, and bureau of civil affairs for a specific city.

*Emergency* includes the services that deal with all kinds of emergencies. These actually are very familiar in our daily life. For example, if someone has a serious disease, it would need to call the *first-aid post* and then the patient would be sent to *hospital*. Other classes of *emergency* include *fire station, police station, emergency call station*, etc.

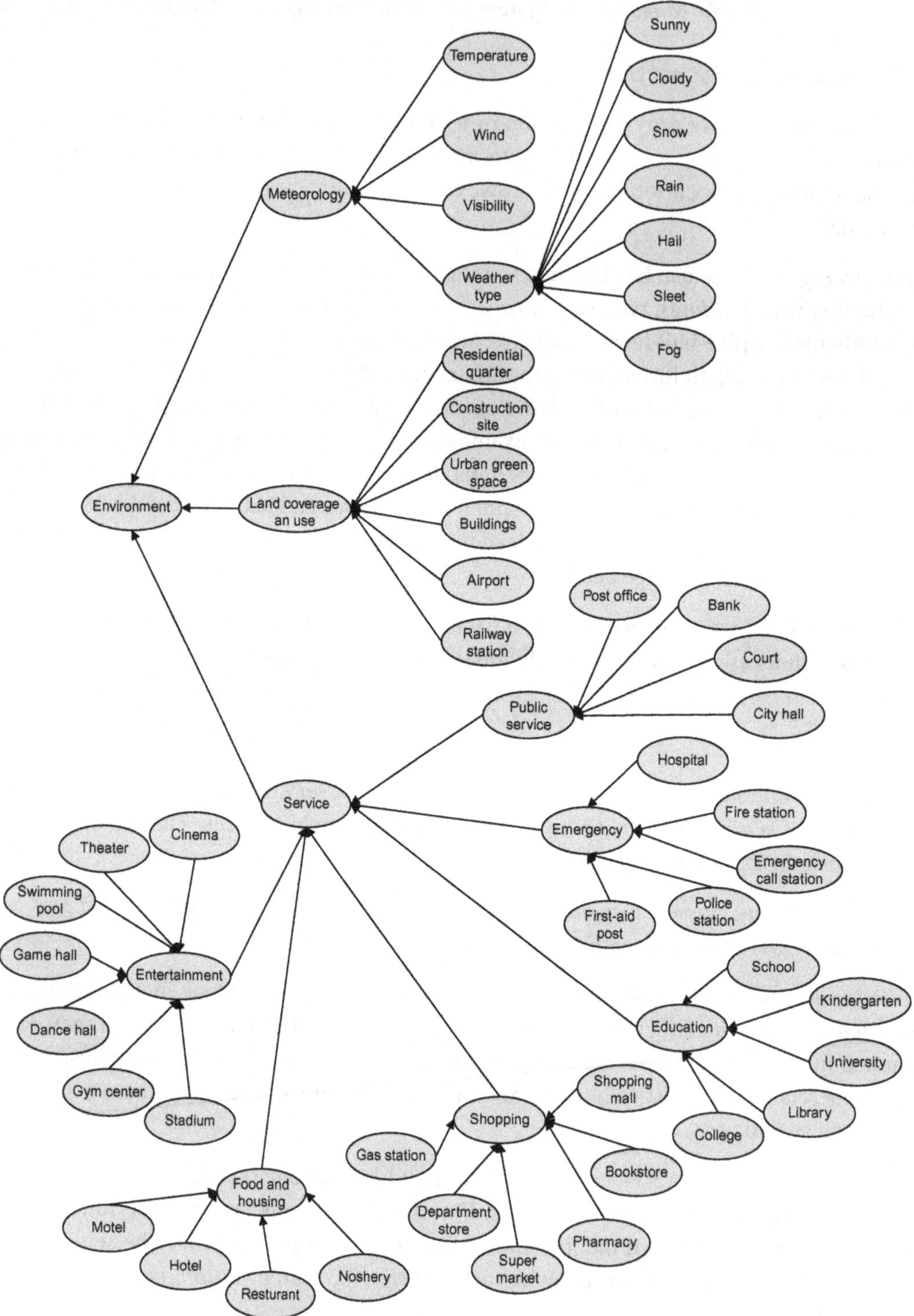


Figure 17.3: Environment.

*Education* describes the service buildings related to education. It is classed into *kindergarten, school, college, university*, and *library* according to different groups of people. Also there may be some other types of educational institution, such as private training schools and community colleges.

*Shopping* is a kind of service for customers, which may be the most common in the city. As we know, shopping service includes *shopping mall, book store, pharmacy, supermarket, department store, gas station*, etc. For an urban traffic management system, the shopping services are typical crowded areas.

*Food and housing* is important for travelers. In our ontology-based system, it includes *motel, hotel, restaurant, noshery*, etc. Here, we put food and housing together because sometimes they are provided at the same time. For example, a hotel usually affords both food and housing.

*Entertainment* means the sites that provide activities for people to enjoy. The modes of entertainment are numerous, thus it leads to many places of entertainment, which in our ontology include *cinema, theater, swimming pool, game hall, dance hall, gym center, stadium*, etc. Each class has general attributes, such as location and charge.

### 17.3.3 Traffic Information

Traffic information provides the basis for management decisions and control strategies for urban traffic management systems. It covers data, policies, rules, and temporary measures. Here we class it into three subontologies as follows (see Fig. 17.4).

#### 17.3.3.1 Parameter

*Parameter* contains most information of traffic control algorithms. Common examples of parameters include *flow volume, average speed, density, occupancy, queue length, period, phase, split*, etc. *Flow volume* is a macroscopic index of traffic conditions, which neglects the complex and nonlinear interactions of vehicles while focusing on the entire phenomena arising from numerous vehicles. *Average speed* is another index that reflects the traffic status. The speed can be further split into two indexes, namely time mean speed and space mean speed, according to different measurement methods. *Density* is defined as the number of vehicles per unit area of the roadway. Sometimes both maximum and minimum densities are required for management systems since jam is actually the density that exceeds a certain degree. *Occupancy* reflects the amount of vehicles on the road, which is defined by the ratio of occupant length of vehicles to road segment length. *Queue length* describes the traffic situation of junctions, where traffic flows from different directions would have to pass through in turn. Several other parameters, i.e., *period, phase*, and *split*, play important roles in traffic signal control.

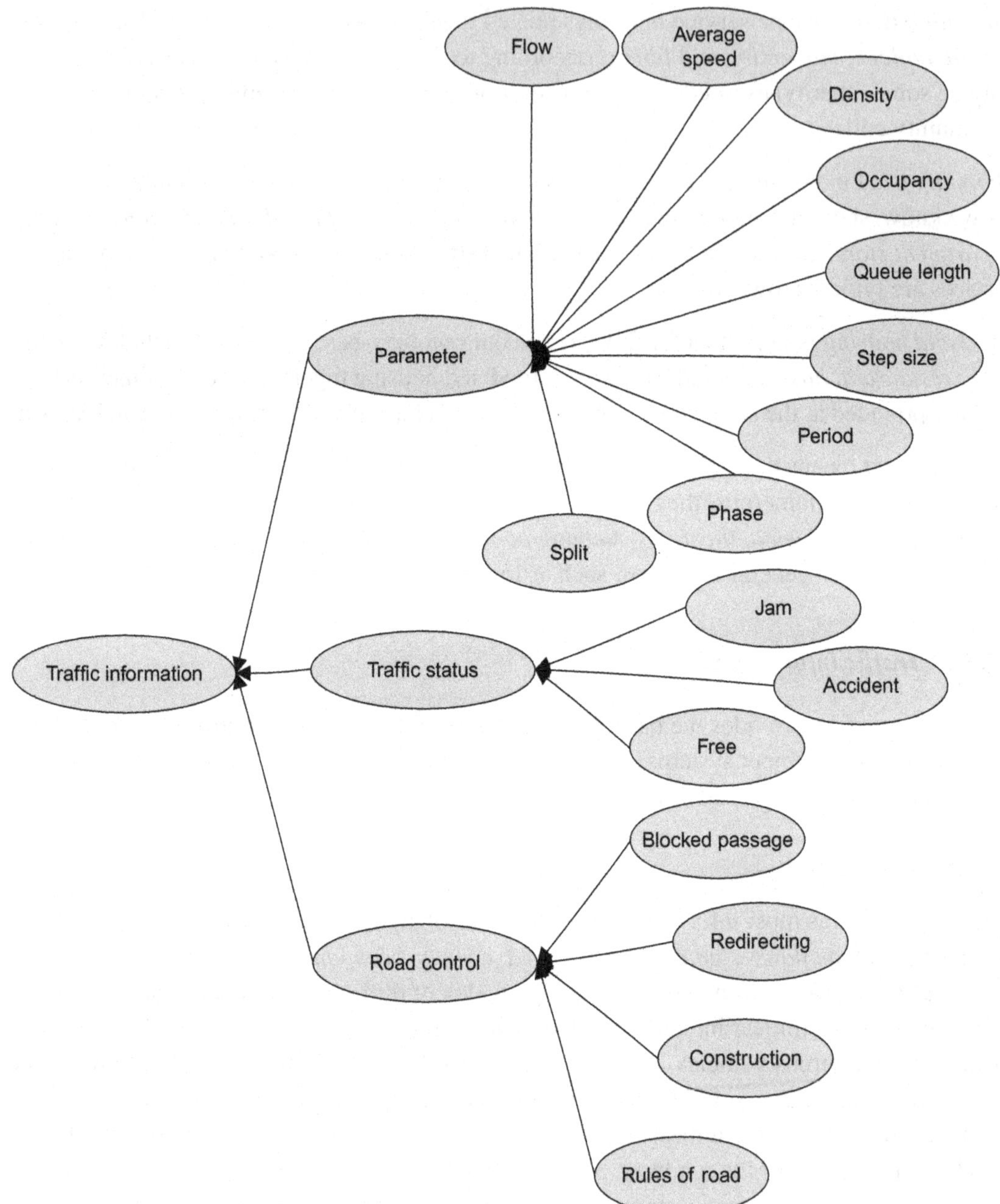


**Figure 17.4: Traffic information.**

According to the application to a specific city, there may be some other parameters. Among different parameters, there are various relations between them. These relations should be clearly predefined; however, this may bring about the problem of coordination and unification of different parameters. For example, there is a basic relationship between flow, average speed, and density as follows:

$$\text{Flow} = \text{Average Speed} \times \text{Density}$$

When the data of these three parameters do not coincide, the management system has to recheck which one is wrong.

#### 17.3.3.2 Traffic Status

*Traffic status* presents the basic knowledge of road traffic. Here the traffic status is preliminarily divided into three classes, namely *free, jam*, and *accident*. By *free*, we mean that the drivers do not have to slow down due to external factors. Note that *free* and *jam* cannot occur simultaneously in the same road segment. Thus in order to confirm the status automatically in the management system, there must be criteria for *free*. The attribute can be defined on the basis of average speed or flow. *Jam* can be graded according to its severity. The jam grade an one important attribute of *jam*. *Accident* has general attributes such as location, grade, etc. The *accident* class can be further divided according to type or object if necessary in applications.

#### 17.3.3.3 Road Control

*Road control* includes related messages on traffic control and management policies, which usually are time dependent and updated by the traffic administrative department. These messages are announced by different media, such as variable message signboard (VMS), website, radio, television, vehicle-carrying terminal, etc. *Road control* subontology is divided into four classes, namely *blocked passage, redirecting, construction*, and *rules of road*.

*Blocked passage* proposes information on the road that is blocked and thus vehicles have to choose alternative routes. It has attributes such as *blocked passage location* and *blocked passage type*. The *blocked passage location* can be valued by the blocked start and the blocked ending. The *blocked passage type* can be temporary or permanent. If it is temporary, there usually exist a start time and a stop time.

*Redirecting* occurs when there is an emergency, construction or traffic control. Its component *redirecting location* is valued by the start junction and ending junction, since usually the redirection is regulated to individual road segments rather than part of a road segment. Also there can be an attribute *redirecting type*, which is temporary or permanent.

*Construction* includes types of information about road constructions and related constructions that may affect road traffic. *Construction* has attribute *construction type*, which may have values such as under construction but passable, under construction and impassable, under construction from start time to end time, etc.

*Rules of road* describes all traffic rules for a specific road segment. As agents may have known some basic rules of traffic, thus the rules of road can be divided into *general rules of road* and *specific rules of road*. The former can be given for all vehicles in advance and the latter can be achieved when a vehicle arrives at the road segment.

### 17.3.4 Traffic Facility

Traffic facilities means physical facilities, which can deliver information, indicate route, administrate traffic, and store data. These facilities are very important in the traffic system as they are the primary information sources for drivers to charge their driving.

#### 17.3.4.1 Variable Message Signboard

*Variable message signboard* is used to display many kinds of informational messages, such as road works, road safety reminders, speed limits, or planned road closures. When an accident happens, warning messages will be displayed on the signboards so that motorists can make alternative travel pans. The VMS has attributes such as *shape, size, location, message display type, message maximum length*, etc. All these attributes are able to define VMS according to practical applications. Note that VMS is the most important means of information transmission, thus it has been integrated with an intelligent transportation system in many research papers and applications.

#### 17.3.4.2 Traffic Light

*Traffic light* is a signal device positioned at road junctions, pedestrian crossings, and other locations to control competing traffic flow. The types of *traffic light* include *three-color light, flash-warning light, direction light*, and *pedestrian crossing light. Three-color light* is the most common traffic light where the three colors are red, green, and yellow, representing no passing, passing, and warning of light switch, respectively. *Flash warning light* is usually a continuously flashing yellow light, which alerts vehicles and pedestrians to take notice of road safety. *Direction light* is a type of special light for vehicle direction command. The direction lights are composed of different arrows. *Pedestrian crossing light* is positioned at the pedestrian crossings of a road segment, giving instructions for pedestrians (Fig. 17.5).

#### 17.3.4.3 Road Sign

*Road sign* can be grouped into several types, such as *warning sign, prohibition sign, mandatory sign, guide sign, construction sign*, and *auxiliary sign* (see Fig. 17.5).

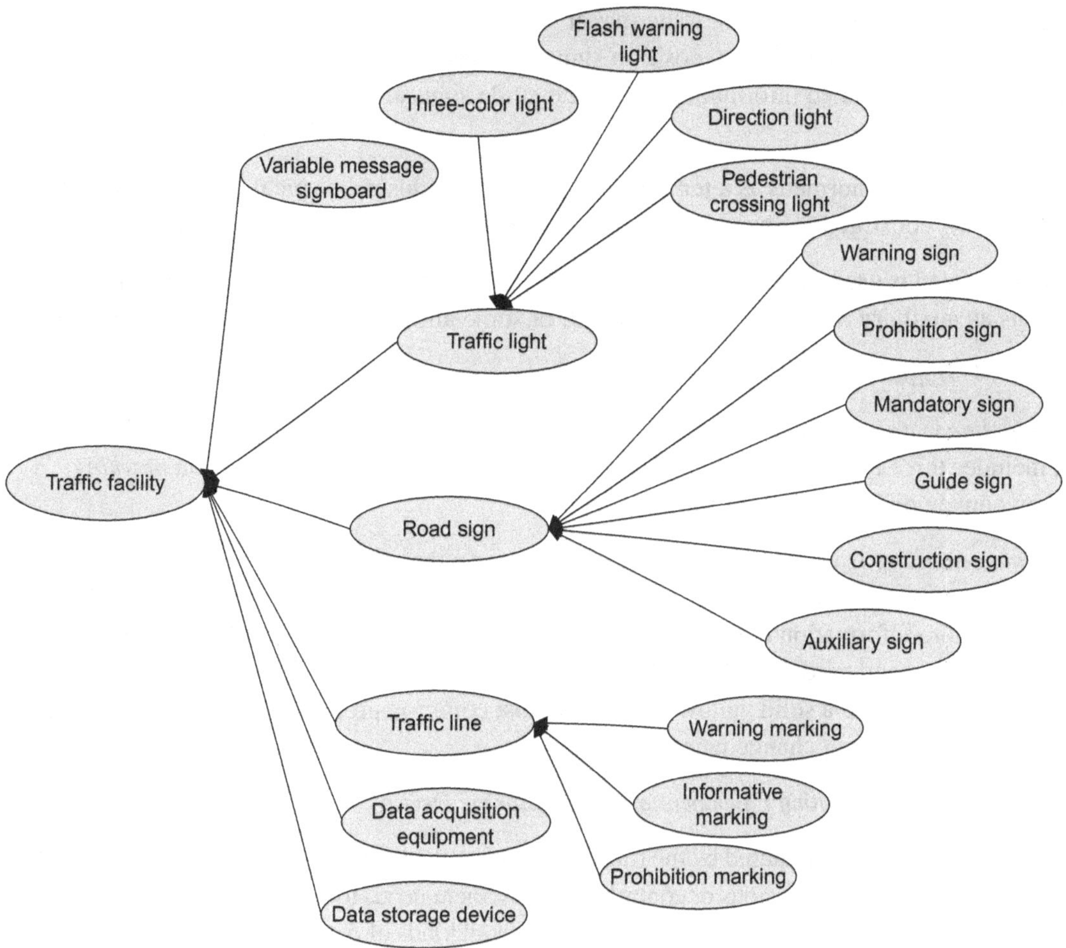


**Figure 17.5: Traffic facility.**

*Warning sign* indicates a hazard ahead on the road that the drivers should pay more attention to. This kind of sign is generally displayed in a yellow color. Some examples are sharp curve ahead sign, winding road ahead sign, intersection ahead sign, right lane ends sign, pedestrian crossing ahead sign, tunnel ahead sign, etc.

*Prohibition sign* is used to prohibit certain types of traffic action. Some typical examples are no entry sign, no parking sign, no overtaking sign, no left sign, no U-turn sign, speed limit sign, etc.

*Mandatory sign* is used to set the obligations of all traffic that use a specific area of road. The difference between prohibition sign and mandatory sign is that the former says what must not be done while the latter says what must be done. For example, pass on the right sign, go straight only sign, and motor vehicles only sign are typical mandatory signs.

*Guide sign* includes street signs, route maker signs, freeway signs, welcome signs, informational signs, place signs, position signs, etc. The guide signs are used to identify road names and other related information. Also, the guide signs are usually labeled with language rather than symbols.

*Construction sign* normally is a temporary sign for road closure. Typical examples include traffic cone, construction ahead sign, road closure sign, right way closed sign, etc.

*Auxiliary sign* is used to show some auxiliary information for drivers. For example, time range sign is an auxiliary to express the time length of some allowable or prohibitive commands.

#### 17.3.4.4 Traffic Line

*Traffic line* is a kind of marking that is used on the road to convey official information. It includes three types: *warning marking, informative marking*, and *prohibition marking*. They convey different kinds of traffic orders.

*Warning marking* is used to alert drivers to changes on the road, such as three-lane changing to two-lane, speed-down marking, and barrier in the road. *Informative marking* is used to provide more information to drivers, such as driving direction sign, pedestrian crossing, road boundary line, and road dotted line. *Prohibition marking* denotes a prohibitive command for drivers, for example a solid yellow line is the most common prohibition marking, meaning that vehicles must not change lanes.

#### 17.3.4.5 Data Acquisition Equipment and Data Storage Device

These two facilities are required by the traffic management system. As traffic real-time information is the basis for making decisions or control strategies, there needs to be data acquisition equipment located at junctions, roads, and other places to collect all kinds of data. On the other hand, historical data are also useful for the system to make good predictions, which further requires many devices to store data as the traffic data are considerable. Usually the data are submitted to the control center and centrally stored, while the control site stores as few data as are necessary.

### 17.3.5 Vehicle

As there are many brands and models of vehicles, it is impractical to name all types. Thus here we classify the vehicles by their type, as shown in Fig. 17.6.

#### 17.3.5.1 Car

*Car* means the type of vehicle that is used to carry not more than nine people. It is the most common type of vehicle on the road, which is because private cars are typical cars. Cars can be further classified by their engine capacities as minicar, popular car, mid-level car, and limousine. Note that for the convenience of traffic management, a car should have an important attribute *car type*, which is valued by private or public, like taxis.

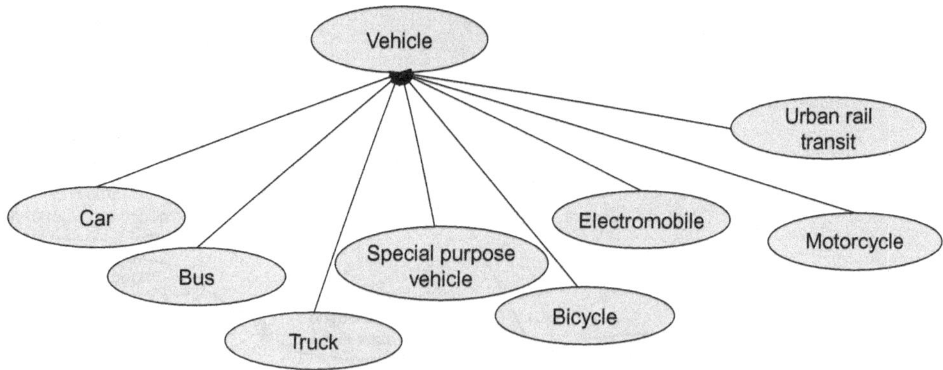


**Figure 17.6: Types of vehicle.**

#### 17.3.5.2 Bus

*Bus* is the type of vehicle that is also used to carry passengers, but more than nine, which distinguishes it from *car*. Typical example of *bus* is bus for citizens, and another example is commuter express. Buses can be classified according to length.

#### 17.3.5.3 Truck

*Truck* denotes the type of vehicle that is designed to transport cargo. It is usually classified by its total mass as light rigid, medium rigid, heavy rigid, heavy combination, and multicombination.

#### 17.3.5.4 Special Purpose Vehicle

*Special purpose vehicle* denotes the summation of other motor vehicles except motorcycles. For a city's daily applications, some familiar examples are as follows: sanitation vehicle, road sprinkler, garbage truck, crane truck, fire engine, etc.

#### 17.3.5.5 Bicycle, Electro-mobile, and Motorcycle

*Bicycle, electro-mobile*, and *motorcycle* include vehicles that are usually on two wheels. These vehicles are used for short trips.

#### 17.3.5.6 Urban Rail Transit

*Urban rail transit* includes the following transit types: *metro, light rail*, and *maglev*. These types of transit may alleviate the current traffic problems efficiently in modern mega cities.

### 17.3.6 Algorithm

Algorithm subontology covers the algorithms that are used in the management system to control urban traffic. It is the algorithm that makes the urban traffic management system intelligent. According to the aims of the algorithms, they are classified into three types: *monitoring algorithm, control algorithm*, and *basic algorithm*. The details are described as follows (Fig. 17.7).

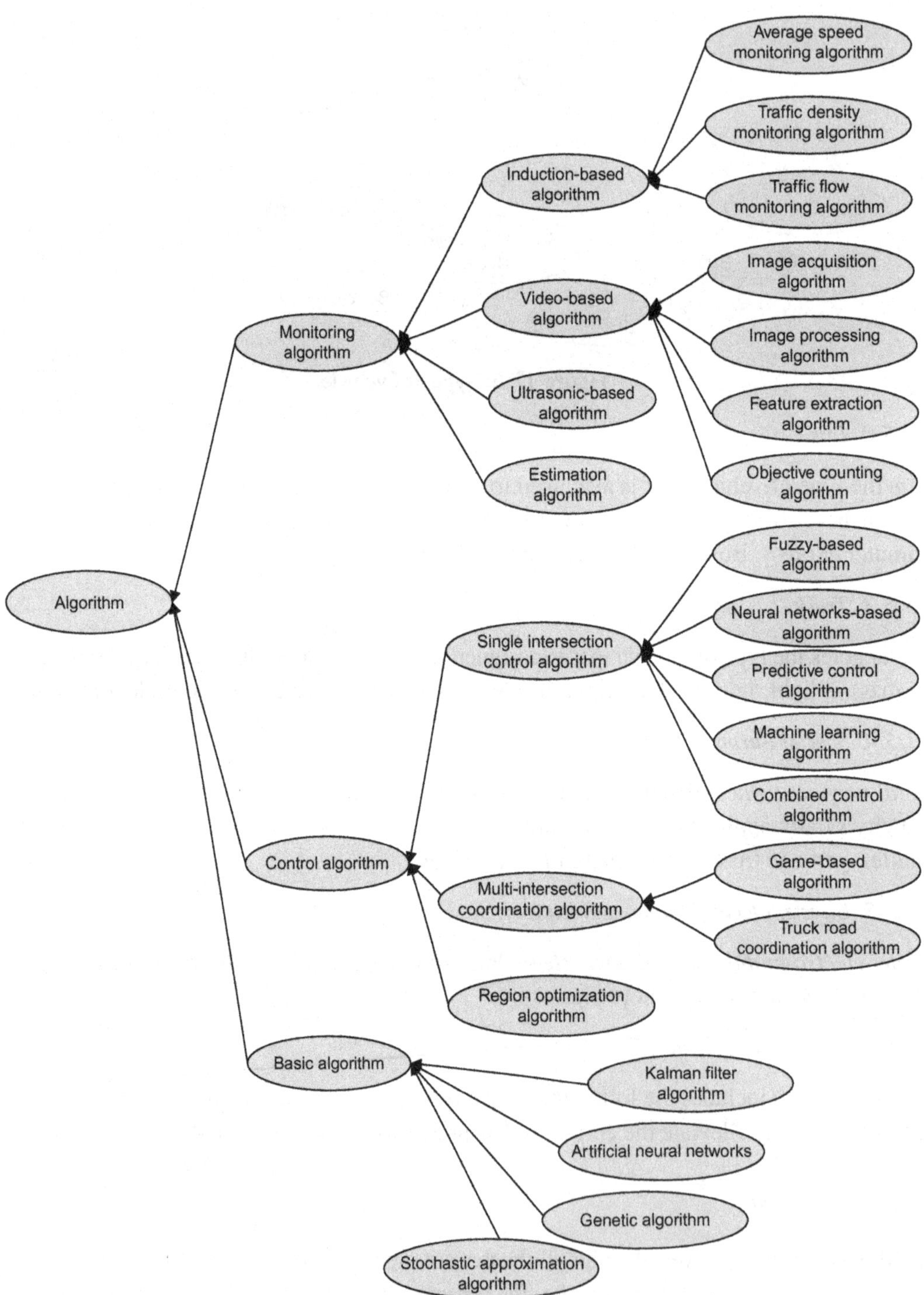


**Figure 17.7: Urban traffic control algorithms.**

#### 17.3.6.1 Monitoring Algorithm

*Monitoring algorithm* is the basis for further control of urban traffic, as it provides traffic information for control, coordination, and optimization algorithms. There are many types of detectors such as circular coil detector, supersonic detector, video detector, magnetic inductive detector, and radar detector. However, the first three of these are most commonly used, therefore we classify monitoring algorithms into the following four types: *induction-based algorithm, video-based algorithm, ultrasonic-based algorithm*, and *estimation algorithm*.

*Induction-based algorithm* mainly focuses on the identification of basic traffic information. Thus it includes *average speed monitoring algorithm, traffic density monitoring algorithm, traffic flow monitoring algorithm*, etc.

*Video-based algorithm* is used for multitasks such as counting, detecting, and capturing. As cameras are widely and densely distributed in developed cities, video-based algorithm is an efficient approach on the basis of digital image recognition technology. According to the processing procedure, *video-based algorithm* can be divided into *image acquisition algorithm, image processing algorithm, feature extraction algorithm*, and *objective counting algorithm*.

*Ultrasonic-based algorithm* is a kind of detecting approach with the operating principle using the Doppler effect. This kind of algorithm is mainly used for expressway monitoring and management, expressway ramp metering control, multilane intersection monitoring, etc.

*Estimation algorithm* is the one that calculates the traffic parameters by some measurement information rather than using directly obtained data from detectors. For example, estimation of queue length based on Kalman filter is an effective method for finding the queue length based on the traffic flow.

#### 17.3.6.2 Control Algorithm

By *control algorithm* we mean the algorithm used to control, coordinate, and optimize urban traffic. As we stated before, our system architecture is divided into three levels, which is convenient in realizing the transformation from control algorithms to control agents. Thus the *control algorithm* falls into three classes, namely *single intersection control algorithm, multi-intersection coordination algorithm*, and *region optimization algorithm*.

The traffic control of a single intersection is the basis of the traffic management and control system. Meanwhile, algorithms designed for single intersections are much more numerous than for multi-intersection coordination and/or regional optimization. The most common algorithms cab be classified as *fuzzy-based algorithm, neural networks-based algorithm, predictive control algorithm, machine learning algorithm*, and *combined control algorithm*. For example, Pappis and Mamdani (1977) developed the first fuzzy-based control algorithm for traffic junctions.

For the *multi-intersection coordination algorithm*, there are two common types, i.e., *game-based algorithm* and *truck road coordination algorithm*. The former is used in cases where the multi-intersections are on an equal footing, so that the coordination algorithm has to provide good planning to optimize the overall index. The latter is mainly used for truck roads since these take the major traffic load. Truck road coordination is called line control in some of the literature.

*Region optimization algorithm* refers to the one used for a large region or the whole city. The larger range the algorithm applies, the less detail is taken into consideration. This kind of algorithm is relatively rare in the published literature. Durfee and Lesser (1991) and Decker and Lesser (1992) propose two feasible ideas.

#### 17.3.6.3 Basic Algorithm

*Basic algorithm* provides the urban traffic management system with basic theory for guidelines. This subontology includes classic monitoring, control and optimization algorithms such as the Kalman filter algorithm, artificial neural networks, genetic algorithm, stochastic approximation algorithm, etc. As we know, there may be more than one form of the algorithm, thus this subontology is rather large.

### 17.3.7 Time

*Time* is one of the most important attributes of any traffic event, so UTMS should include an expression for time. According to users' customs, a complete state of time should have the following elements: *year, month, day, hour, minute*, and *second*. Sometimes *week* is also used, which includes *Monday, Tuesday, Wednesday, Thursday, Friday, Saturday*, and *Sunday*.

In general, the usual usages of time are specific moment and duration. A specific moment often is used to denote the occurrence of a traffic event or the start and end times of its duration, while duration indicates a sustained traffic event such as a jam.

## 17.4 Multiagent System Architecture

With the development of Internet communication technology, the world is stepping into Internet era. The network now has become an inseparable part of our lives and work. Based on this, it is believed that the network can produce better performance, higher intelligence, and more convenience. However, how to realize it remains a fundamental problem. As we know, the control algorithm is the core in the age of electricity, which requires corresponding storage space and computational power. In the Internet era, keeping this traditional approach would cause great waste and complexity. Wang (2000, 2005) presents the agent-based control method in his studies.

As explained by Wang (2005), the transfer from control algorithms to control agents would make control become an independent entity instead of an affiliated function in system design.

In the new approach, the control algorithms are divided into many task-oriented control agents that are distributed over networks. Various control and/or management tasks can be accomplished by the cooperation of these agents. It is obvious that one individual agent cannot solve any complex problem; therefore more research is required on the cooperation of many different kinds of agents. To ensure a coherent control and communication mechanism among different agents, we could adapt the hierarchical architecture from the works of Wang (2005) for UTMS, as shown in Fig. 17.8. Here the system's structure is divided into three levels: *global organization center level, regional coordination center level*, and *local execution device level.*

In a traffic control and management system, the function of the *global organization center level* is task planning and decomposition, reasoning based on the traffic objectives and current traffic status, and scheduling. The global organization center has great computational power, which makes it able to undertake complex reasoning and learning, carry out online simulation and prediction, and provide communication protocols and knowledge for agents. Thus the center should include *naming agent, creating agent, storage agent, scheduling agent, monitoring agent, decomposition agent, resource allocation agent*, etc. Each agent has a main purpose. For example, the *naming agent* can provide a name for any agent; the *creating agent* can create new agent for specific requirement; the *storage agent* can store specific information and send real-time traffic data back from local sites; the *scheduling agent* can choose and allocate suitable agents for different traffic devices; the *monitoring agent* is used to receive information fed back from regional coordination centers; the *decomposition agent* is used to divide complex traffic control and management tasks into many subtasks and carry out basic analysis; the *resource allocation agent* is used to uniformly manage all the traffic resources such as traffic devices.

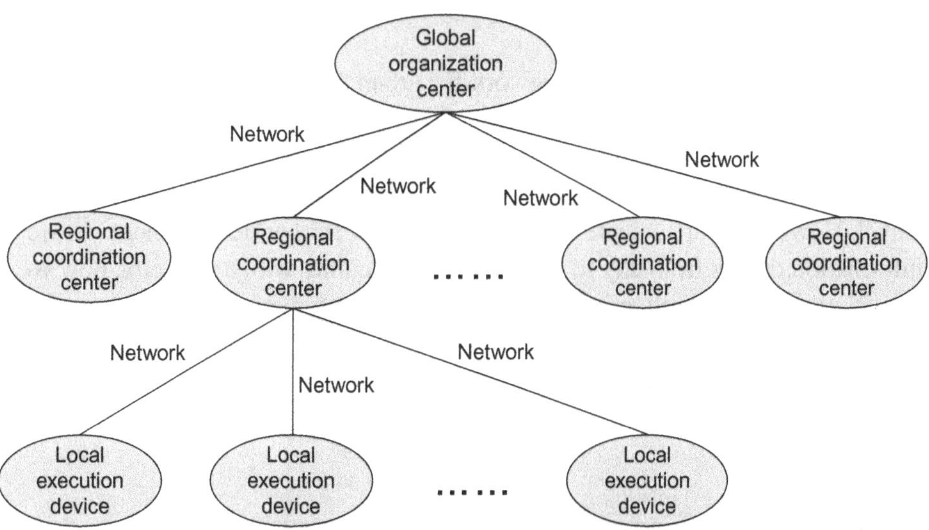


**Figure 17.8: Multiagent framework.**

The *regional coordination center level* is the interface between the organization and execution levels. The function of this level is receiving agents from the organization level and deploying the agents to the appropriate devices. Thus, in general, this level should consist of dispatchers and coordinators. The regional coordination center includes *receiving agent, resolution agent, dispatching agent, transmission agent*, etc. The *receiving agent* takes charge of receiving agents from the organization center; the *resolution agent* is used to resolve address and other information integrated in the agents; the *dispatching agent* mainly deploys agents to corresponding traffic devices and sends feedback information to the organization level; the *transmission agent* is responsible for data transmission from local sites to the organization level.

The *local execution device level* consists of different kinds of traffic devices, such as crossroad signal controllers, video detectors, snapshot cameras, and variable message boards. These devices directly execute all control and management commands and obtain all kinds of traffic information. In general, the local execution device should have *downloading agent, status assessment agent, switching agent, information agent, deleting agent*, etc. The *downloading agent* is responsible for receiving agents from the coordination level and sending acknowledgement messages to the coordination level. The *status assessment agent* then analyzes the current traffic conditions and internal state of the traffic device and makes a decision based on rules. If there is a need to change agents, the *switching agent* would make the switchover of coming agent and original agent. If the agent replaced does not have to return to the center, then the *deleting agent* would delete it and send an acknowledgement message to the coordination level. The function of the *information agent* is to collect all kinds of traffic information and send it to the coordination level.

## 17.5 Conclusions

In this chapter, we have proposed a framework for urban traffic management systems based on ontology and multiagent system technology. Specifically, we construct a basic traffic ontology, which includes seven subontologies. Then the three-leveled architecture for an agent-based distributed traffic control system is proposed with an explanation of the functions at each level. However, there are still many issues left for further development. As we know, traffic information differs from city to city, thus when dealing with a new city, how we can merge the original traffic ontology with the new concept automatically is an important topic. On the other hand, we have to work on the goals, for example study and implement the agent-based distributed traffic control system proposed.

## References

Adler, J. L. & Blue, V. J. (2002). A cooperative multi-agent transportation management and route guidance system. *Transportation Research Part C, 10*(5/6), 433–454.

Bermejo-Alonso, J. & Sanz, R. (2006). A survey on ontologies for agents. UPM Autonomous Systems Laboratory, ASLab-ICEA-R-2006-002, v 1.0 Draft. Retrieved from http://tierra.aslab.upm.es/documents/controlled/ASLAB-R-2006-05-v1-Draft-JB.pdf.

Borst, W. N. (1997). *Construction of engineering ontologies for knowledge sharing and reuse* (Doctoral thesis). University of Twente, Enschede.

Calero, C., Ruiz, F., & Piattini, M. (2006). *Ontologies for software engineering and software technology*. Berlin: Springer.

Chandrasekaran, B., Josephson, J. R., & Benjamins, V. R. (1999). What are ontologies, and why do we need them? *IEEE Intelligent Systems, 14*(1), 20–26.

Chen, B. & Cheng, H. H. (2010). A review of the applications of agent technology in traffic and transportation systems. *IEEE Transactions on Intelligent Transportation Systems, 11*(2), 485–496.

Chen, B., Cheng, H. H., & Palen, J. (2009). Integrating mobile agent technology with multi-agent systems for distributed traffic detection and management systems. *Transportation Research Part C, 17*(1), 1–10.

Chen, R. S. & Chen, D. K. (2008). Apply ontology and agent technology to construct virtual observatory. *Expert Systems with Applications, 34*(3), 2019–2028.

Choy, M. C., Srinivasan, D., & Cheu, R. L. (2003). Cooperative, hybrid agent architecture for real-time traffic signal control. *IEEE Transactions on Systems, Man and Cybernetics-Part A, 33*(5), 597–607.

Davy, S., Barrett, K., Serrano, M., Strassner, J., Jennings, B., & van der Meer, S. (2007). Policy interactions and management of traffic engineering services based on ontologies. In *Proceedings of Network Operations and Management Symposium (LANOMS 2007)*, Rio de Janeiro, Sept. 10–12, pp. 95–105.

Decker, K. S. & Lesser, V. R. (1992). Generalizing the partial global planning algorithm. *International Journal of Intelligence and Cooperative Information System, 1*(2), 319–346.

Dolan, M. E. & Blake, J. A. (2009). Using ontology visualization to facilitate access to knowledge about human disease genes. *Applied Ontology, 4*(1), 35–49.

Durfee, E. H. & Lesser, V. R. (1991). Partial global planning: A coordination framework for distributed hypothesis formation. *IEEE Transactions on System, Man, and Cybernetics, 21*(5), 1167–1183.

Eschenbach, C. & Grüninger, M. (2008). Formal ontology in information systems. In *Proceedings of the Fifth International Conference* (*FOIS 2008*), Saarbrücken, Germany, Oct. 31–Nov. 3. IOS Press.

Fernández-López, M., Gómez-Pérez, A., Pazos-Sierra, A., & Pazos-Sierra, J. (1999). Building a chemical ontology using methontology and the ontology design environment. *IEEE Intelligent Systems, 14*(1), 37–46.

Fonseca, F. T. & Egenhofer, M. J. (1999). Ontology-driven geographic information systems. In *Proceedings of ACM-GIS*, Kansas City, USA, 2–6 Nov., pp. 14–19.

Freitas, F., Stuckenschmidt, H., Malucelli, A., & Pinto, H. S. (2007). Special issue on ontologies and their applications. *Journal of Universal Computer Science, 13*(12), 1801–1969.

Gokulan, B. P. & Srinivasan, D. (2010). Distributed geometric fuzzy multiagent urban traffic signal control. *IEEE Transactions on Intelligent Transportation Systems, 11*(3), 714–727.

Gómez-Pérez, A., Fernández-López, M., & Corcho, O. (2004). *Ontological engineering: With examples from the areas of knowledge management, ecommerce and the semantic web*. London, UK: Springer.

Gruber, T. R. (1993). A translation approach to portable ontology specifications. *Knowledge Acquisition, 5*(2), 199–220.

Gruber, T. R. (1995). Toward principles for the design of ontologies used for knowledge sharing. *International Journal Human-Computer Studies, 43*(5–6), 907–928.

Happel, H. & Seedorf, S. (2006). Applications of ontologies in software engineering. In *International Workshop on Semantic Web Enabled Software Engineering (SWESE'06)*, Athens, GA, 6 Nov., pp. 1–14. Citeseer.

He, F., Miao, Q., Li, Y., Wang, F. Y., & Tang, S. (2006). Modeling and analysis of artificial transportation system based on multi-agent technology. In *Proceedings of the IEEE International Conference on Intelligent Transportation Systems (ITSC 2006)*, Toronto, Sept. 17–20, pp. 1120–1124.

Huhns, M. N. & Singh, M. P. (1997). Ontologies for agents. *IEEE Internet Computing, 1*(6), 81–83.

Jennings, N. R. & Wooldridge, M. (1998). Applications of intelligent agents. In: N. R. Jennings & M. Wooldridge (Eds.), *Agent technology: Foundations, applications, and markets* (pp. 3–28). Berlin: Springer.

Laclavík, M. (2005). *Ontology and agent based approach for knowledge management* (Doctoral thesis). Institute of Informatics, Slovak Academy of Sciences.

Li, Z. (2007). *A study of urban traffic signal control system based on mobile multi-agent* (Doctoral thesis). Chinese Academy of Sciences.

Liu, X. M. & Wang, F. Y. (2002). Study of city area traffic coordination control on the basis of agent. In *IEEE 5th International Conference on Intelligent Transportation Systems*, Singapore, Sept. 3–6, pp. 758–761.

Loia, V. & Lee, C. S. (2010). Special issue on new trends for ontology-based knowledge discovery. *International Journal of Intelligent Systems, 12*(12), 1141–1264.

Merdan, M., Koppensteiner, G., Hegny, I., & Favre-Bulle, B. (2008). Application of an ontology in a transport domain. In *IEEE International Conference on Industrial Technology (IEEE ICIT2008)*, Chengdu, China, April 21–24, pp. 1–6.

Merdan, M., Vittori, L., Koppensteiner, G., Vrba, P., & Favre-Bulle, B. (2008). Simulation of an ontology-based multi-agent transport system. In *SICE Annual Conference 2008*, Tokyo, Japan, Aug. 20–22, pp. 3339–3343.

Neches, R., Fikes, R., Finin, T., Gruber, T. R., Ratil, R., Senator, T., & Swartout, W. R. (1991). Enabling technology for knowledge sharing. *AI Magazine, 12*(3), 36–56.

Pappis, C. P. & Mamdani, E. H. (1977). A fuzzy logic controller for a traffic junction. *IEEE Transactions on Systems, Man, and Cybernetics, 7*(10), 707–717.

Pérez A. G. & Benjamins, V. R. (1991). Overview of knowledge sharing and reuse components: Ontologies and problem-solving methods. In *Proceedings of the IJCAI-99 Workshop on Ontologies and Problem-Solving Methods*, Stockholm, Sweden, Aug. 2, pp. 1–15.

Pinto, H. S. & Martins, J. P. (2004). Ontologies: How can they be built? *Knowledge and Information Systems, 6*(4), 441–464.

Roozemond, D. A. (1999). Using autonomous intelligent agents for urban traffic control systems. In *Proceedings of the 6th World Congress on Intelligent Transport Systems*, Toronto, Canada, Nov. 8–12, pp. 69–79.

Saridis, G. N. (1979). Toward the realization of intelligent controls. *Proceedings of the IEEE, 67*(8), 1115–1133.

Studer, R., Benjamins, V. R., & Fensel, D. (1998). Knowledge engineering: Principles and methods. *Data and Knowledge Engineering, 25*(1–2), 161–197.

Su, X., Matskin, M., & Rao, J. (2003). Implementing explanation ontology for agent system. In *Proceedings of the 2003 IEEE/WIC International Conference on Web Intelligence*, Halifax, Canada, Oct. 13–17, pp. 1–7.

Tamma, V., Cranefield, S., Finin, T. W., & Willmott, S. (2005). *Ontologies for agents: Theory and experiences. Basel*, Switzerland: Birkhäuser.

Tomás, V. R. & García, L. A. (2005). Agent-based management of non-urban road meteorological incidents. In *Proceedings of Multi-Agent Systems and Applications IV*, Budapest, Hungary, Sept. 15–17, Vol. 3690, pp. 213–222.

Tran, Q. N. & Low, G. (2008). MOBMAS: A methodology for ontology-based multi-agent systems development. *Information and Software Technology, 50*(7–8), 697–722.

Uschold, M. & Gruninger, M. (1996). Ontologies: Principles, methods, and applications. *Knowledge Engineering Review, 11*(2), 93–155.

Uschold, M. & Jasper, R. (1999). A framework for understanding and classifying ontology applications. In *Proceedings of the IJCAI-99 Workshop on Ontologies and Problem-Solving Methods*, Stockholm, Sweden, Aug. 2, pp. 16–21.

Wang, F. Y. (2000). ABCS: *Agent-based control systems* (SIE Working Paper). Tucson: University of Arizona.

Wang, F. Y. (2005). Agent-based control for networked traffic management systems. *IEEE Intelligent Systems, 20*(5), 92–96.

Wang, F. Y. (2010). Parallel control and management for intelligent transportation systems: Concepts, architectures, and applications. *IEEE Transactions on Intelligent Transportation Systems, 11*(3), 630–638.

Warden, T., Porzel, R., Gehrke, J. D., Herzog, O., Langer, H., & Malaka, R. (2010). Towards ontology-based multiagent simulations: The PlaSMA approach. In *24th European Conference on Modelling and Simulation (ECMS 2010)*, Kuala Lumpur, Malaysia, June 1–4, pp. 50–56.

Wooldridge, M. (2009). *An introduction to multi agent systems*. Chichester, UK: John Wiley & Sons.

Wooldridge, M. & Jennings, N. R. (1995). Intelligent agents: Theory and practice. *The Knowledge Engineering Review, 10*(2), 115–152.

Zhu, F., Li, G., Li, Z., Chen, C., & Wen, D. (2011). A case study of evaluating traffic signal control systems using computational experiments. *IEEE Transactions on Intelligent Transportation Systems, 12*(4), 1220–1226.

# Index

Page numbers followed by *f* indicates a figure and *t* indicates a table.

## N



## O



## P

Printed in the United States
By Bookmasters